Novel and Alternative Methods in Food Processing

Biotechnological, Physicochemical, and Mathematical Approaches

Innovations in Agricultural and Biological Engineering

Novel and Alternative Methods in Food Processing

Biotechnological, Physicochemical, and Mathematical Approaches

Edited by

N. Veena, PhD

Megh R. Goyal, PhD

Ritesh B. Watharkar, PhD

AAP | APPLE ACADEMIC PRESS

First edition published 2023

Apple Academic Press Inc.
1265 Goldenrod Circle, NE,
Palm Bay, FL 32905 USA

760 Laurentian Drive, Unit 19,
Burlington, ON L7N 0A4, CANADA

CRC Press
6000 Broken Sound Parkway NW,
Suite 300, Boca Raton, FL 33487-2742 USA

4 Park Square, Milton Park,
Abingdon, Oxon, OX14 4RN UK

Library and Archives Canada Cataloguing in Publication

Title: Novel and alternative methods in food processing : biotechnological, physicochemical, and mathematical approaches / edited by N. Veena, PhD, Megh R. Goyal, PhD, Ritesh B. Watharkar, PhD.

Names: Veena, N., editor. | Goyal, Megh R., editor. | Watharkar, Ritesh B., editor.

Series: Innovations in agricultural and biological engineering.

Description: First edition. | Series statement: Innovations in agricultural and biological engineering | Includes bibliographical references and index.

Identifiers: Canadiana (print) 20220485917 | Canadiana (ebook) 20220485984 | ISBN 9781774911624 (hardcover) | ISBN 9781774911631 (softcover) | ISBN 9781003328605 (ebook)

Subjects: LCSH: Food industry and trade—Technological innovations. | LCSH: Food industry and trade—Mathematical models. | LCSH: Food—Biotechnology. | LCSH: Food—Composition. | LCSH: Functional foods.

Classification: LCC TP370 .N68 2023 | DDC 664—dc23

Library of Congress Cataloging-in-Publication Data

Names: Veena, N., editor. | Goyal, Megh R., editor. | Watharkar, Ritesh B., editor.

Title: Novel and alternative methods in food processing : biotechnological, physicochemical, and mathematical approaches / edited by N. Veena, PhD, Megh R. Goyal, PhD, Ritesh B. Watharkar, PhD.

Description: First edition. | Palm Bay, FL : Apple Academic Press, [2023] | Series: Innovations in agricultural and biological engineering | Includes bibliographical references and index. | Summary: "This new volume explores an array of emerging and advanced techniques in the food processing sector. Novel food processing methods such as ultrasound processing, microwave heating, advanced drying methods, and nonthermal technologies are discussed in detail. The volume also covers the application of irradiation in food and food products and highlights the potential of nutraceutical foods and application of encapsulation methods in the food industry. It illustrates the various applications of these new novel food processing, preservation, and encapsulation methods. Apart from new processing technologies, mathematical modeling concepts and case studies are included to illustrate applications of modeling techniques in food processing. The volume promotes the understanding of the thermodynamics of food polymers, structural design principles, structural hierarchy, and the steps involved in food structuring and structure measurement techniques. Also provided is an overview of the scope and significance of food biotechnology, the benefits, and concerns of using biotechnology in food processing, as well as microbial valorizing approach as a novel technique and a source for nutrition and nutraceuticals. Given the book's vast coverage, this volume has tremendous applications in different areas of the food industry, including in food processing, preservation, safety, and quality evaluation of processed foods"-- Provided by publisher.

Identifiers: LCCN 2022057977 (print) | LCCN 2022057978 (ebook) | ISBN 9781774911624 (hardcover) | ISBN 9781774911631 (paperback) | ISBN 9781003328605 (ebook)

Subjects: LCSH: Food science. | Food industry and trade.

Classification: LCC TP370 .N58 2023 (print) | LCC TP370 (ebook) | DDC 664--dc23/eng/20230111

LC record available at https://lccn.loc.gov/2022057977

LC ebook record available at https://lccn.loc.gov/2022057978

ISBN: 978-1-77491-162-4 (hbk)
ISBN: 978-1-77491-163-1 (pbk)
ISBN: 978-1-00332-860-5 (ebk)

ABOUT THE BOOK SERIES: INNOVATIONS IN AGRICULTURAL AND BIOLOGICAL ENGINEERING

Under this book series, Apple Academic Press Inc. is publishing book volumes over a span of 8–10 years in the specialty areas defined by the American Society of Agricultural and Biological Engineers (<asabe.org>). Apple Academic Press Inc. aims to be a principal source of books in agricultural and biological engineering. We welcome book proposals from readers in areas of their expertise.

The mission of this series is to provide knowledge and techniques for agricultural and biological engineers (ABEs). The book series offers high-quality reference and academic content on agricultural and biological engineering (ABE) that is accessible to academicians, researchers, scientists, university faculty and university-level students, and professionals around the world.

Agricultural and biological engineers ensure that the world has the necessities of life, including safe and plentiful food, clean air and water, renewable fuel and energy, safe working conditions, and a healthy environment by employing knowledge and expertise of the sciences, both pure and applied, and engineering principles. Biological engineering applies engineering practices to problems and opportunities presented by living things and the natural environment in agriculture.

ABE embraces a variety of the following specialty areas (<asabe.org>): aquaculture engineering, biological engineering, energy, farm machinery and power engineering, food, and process engineering, forest engineering, information, and electrical technologies, soil, and water conservation engineering, natural resources engineering, nursery, and greenhouse engineering, safety, and health, and structures and environment.

For this book series, we welcome chapters on the following specialty areas (but not limited to):

- Academia to industry to end-user loop in agricultural engineering;
- Agricultural mechanization;
- Aquaculture engineering;
- Biological engineering in agriculture;

- Biotechnology applications in agricultural engineering;
- Energy source engineering;
- Farm to fork technologies in agriculture;
- Food and bioprocess engineering;
- Forest engineering;
- GPS and remote sensing potential in agricultural engineering;
- Hill land agriculture;
- Human factors in engineering;
- Impact of global warming and climatic change on agriculture economy;
- Information and electrical technologies;
- Irrigation and drainage engineering;
- Nanotechnology applications in agricultural engineering;
- Natural resources engineering;
- Nursery and greenhouse engineering;
- Potential of phytochemicals from agricultural and wild plants for human health;
- Power systems and machinery design;
- Robot engineering and drones in agriculture;
- Rural electrification;
- Sanitary engineering;
- Simulation and computer modeling;
- Smart engineering applications in agriculture;
- Soil and water engineering;
- Micro-irrigation engineering;
- Structures and environment engineering;
- Waste management and recycling;
- Any other focus areas.

For more information on this series, readers may contact:

Megh R. Goyal, PhD, PE
Book Series Senior Editor-in-Chief: Innovations in Agricultural and Biological Engineering
E-mail: goyalmegh@gmail.com

OTHER BOOKS ON AGRICULTURAL AND BIOLOGICAL ENGINEERING BY APPLE ACADEMIC PRESS, INC.

Management of Drip/Trickle or Micro Irrigation
Megh R. Goyal, PhD, PE, Senior Editor-in-Chief
Evapotranspiration: Principles and Applications for Water Management
Megh R. Goyal, PhD, PE and Eric W. Harmsen, PhD Editors

Book Series: Research Advances in Sustainable Micro Irrigation
Senior Editor-in-Chief: Megh R. Goyal, PhD, PE

- **Volume 1:** Sustainable Micro Irrigation: Principles and Practices
- **Volume 2:** Sustainable Practices in Surface and Subsurface Micro Irrigation
- **Volume 3:** Sustainable Micro Irrigation Management for Trees and Vines
- **Volume 4:** Management, Performance, and Applications of Micro Irrigation Systems
- **Volume 5:** Applications of Furrow and Micro Irrigation in Arid and Semi-Arid Regions
- **Volume 6:** Best Management Practices for Drip Irrigated Crops
- **Volume 7:** Closed Circuit Micro Irrigation Design: Theory and Applications
- **Volume 8:** Wastewater Management for Irrigation: Principles and Practices
- **Volume 9:** Water and Fertigation Management in Micro Irrigation
- **Volume 10:** Innovation in Micro Irrigation Technology

Book Series: Innovations and Challenges in Micro Irrigation
Senior Editor-in-Chief: Megh R. Goyal, PhD, PE

- Engineering Interventions in Sustainable Trickle Irrigation: Water Requirements, Uniformity, Fertigation, and Crop Performance
- Management Strategies for Water Use Efficiency and Micro Irrigated Crops: Principles, Practices, and Performance

- Micro-Irrigation Engineering for Horticultural Crops: Policy Options, Scheduling, and Design
- Micro-Irrigation Management: Technological Advances and Their Applications
- Micro-Irrigation Scheduling and Practices
- Performance Evaluation of Micro-Irrigation Management: Principles and Practices
- Potential of Solar Energy and Emerging Technologies in Sustainable Micro-Irrigation
- Principles and Management of Clogging in Micro-Irrigation
- Sustainable Micro-Irrigation Design Systems for Agricultural Crops: Methods and Practices

Book Series: Innovations in Agricultural and Biological Engineering
Senior Editor-in-Chief: Megh R. Goyal, PhD, PE

- Advanced Research Methods in Food Processing Technologies
- Advances in Food Process Engineering: Novel Processing, Preservation and Decontamination of Foods
- Advances in Green and Sustainable Nanomaterials: Applications in Energy, Biomedicine, Agriculture, and Environmental Science
- Advances in Sustainable Food Packaging Technology
- Analytical Methods for Milk and Milk Products, 2-volume set:
    - Volume 1: Sampling Methods, Chemical and Compositional Analysis
    - Volume 2: Physicochemical Analysis of Concentrated, Coagulated and Fermented Products
- Biological and Chemical Hazards in Food and Food Products: Prevention, Practices, and Management
- Bioremediation and Phytoremediation Technologies in Sustainable Soil Management, 4-volume set:
    - Volume 1: Fundamental Aspects and Contaminated Sites
    - Volume 2: Microbial Approaches and Recent Trends
    - Volume 3: Inventive Techniques, Research Methods, and Case Studies
    - Volume 4: Degradation of Pesticides and Polychlorinated Biphenyls
- Dairy Engineering: Advanced Technologies and Their Applications
- Developing Technologies in Food Science: Status, Applications, and Challenges

- Emerging Technologies in Agricultural Engineering
- Engineering Interventions in Agricultural Processing
- Engineering Interventions in Foods and Plants
- Engineering Practices for Agricultural Production and Water Conservation: An Interdisciplinary Approach
- Engineering Practices for Management of Soil Salinity: Agricultural, Physiological, and Adaptive Approaches
- Engineering Practices for Milk Products: Dairyceuticals, Novel Technologies, and Quality
- Enzyme Inactivation in Food Processing: Technologies, Materials, and Applications
- Field Practices for Wastewater Use in Agriculture: Future Trends and Use of Biological Systems
- Flood Assessment: Modeling and Parameterization
- Food Engineering: Emerging Issues, Modeling, and Applications
- Food Process Engineering: Emerging Trends in Research and Their Applications
- Food Processing and Preservation Technology: Advances, Methods, and Applications
- Food Technology: Applied Research and Production Techniques
- Functional Dairy Ingredients and Nutraceuticals: Physicochemical, Technological, and Therapeutic Aspects
- Handbook of Research on Food Processing and Preservation Technologies, 5-volume set:
    - Volume 1: Nonthermal and Innovative Food Processing Methods
    - Volume 2: Nonthermal Food Preservation and Novel Processing Strategies
    - Volume 3: Computer-Aided Food Processing and Quality Evaluation Techniques
    - Volume 4: Design and Development of Specific Foods, Packaging Systems, and Food Safety
    - Volume 5: Emerging Techniques for Food Processing, Quality, and Safety Assurance
- Modeling Methods and Practices in Soil and Water Engineering
- Nanotechnology and Nanomaterial Applications in Food, Health, and Biomedical Sciences
- Nanotechnology Applications in Agricultural and Bioprocess Engineering: Farm to Table

- Nanotechnology Applications in Dairy Science: Packaging, Processing, and Preservation
- Nanotechnology Horizons in Food Process Engineering, 3-volume set:
    - Volume 1: Food Preservation, Food Packaging and Sustainable Agriculture
    - Volume 2: Scope, Biomaterials, and Human Health
    - Volume 3: Trends, Nanomaterials, and Food Delivery
- Novel and Alternative Methods in Food Processing: Biotechnological, Physicochemical, and Mathematical Approaches
- Novel Dairy Processing Technologies: Techniques, Management, and Energy Conservation
- Novel Processing Methods for Plant-Based Health Foods: Extraction, Encapsulation and Health Benefits of Bioactive Compounds
- Novel Strategies to Improve Shelf-Life and Quality of Foods: Quality, Safety, and Health Aspects
- Phytochemicals and Medicinal Plants in Food Design: Strategies and Technologies for Improved Healthcare
- Processing of Fruits and Vegetables: From Farm to Fork
- Processing Technologies for Milk and Milk Products: Methods, Applications, and Energy Usage
- Quality Control in Fruit and Vegetable Processing: Methods and Strategies
- Scientific and Technical Terms in Bioengineering and Biological Engineering
- Soil and Water Engineering: Principles and Applications of Modeling
- Soil Salinity Management in Agriculture: Technological Advances and Applications
- State-of-the-Art Technologies in Food Science: Human Health, Emerging Issues and Specialty Topics
- Sustainable and Functional Foods from Plants
- Sustainable Biological Systems for Agriculture: Emerging Issues in Nanotechnology, Biofertilizers, Wastewater, and Farm Machines
- Sustainable Nanomaterials for Biomedical Engineering: Impacts, Challenges, and Future Prospects
- Sustainable Nanomaterials for Biosystems Engineering: Trends in Renewable Energy, Environment, and Agriculture
- Technological Interventions in Dairy Science: Innovative Approaches in Processing, Preservation, and Analysis of Milk Products

- Technological Interventions in Management of Irrigated Agriculture
- Technological Interventions in the Processing of Fruits and Vegetables
- Technological Processes for Marine Foods, From Water to Fork: Bioactive Compounds, Industrial Applications, and Genomics
- The Chemistry of Milk and Milk Products: Physicochemical Properties, Therapeutic Characteristics, and Processing Methods

ABOUT SENIOR-EDITOR-IN-CHIEF

Megh R. Goyal, PhD, PE

Retired Professor in Agricultural and Biomedical Engineering, University of Puerto Rico, Mayaguez Campus; Senior Acquisitions Editor, Biomedical Engineering and Agricultural Science, Apple Academic Press, Inc.

Megh R. Goyal, PhD, PE, is a Retired Professor in Agricultural and Biomedical Engineering from the General Engineering Department in the College of Engineering at the University of Puerto Rico–Mayaguez Campus. During his professional career, he has worked as a Soil Conservation Inspector; Research Assistant at Haryana Agricultural University and Ohio State University; Research Agricultural Engineer/Professor at the Department of Agricultural Engineering of UPRM; and Professor of Agricultural and Biomedical Engineering in the General Engineering Department of UPRM. He spent a one-year sabbatical leave in 2002–2003 at the Biomedical Engineering Department of Florida International University, Miami, USA.

Dr. Goyal was the first agricultural engineer to receive a professional license in agricultural engineering from the College of Engineers and Surveyors of Puerto Rico. In 2005, he was proclaimed the "Father of Irrigation Engineering in Puerto Rico for the Twentieth Century" by the American Society of Agricultural and Biological Engineers, Puerto Rico Section, for his pioneering work on micro irrigation, evapotranspiration, agroclimatology, and soil and water engineering.

During his professional career of 52 years, he has received many awards, including Scientist of the Year, Membership Grand Prize for the American Society of Agricultural Engineers Campaign, Felix Castro Rodriguez Academic Excellence Award, Man of Drip Irrigation by the Mayor of Municipalities of Mayaguez/Caguas/Ponce and Senate/Secretary of Agriculture of ELA, Puerto Rico, and many others. He has been recognized as one of the experts "who rendered meritorious service for the development of [the] irrigation sector in India" by the Water Technology Centre of Tamil Nadu Agricultural University in Coimbatore, India, and ASABE who bestowed on him the 2018 Netafim Microirrigation Award.

Dr. Goyal has authored more than 200 journal articles and edited more than 100 books. AAP has published many of his books, including *Management of Drip/Trickle or Micro Irrigation; Evapotranspiration: Principles and Applications for Water Management;* ten-volume set on *Research Advances in Sustainable Micro Irrigation.* Dr. Goyal has also developed several book series with AAP, including Innovations in Agricultural & Biological Engineering (with over 60 titles in the series to date); Innovations and Challenges in Micro Irrigation; and Innovations in Plant Science for Better Health: From Soil to Fork.

Dr. Goyal received his BSc degree in Engineering from Punjab Agricultural University, Ludhiana, India, and his MSc and PhD degrees from the Ohio State University, Columbus, Ohio, USA. He also earned a Master of Divinity degree from the Puerto Rico Evangelical Seminary, Hato Rey, Puerto Rico, USA.

Readers may contact him at goyalmegh@gmail.com.

ABOUT THE EDITORS

N. Veena, PhD
Assistant Professor,
Department of Dairy Chemistry,
College of Dairy Science and Technology,
Guru Angad Dev Veterinary and
Animal Sciences University, Ludhiana–141004,
Punjab, India

N. Veena, PhD, is working as an Assistant Professor in the Department of Dairy Chemistry, College of Dairy Science and Technology, Guru Angad Dev Veterinary and Animal Sciences University, Ludhiana, Punjab, India.

She received her BTech degree in Dairy Technology from Karnataka Veterinary, Animal, and Fisheries Sciences University, Bidar, Karnataka, India, and her MTech and PhD degrees in Dairy Chemistry from ICAR-National Dairy Research Institute, Karnal, Haryana, India.

Her area of expertise includes fortification of milk, development and characterization of functional dairy foods, and nanoencapsulation of nutraceuticals and their delivery. She is the recipient of junior research and institutional fellowships during her master's and doctoral programs from the ICAR-National Dairy Research Institute, Karnal, Haryana, India. She has been involved in different institutional and externally funded research projects such as Milkfed (Punjab), ICAR, DST, and RKVY as PI and Co-PI.

She has published 17 research papers in peer-reviewed journals, one edited book, 11 book chapters, 15 popular articles, and seven practical manuals. She has developed the e-learning content for post-graduate dairy chemistry courses under an NAHEP project. She has presented her research work at various national and international seminars and conferences and received several awards for best oral and poster presentations. She holds membership in various professional bodies such as the Indian Dairy Association, Association of Food Scientists and Technologists (India) and Dairy Technology Society of India, etc.

Readers may contact her at: veena.ndri@gmail.com.

Ritesh B. Watharkar, PhD
Assistant Professor, College of Agriculture and Food Business Management,
MIT-ADT University, Rajbaugh,
Solapur–Pune Hwy, Near Bharat Petrol Pump,
Loni Kalbhor 412201, Maharashtra–India

Ritesh B. Watharkar, PhD, is currently working as an Assistant Professor in the College of Agriculture and Food Business Management at MIT-ADT University, Maharashtra, India.

He received his BTech degree in Food Technology from Dr. Balasaheb Sawant Konkan Agriculture University, Dapoli, Maharashtra, India; his MTech in Food Processing and Engineering degree is from the Karunya University, India; his PhD in Food Engineering and Technology degree from Tezpur University (A Central University), Assam, India.

He was formerly a Senior Research Fellow under a funded project by the Ministry of Food Processing Industries, New Delhi (2013–2015), and Assistant Professor at K.K. Wagh Food Tech College (2011–2013); at Shramshakti College of Food Technology, affiliated by Mahatma Phule Agricultural University (2019–2020).

During his professional career, he received awards such as senior research fellow under the UNESCO/China Great Wall Fellowship Program and worked for one year in the School of Biosystem Engineering and Food Science at Zhejiang University, Hangzhou, China. He was also awarded for best popular article from *Krishidut Magazine,* Maharashtra, India. Currently, he is a life member of the Association of Food Scientists and Technologists, India. He has published more than 10 journal articles and attended more than 10 national and international conferences. He has also published more than 10 popular articles in the *Marathi Literacy Magazine*.

Readers may contact him at: watharkarritesh2019@gmail.com.

CONTENTS

Contributors .... *xix*

Abbreviations .... *xxiii*

Preface .... *xxix*

PART I: Novel Methods for Foods Processing and Preservation 1

1. **Ultrasound Processing of Foods and Food Products 3**
   Deepika Kohli, Mahipal S. Tomar, and Rishi Richa

2. **Microwave Heating of Food Products 23**
   Gazia Nasir, Sadaf Zaidi, and Farhana M. Allai

3. **Potential of Nonthermal Technologies in the Food Industry 47**
   R. Kaavya, R. Pandiselvam, Sergio I. Martinez-Monteagudo, Anjineyulu Kothakota, N. U. Sruthi, and Kali Ramesh

4. **Thermophysical and Physicochemical Technologies for Drying of Food Products 69**
   Namrata A. Giri, T. Krishnakumar, G. Jeevarathinam, and M. S. Sajeev

5. **Irradiation of Foods and Food Products 93**
   Pavuluri Pavani, S. Subhashini, and S. Niveadhitha

PART II: Alternate Methods in Food Processing 113

6. **Mathematical Modeling in Food Processing 115**
   Sagar V. Gokhale, Jyoti S. Gokhale, and Smita S. Lele

7. **Structural Design of Foods Based on Laws of Physics 149**
   Palak Mahajan, Shubham Saraswat, and Manab B. Bera

8. **Applications of Biotechnological Approaches in the Food Industry: Review 177**
   Sreejit Valiathan and Preetha Radhakrishnan

9. **Valorization of Agro-Industrial Wastes Using Fermentation Technology 205**
   Mandira Kapri, Prem P. Srivastav, and Satyawati Sharma

PART III: Nutraceuticals and Encapsulation Methods in the Food Industry 235

10. Prospects of Citrus as a Nutraceutical Food 237
Upasana Deb and Sheena Haorongbam

11. Potential of Functional Lipids: Production, Properties, and Applications 253
Ankita Kataria, Bobade Hanuman, and Rajan Sharma

12. Encapsulation Technologies in The Food Industry 287
Nadira Anjum, Insha Zahoor, Meenakshi Trilokia, and Anju Bhat

13. Technical Characteristics for Encapsulation in the Food Industry 303
Carolina E. Demaman Oro, Marcelo L. Mignoni, Giovani L. Zabot, Geciane T. Backes, Rogério M. Dallago, and Marcus V. Tres

Index 321

CONTRIBUTORS

Farhana M. Allai
Department of Post-Harvest Engineering and Technology, Faculty of Agricultural Sciences, Aligarh Muslim University, Aligarh 202002, Uttar Pradesh, India
Department of Food Technology, Islamic University of Science and Technology, Awantipora 192122, Jammu & Kashmir, India; E-mail: faruallai@gmail.com

Nadira Anjum
Division of Food Science and Technology, SKUAST, Jammu, Chatha 180009, Jammu & Kashmir, India; E-mail: Nadira Anjum1736@gmail.com

Geciane T. Backes
Department of Food Engineering, URI Erechim, 1621 Sete de Setembro Av., Fátima, Erechi, RS 99709-910, Brazil; E-mail: gtoniazzo@uricer.edu.br

Manab B. Bera
Department of Food Engineering and Technology, Sant Longowal Institute of Engineering and Technology, Longowal 148106, Sangrur, Punjab, India; E-mail: drmbbera@yahoo.com

Anju Bhat
Division of Food Science and Technology, SKUAST, Jammu, Chatha 180009, Jammu & Kashmir, India; E-mail: anjubhat.skuastj@gmail.com

Rogério M. Dallago
Department of Food Engineering, URI Erechim, 1621 Sete de Setembro Av., Fátima, Erechim, RS 99709-910, Brazil; E-mail: dallago@uricer.edu.br

Upasana Deb
Department of Agribusiness Management and Food Technology, North Eastern Hill University, Tura Campus, Tura 794002, Meghalaya, India; E-mail: upasanadeb123@gmail.com

Namrata A. Giri
Division of Crop Utilization, ICAR-Central Tuber Crops Research Institute (CTCRI), Sreekariyam 695017, Thiruvananthapuram, Kerala, India; E-mail: namrata.AG@icar.gov.in

Jyoti S. Gokhale
Department of Food Engineering and Technology, Institute of Chemical Technology, Nathalal Parekh Marg, Matunga (East), Mumbai 400019, Maharashtra, India;
E-mail: js.gokhale@ictmumbai.edu.in

Sagar V. Gokhale
Ojman Foodbio, Survey No. 77/5 & 6, Behind Vishnu-Malati Industrial Estate, Shivane 411023, Pune, Maharashtra, India; E-mail: sagar.gokhale@ojmanfoodbio.in

Megh R. Goyal
Agricultural and Biomedical Engineering from College of Engineering at University of Puerto Rico Mayaguez Campus, Mayagüez, Puerto Rico
Agricultural and Biomedical Engineering for Apple Academic Press Inc.; PO Box 86, Rincon, Puerto Rico 00677-0086, USA; E-mail: goyalmegh@gmail.com

Sheena Haorongbam
Department of Agribusiness Management and Food Technology, North Eastern Hill University, Tura Campus, Tura 794002, Meghalaya, India; E-mail: sheenahaorongbam@gmail.com

Bobade Hanuman
Department of Food Science and Technology, Punjab Agricultural University, Ludhiana 141004, Punjab, India; E-mail: hanu.chk@gmail.com

G. Jeevarathinam
Department of Food Technology, Hindusthan College of Engineering and Technology, Coimbatore 641032, Tamil Nadu, India; E-mail: jeevathefoodengineer@gmail.com

R. Kaavya
Dairy and Food Science Department, South Dakota State University, Alfred Dairy Science Hall, Brookings 57007, South Dakota, USA; E-mail: kaavya.rathnakumar@sdstate.edu

Mandira Kapri
Indian Institute of Technology (IIT), Hauz Khas 110016, New Delhi, India; E-mail: kaprimandira@gmail.com

Ankita Kataria
Department of Food Science and Technology, Punjab Agricultural University, Ludhiana 141004, Punjab, India; E-mail: ankitakataria92@gmail.com

Deepika Kohli
Department of Processing and Food Engineering, CTAE, MPUAT, Udaipur 313004, Rajasthan, India; E-mail: deepikakohli27@gmail.com

Anjineyulu Kothakota
Agro-Processing & Technology Division, CSIR - National Institute for Interdisciplinary Science and Technology (NIIST), Trivandrum 695019, Kerala, India; E-mail: anjineyuluk@niist.res.in

T. Krishnakumar
Division of Crop Utilization, ICAR-Central Tuber Crops Research Institute (CTCRI), Sreekariyam 695017, Thiruvananthapuram, Kerala, India; E-mail: krishnakumar.T@icar.gov.in

Smita S. Lele
Department of Food Engineering and Technology, Institute of Chemical Technology, Nathalal Parekh Marg, Matunga (East), Mumbai 400019, Maharashtra, India; E-mail: ss.lele@ictmumbai.edu.in

Palak Mahajan
Department of Food Engineering and Technology, Sant Longowal Institute of Engineering and Technology, Longowal 148106, Sangrur, Punjab, India; E-mail: mahajanplk1995@gmail.com

Marcelo L. Mignoni
Department of Food Engineering, URI Erechim, 1621 Sete de Setembro Av., Fátima, Erechim, RS 99709-910, Brazil; E-mail: mignoni@uricer.edu.br

Sergio I. Martinez-Monteagudo
Dairy and Food Science Department, South Dakota State University, Alfred Dairy Science Hall, Brookings 57007, South Dakota, USA; E-mail: Sergio.MartinezMonteagudo@sdstate.edu

Gazia Nasir
Department of Post-Harvest Engineering and Technology, Faculty of Agricultural Sciences, Aligarh Muslim University, Aligarh 202002, Uttar Pradesh, India; E-mail: gazianasir@gmail.com

S. Niveadhitha
Department of Food Technology, Rajalakshmi Engineering College, Rajalakshmi Nagar, Thandalam 602105, Chennai, Tamil Nadu, India; E-mail: s.nivi333@gmail.com

Carolina E. Demaman Oro
Department of Food Engineering, URI Erechim, 1621 Sete de Setembro Av., Fátima, Erechim, RS 99709-910, Brazil; E-mail: carolinae.oro@hotmail.com

R. Pandiselvam
Biochemistry and Post-Harvest Technology Division, ICAR—Central Plantation Crops Research Institute, Kasaragod 671124, Kerala, India; E-mail: anbupandi1989@yahoo.co.in

Pavuluri Pavani
Department of Food Technology, FET, Jain (Deemed-to-be- University), Jain Global Campus, Jakkasandra, Kanakapura 562112, Ramanagara, Karnataka, India; E-mail: pavanipavuluri56@gmail.com

Preetha Radhakrishnan
Department of Food Process Engineering, School of Bioengineering,
College of Engineering and Technology, SRM Institute of Science and Technology, SRM Nagar, Kattankulathur 603203, Chengalpattu, Chennai, Tamil Nadu, India; E-mail: preethar@srmist.edu.in

Kali Ramesh
Department of Grain Science & Industry, Kansas State University, Manhattan 66502, Kansas, USA; E-mail: kaliramesh@ksu.edu

Rishi Richa
College of Agricultural Engineering & Technology, SKUAST, Kashmir 190025, Jammu & Kashmir, India; E-mail: rishiricha24@gmail.com

M. S. Sajeev
Division of Crop Utilization, ICAR-Central Tuber Crops Research Institute (CTCRI), Sreekariyam 695017, Thiruvananthapuram, Kerala, India; E-mail: ms_sajeev@rediffmail.com

Shubham Saraswat
Department of Food Engineering and Technology, Sant Longowal Institute of Engineering and Technology, Longowal 148106, Sangrur, Punjab, India; E-mail: sshubhamsaraswat@gmail.com

Rajan Sharma
Department of Food Science and Technology, Punjab Agricultural University, Ludhiana 141004, Punjab, India; E-mail: rajan-fst@pau.edu

Satyawati Sharma
Indian Institute of Technology, Hauz Khas 110016, New Delhi, India; E-mail: satyawatis@hotmail.com

Prem P. Srivastav
Indian Institute of Technology, Kharagpur 721302, West Bengal, India; E-mail: pps@agfe.iitkgp.ac.in

N. U. Sruthi
Department of Agricultural and Food Engineering, Indian Institute of Technology (IIT), Kharagpur 721302, West Bengal, India; E-mail: sruthisparks@gmail.com

S. Subhashini
Department of Food Technology, FET, Jain (Deemed-to-be- University), Jain Global Campus, Jakkasandra, Kanakapura 562112, Ramanagara, Karnataka, India; E-mail: s.subhashini10@gmail.com

Mahipal S. Tomar
Department of Food Processing Engineering, National Institute of Technology, Rourkela 769001, Odisha, India; E-mail: mpstomar95@gmail.com

Marcus V. Tres
Laboratory of Agroindustrial Processes Engineering (LAPE), Federal University of Santa Maria, 1040 Sete de Setembro St., Center DC, Cachoeira do Sul, RS 96508-010, Brazil;
Telephone: +55-5137248417; E-mail: marcus.tres@ufsm.br

Meenakshi Trilokia
Division of Food Science and Technology, SKUAST, Jammu, Chatha 180009, Jammu & Kashmir, India; E-mail: meenakshi.trilokia@gmail.com

Sreejit Valiathan
Department of Food Process Engineering, School of Bioengineering, College of Engineering and Technology, SRM Institute of Science and Technology, SRM Nagar, Kattankulathur 603203, Chengalpattu, Chennai, Tamil Nadu, India; E-mail: sreejitv@srmist.edu.in

N. Veena
Department of Dairy Chemistry, College of Dairy Science and Technology, Guru Angad Dev Veterinary and Animal Sciences University, Ludhiana 141004, Punjab, India; E-mail: veena.ndri@gmail.com

Ritesh B. Watharkar
College of Management, Food and Agri. Management Department, MIT Art, Design and Technology University, Loni Kalbhor, Pune 412201, Maharashtra, India; E-mail: watharkarritesh2019@gmail.com

Giovani L. Zabot
Laboratory of Agroindustrial Processes Engineering (LAPE), Federal University of Santa Maria, 1040 Sete de Setembro St., Center DC, Cachoeira do Sul, RS 96508-010, Brazil; Telephone: +55-5137248419; E-mail: giovani.zabot@ufsm.br

Insha Zahoor
Department of Post-Harvest Engineering & Technology, Faculty of Agricultural Sciences, Aligarh Muslim University, Aligarh 202002, Uttar Pradesh, India; E-mail: insha.zahor@gmail.com

Sadaf Zaidi
Department of Post-Harvest Engineering and Technology, Faculty of Agricultural Sciences, Aligarh Muslim University, Aligarh 202002, Uttar Pradesh, India; E-mail: sadaf63in@yahoo.com

ABBREVIATIONS

AA	arachidonic acid
ABTS	2,2′-azino-bis(3-ethylbenzothiazoline-6-sulfonic acid)
AFM	atomic force microscopy
AIDS	acquired immunodeficiency syndrome
ALA	α-Linolenic acid
ALC	amylose-lipid complex
ANN	artificial neural network
BV	biological value
CAC	cholesterol absorbing capacity
CFM	confocal microscopy
CFU	colony forming unit
CHD	coronary heart diseases
CLA	conjugated linoleic acid
Co-60	cobalt-60
CP	conformational potential
Cs-137	caesium-137
CTD	cast tape drying
CWW	cheese whey wastewater
DHA	docosahexaenoic acid
DHGLA	di-homo-γ-linolenic acid
DNA	deoxyribonucleic acid
DPPH	2,2-diphenyl-1-picrylhydrazyl
DSC	differential scanning calorimetry
DW	dextrose water
Ea	activation energy
EBI	electron beam irradiation
EF	electric field
EFSA	European Food Safety Authority
EHD	electrohydrodynamic
EMR	electromagnetic radiation
EPA	eicosapentaenoic acid
ESEM	environmental scanning electron microscopy
ETA	eicosatetraenoic acid
EU	European Union

EV	excluded volume
FA	fatty acid
FAO	Food and Agriculture Organization
FD	freeze drying
FDA	Food and Drug Administration
FFA	free fatty acids
FHIP	Flory-Huggins interaction parameter
FHT	Flory-Huggins theory
FIR	far-infrared
FRAP	Ferric Reducing Antioxidant Power
FRB	fermented rice bran
FSD	food structural designing
FV	free volume
GA	genetic algorithm
GAIN	global alliance for improved nutrition
GC-MS	gas chromatography–mass spectrometry
GDRI	glucose dialysis retardation index
GHz	gigahertz
GI	geographical indicators
GI	gastrointestinal
GLA	gamma (γ)-linolenic acid
GM	genetically modified
GMP	good manufacturing practices
GOFA	4'-geranyloxyferulic
GRAS	generally recognized as safe
GT	Gordon Taylor
GTT	glass transition temperature
HDL	high-density lipoprotein
HIV	human immunodeficiency virus
HMF	5-hydroxymethylfurfural
HPD	heat pump dryer
HPP	high-pressure processing
HVED	high-voltage electric discharge
Hz	Hertz
IAEA	International Atomic Energy Agency
ICFI	International Council on Food Irradiation
ICGFI	International Consultative Group on Food Irradiation
IFFIT	International Facility for Food Irradiation Technology
IFIP	International Food Irradiation Project

IP	interaction potential
IR	infrared
J	Joule
JECFI	Joint Expert Committee on Food Irradiation
kGy	kilogray
KHz	kilohertz
kV	kilovolt
LA	linoleic acid
LAB	lactic acid bacteria
LCST	lower critical solution temperature
LCTG	long chain triglyceride
LDL	low-density lipoprotein
LLC	Lewis lung carcinoma cells
MAE	microwave-assisted extraction
MAFD	microwave-assisted freeze drying
MATS	microwave-assisted thermal sterilization
MAVD	microwave-assisted vacuum drying
MCFA	medium chain fatty acid
MCTG	medium chain triglyceride
MDS	molecular dynamic simulation
MeV	mega electronvolt
MF	microfiltration
mg	milligram
MHz	megahertz
ml	milliliter
MPa	megapascal
MRI	magnetic resonance imaging
MS	manosonication
MTS	manothermosonication
MW	molecular weight
$NaHCO_3$	sodium bicarbonate
NEFA	nonesterified fatty acids
NIR	near-infrared
nm	nanometer
NPU	net protein utilization
O/W	oil-in-water
O/W/O	oil-in-water-in-oil
OAC	oil absorption capacity
OD	osmotic dehydration

OECD	Organization for Economic Co-Operation and Development
OH	ohmic heating
OM	optical microscopy
PALS	positron annihilation lifetime spectroscopy
PATP	pressure-assisted thermal processing
PD	phase diagram
PEF	pulsed electric field
PEM	protein-energy-malnutrition
PLGA	poly(lactic-co-glycolic acid)
POTP	pressure-ohmic thermal processing
PPAR	peroxisome proliferator-activated receptor
PPO	polyphenol oxidase
PPP	pineapple peel powder
PT	phase transition
PUFA	polyunsaturated fatty acid
PVA	polyvinyl alcohol
PVP	polyvinyl pyrrolidone
R^2	correlation regression coefficient
RF	radio frequency
RFD	radio frequency drying
RTD	residence time distribution
RTE	ready-to-eat
RW	Refractance window
SARS-CoV-2	severe acute respiratory syndrome coronavirus-2
SAS	supercritical antisolvent
SC	swelling capacity
SCAC	sodium cholate adsorption capacity
SCM	soft condensed material
SCP	single cell protein
SD	state diagram
SDA	stearidonic acid
SDE	structural design engineering
SEM	scanning electron microscopy
SF	submerged fermentation
SG	starch and extracted gluten
SI	solubility index
SL	structured lipid
SM	soft matter

SPI	soy protein isolate
SSD	superheated steam dryer
SSF	solid state fermentation
TAG	triacylglycerol
TEM	transmission electron microscopy
TFC	total flavonoid content
TPC	total phenol content
TS	thermosonication
TS	transverse section
TVA	trans-vaccenic acid
UCST	upper critical solution temperature
UHVEF	ultra high-voltage electric fields
UNESCO	United Nations Educational, Scientific, and Cultural Organization
US	ultrasonication
UV	ultraviolet
VLDL	very low-density lipoprotein
W	watt
W/cm^2	watt per centimeter square
W/O	water-in-oil
W/O/W	water-in-oil-in-water
WAC	water absorption capacity
WHC	water holding capacity
WHO	World Health Organization
WI	Widom insertion
XRT	x-ray tomography
η	viscosity

PREFACE

Modern food processing is used to increase efficiency, improve food safety, increase shelf life, and enhance nutritional and sensory quality of the final product without compromising on the natural quality of foods. Increase in consumer demand for such foods has led to the exploration of novel and alternate food-processing applications in the food business.

This book attempts to illustrate various applications of novel and alternate food processing, preservation, and encapsulation methods. This book consists of three main sections. Section I—Novel Methods for Food Processing and Preservation explores the principles, operations of the process, benefits and techniques used, and recent developments in the applications of ultrasound processing, microwave heating, and advanced drying (like refractance window, heat pump, infrared, hybrid, superheated steam, and radio-frequency drying). It also discusses innovative nonthermal technologies (like pulsed electric field, membrane filtration, high-pressure processing, etc.) and irradiation as an aid to processing and preservation of foods.

Section II—Alternate Methods in Food Processing highlights the fundamental principles applicable in food process modeling. In this section, case studies are also discussed to illustrate applications of mathematical-modeling techniques. The concepts of food physics based on thermodynamics, food structural designing, structure measurement technique, method of construction, and understanding of state diagrams for studying the structural changes in food products are elaborated. This section also provides an overview of the scope and significance of food biotechnology, the benefits, and concerns of using biotechnology in food processing as well as microbial valorizing approach as a novel technique and its potential environmental and health benefits. The last section—Nutraceuticals and Encapsulation Methods in Food Industry overviews the nutraceutical components of citrus as well as production, techno-functional attributes, and application of functional lipids. It also explores the encapsulation techniques, characteristics of encapsulated materials, advantages, and application of various encapsulation methods in food processing.

This book has exceeded expectation due to the support of the contributing authors to this book who have been most valuable in this compilation. The names of the contributing authors are mentioned with their respective chapters. The editorial team appreciates the authors for their proficiency, pledge, and perseverance.

We thank the editorial staff at Apple Academic Press, Inc. for their valuable assistance, great support, and guidance throughout this project.

We appeal to readers to provide feedback that could improve the subsequent editions of this book.

We also thank our friends and families for their unlimited support, encouragement, love, and affection during the course of editing this volume. Finally and most importantly, we would like to commend our spouses, Mohan Kumar, Subhadra and Aishwarya, for their understanding and patience throughout this project.

—**Editors**

PART I

NOVEL METHODS FOR FOODS PROCESSING AND PRESERVATION

CHAPTER 1

ULTRASOUND PROCESSING OF FOODS AND FOOD PRODUCTS

DEEPIKA KOHLI, MAHIPAL S. TOMAR, and RISHI RICHA

ABSTRACT

Food standardization is increasing with an increase in the demand for ready-to-eat food. The conventional technologies accelerate the deterioration rate of food quality. Ultrasound technology is a novel method which is environment friendly and hardly damages the raw material. In this chapter, ultrasound processing of food products has focused on the history of ultrasound in the food processing sector, principle, generation of ultrasound, system used for ultrasound, combined treatment of ultrasound with other treatment, and application of ultrasound in the different unit operation of food processing like inactivation of microorganism and enzymes, mixing, homogenization, freezing, extraction, modification of food properties, cleaning and decontamination, cutting, filtration, drying, emulsification, importance of ultrasound processing in comparison with other processing methods.

1.1 INTRODUCTION

Nowadays, increased customer awareness of the nutritional value, quality, and safety became a key concern. So, modern food processing is needed with great advances to increase efficiency, enhance the quality (nutritional and sensory) of the final product, and improve food safety without compromising the natural quality to achieve a long shelf life. Heat treatment is the oldest method of food processing and preservation by deactivating the

Novel and Alternative Methods in Food Processing: Biotechnological, Physicochemical, and Mathematical Approaches. N. Veena, Megh R. Goyal, & Ritesh B. Watharkar (Eds.)

microorganisms but the flavor and nutritional profile of the product may also get degraded by this approach.[36]

The sound wave within frequency from 20 kHz to several MHz or above the human hearing limit is termed as ultrasound. In food processing, based on the few parameters like intensities and frequency range, the ultrasound application can be divided into two divisions: low energy (high frequency and low intensity) and high energy (low frequency and high intensity). The first type (frequency >100 kHz and intensity <1 W/cm^2) is commonly used for nondestructive analysis of food composition and the later type (intensity > 1 W/cm^2 and frequency between 20 and 100 kHz) is used to promote the chemical reaction, clean equipment, emulsion creation, and to physically disrupt the tissues.[5,63] This high intensity and low frequency (20–100 kHz) ultrasound is termed as the power ultrasound which is applied to food processing to perform the process such as drying, filtration, degassing, extraction, sterilization, and emulsification.[49] There is a low degree of degradation of nutritional quality by ultrasound processing which increases its application area to the variety of products like juices of strawberry, apple, tomato, orange, and many more.[11]

In this chapter, ultrasound for food processing is highlighted and discussed in detail as a technique to process and preserve the food items by minimally affecting the quality (nutritional and sensory) of a product.

1.2 HISTORY

In 1880, the piezoelectric effect phenomenon was discovered by Pierre and Jacques Curie which means that under the applied mechanical pressure, a plate cut from Rochelle salt crystal might penetrate the electrical potential. The reverse effect occurs on applying an alternate voltage across the particular plate of crystal. The linear dimension changes due to alternate voltage which results in mechanical vibration. Firstly, crystals were then used to generate and to detect the sound waves in 1881. The ceramic materials used in the transducer to produce ultrasound can produce piezo-electricity are lead titanate, lead zirconate, and barium titanate. The whistling, magnetostricting, and spark discharging in water are the other methods for mechanical vibration to generate ultrasound.[3]

Scientists in 1927 found possible use of power ultrasound, that is, high power and low-frequency form of the ultrasound. The developed power ultrasound was first used only in the process like cleaning of the surface and emulsification.[55] After the 1960s, the power ultrasounds were then accepted to use in other processing operations in the industry.[42] Therefore, it is now considered a promising and novel method in the industries of food processing. In both industry and academy viewpoints, it still has a wide range of possible scope for further development.[17]

1.3 GENERATION OF ULTRASOUND WAVES

The transducer is the main sole device to convert the electrical energy or mechanical energy to the sound energy at an ultrasonic frequency.[7] The transducers are mainly of three types:

1.3.1 PIEZOELECTRIC TRANSDUCER

It is the most common transducer as an electrostrictive device that uses a ceramic material like lead metaniobate, barium titanate, and lead zirconate titanate to generate ultrasound. It is 80–95% efficient to transfer raw energy into acoustic energy. It is less rugged in comparison to the magnetostrictive type of transducer device, but it cannot withstand for a long period of time to expose at a high temperature greater than 85°C.[7]

1.3.2 MAGNETOSTRICTIVE TRANSDUCER

It is electromechanical devices, uses ferromagnetic material that has magnetostriction property, that is, having the ability to change its dimension because of expansion and contraction in an applied magnetic field. The frequency is usually limited to less than 100 kHz. It is not that much more efficient as piezoelectric transducer due to the losses which are mainly because of the heat. It is only 60% efficient to transfer raw energy into acoustic energy but can withstand for a long time to expose at the high temperature.[7] Since it has the additional advantage of robust constructional design, so it can be used in the heavy-duty application.[63]

1.3.3 LIQUID DRIVEN TRANSDUCER

It is a system of liquid whistles in which cavitation is generated by the venturi effect. In this, liquid in high-pressure is forced to pass by a small orifice and while liquid emerges from the orifice, it is at the ambient pressure which causes the decompression and produces cavitation bubbles. This quick alternating pressure in the liquid and cavitation effect increases the degree of mixing of liquid and makes it an appropriate method to be used in operations like emulsification, homogenization, and mixing in the food processing industry.[7,63]

1.4 PRINCIPLE OF ULTRASOUND TECHNOLOGY IN FOOD PROCESSING

Similar to any sound waves, ultrasound also transmits as a sequence of cycle, that is, positive pressure compression and negative pressure rarefaction cycle. The liquid molecules get affected because of the difference in pressures. Voids are created if the force between liquid molecules is lower than the negative pressure (rarefaction cycle). The cavities or voids did not collapse completely because of the presence of vapors in the voids. These series of cycles increase the size of voids or cavity and forms an acoustic cavitation bubble.[29,51,63] Therefore, the main driving force is cavitation for ultrasound processing technology but uncontrolled cavitation should be avoided. This controlled cavitation is also already been used in medical science to deliver the drugs to the targeted organ to cure the disease of tissues.[24,48] Depending on the generation mode, it is categorized into four types, that is, hydro-dynamic, particle, optic, and acoustic.

In the processing of food products, physical or chemical changes are only efficiently resulted by hydrodynamic and acoustic cavitation.[27] On the basis of stability of generated cavitation bubble, it is classified into two types: stable and transient. The first type exists for a longer time, so have much time to come together for the formation of large bubbles while the latter type exists for a very little time so collapses frequently.[59] The thousands of stable and unstable bubbles formed in the targeted liquid. Because of the violent collapse of unstable bubbles, the temperature and pressure of the liquid are generated to about 5000 K and 50 MPa, respectively, and hence generate high-speed liquid jets. This high pressure and temperature and high-speed microjets are mainly the reason for the disruption of microbial cells and other processing applications in food.[51,63]

1.5 ULTRASOUND SYSTEM

The basic parts of this system to apply power ultrasound in foods are generator, transducer, and coupler. The generator is a mechanical or electronic oscillator.

The transducer is the main device to convert electrical energy or mechanical energy to the sound energy at the ultrasonic frequency. The coupler is the end part of the system, which transfers the generated ultrasonic vibrations to the target material or sample. Therefore, the available system only differs on the basis of generator design, transducer type, and reactor or coupler. The five ultrasonic systems are known which are explained in this section.[7]

1.5.1 ULTRASONIC BATHS

It is basically used in the laboratory, as it is cheaper as compared to the probe type and also readily available. In this, the transducers to convert the energy are usually underside of the vessel. It is operated at around the frequency of 40 kHz. The high intensities can be produced only as a result of the reflection of sound waves in the interface of the liquid or air which develops standing waves. To maintain the high intensity, the depth of liquid is one of the important criteria. So, it should be maintained equivalent to the half of wavelength of sound in liquid. If the level of water in the vessel is below the mentioned limit, then the generation of high intensities is not possible. So, if these conditions are maintained properly then the high activity zone can be created by the bath system of the ultrasonic.

1.5.2 ULTRASONIC PROBES

In this type of system to generate energy, the detachable "horn" is attached to the transducers. The length of the horn or probe is generally half (or multiples) wavelength. The gain of amplitude is dependent on shape and diameter difference between the driven and emitting face of the horn. If the probe diameter is uniform to length, then there is no gain of amplitude and it will only transfer the acoustic energy to media.

1.5.3 PARALLEL VIBRATING PLATES

The mechanical effect can be maximized by two opposing vibrating plates instead to single surface. The vibration energy is reverberated by opposite plates before absorbed by liquid. The plates vibrate at the different (20 kHz and 16 kHz) beat frequency which creates large numbers of different stable and unstable cavitation bubbles.

1.5.4 RADIAL VIBRATING SYSTEM

In this, the liquid-flowing pipe itself is used as a vibrating source. This system can handle high viscous fluids and flow rates. The cylindrical resonating pipe helps to focus the ultrasound at the center of tube which results in high energy at the center. This helps to emit low power at the surfaces which finally reduces the chances of erosion problems at the emitter surface.

1.5.5 AIRBORNE POWER ULTRASOUND TECHNOLOGY

Air or gas has high acoustic absorption capacity and low density which create problem to efficiently generate and transmit the ultrasound energy. So, it requires high-power capacity, highly focused or directional radiation, vibration of large amplitude, and good impedance match. Sirens and whistles were the ultrasonic devices used in the air before the late 1980s and after that stepped-plated transducer was then developed. In stepped-plate transducers, a flexible radiating plate of large diameter with stepped-profile is driven at the center by a piezoelectric device. This device is used in operations like dehydration, defoaming of cans liquid, agglomeration of particles, mass transfer, and gas sterilization.

1.6 ULTRASOUND-BASED COMBINED TECHNOLOGIES

The efficiency of ultrasound processing can be improved by combining it with other food preservation treatments. Many studies have been therein this approach to combine ultrasound with either temperature, pressure, or temperature and pressure.[12,29]

1.6.1 ULTRASONICATION (US)

It is the technique in which ultrasound is combined with low temperature. So, it is mostly applied to process temperature-sensitive product. It minimally affects nutrients like ascorbic acid, nonenzymatic browning, and denaturation of protein. It requires a long period exposure to kill the microorganisms or inactivate the enzyme.

1.6.2 THERMOSONICATION (TS)

It is a technique to combine low-frequency ultrasound with the moderate heat. It gives greater efficiency of inactivation of microorganism in comparison to heat alone and also helps to reduce the processing time by 55% and processing temperature by 16%. The sensory quality of final product is good in this combined technique.

1.6.3 MANOSONICATION

It is a technique to combine ultrasound with pressure at low temperature to inactivate the microoragnisms or enzymes. At the same temperature, it gives inactivation efficiency higher than the ultrasound alone.

1.6.4 MANOTHERMOSONICATION

It is the combination of ultrasound with heat and pressure. It gives higher efficiency to inactivate the enzymes and to kill the microorganisms. This treatment can inactivate some enzymes in short time and at low temperature than conventional thermal treatments. The high thermos tolerance microorganisms and thermoresistant enzymes can also be inactivated by this treatment.

1.7 APPLICATIONS OF ULTRASOUND TECHNOLOGY

Figure 1.1 shows the various applications of the ultrasound processing, which are also discussed here in brief. The application of ultrasound can be first categorized on the basis of application mode which includes by submerging in an ultrasonic bath, by coupling with the device, and by direct application to the product. Classification based on the sound waves intensity includes low- and high-power ultrasound.[12]

- The low-power ultrasound waves have high frequency and low intensity (>100 kHz and <1 W/cm^2), which is basically used for nondestructive analytical measurement to monitor the compositions and for analysis of physicochemical properties.
- The high-power ultrasound waves have low frequency and high intensity (from 20 Hz to 100 kHz and >1 W/cm^2) which have the ability to

create cavitation bubbles to safely preserve and process the food by impacting to biological, chemical, and physical properties of food.

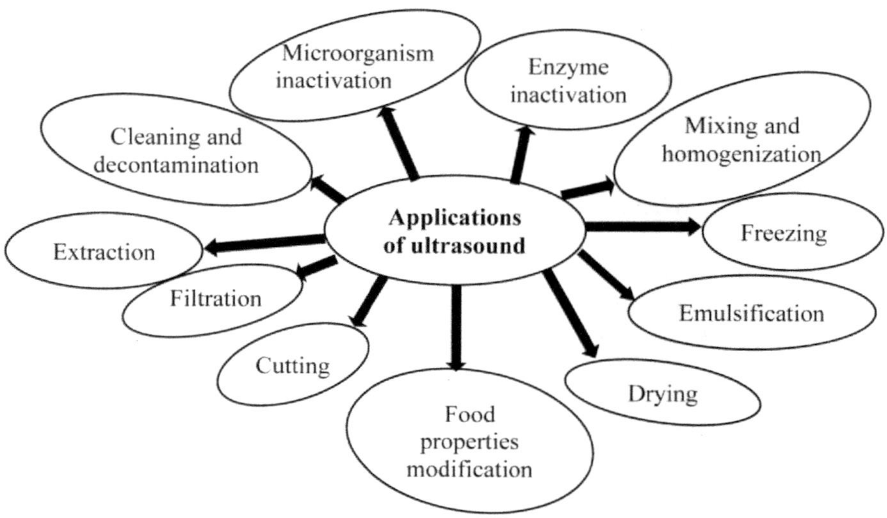

FIGURE 1.1 Applications of the ultrasound processing.

1.7.1 MICROBIAL AND ENZYMATIC ACTIVITY

Two factors, that is, microbial and enzymatic activities are mostly responsible for the deterioration of the food. By reducing these activities, the preservation of food is possible. Besides microbial safety, now food industries are more focusing on the increase in shelf life without or less affecting the quality, color, and flavor of foods. Ultrasound is a novel and nonthermal processing technique to change the microbial and enzymatic activities of food material which affects the various quality of the food, that is, color changes, texture changes, and mostly affect the chemical changes.[13,47] When ultrasound waves are passed through the liquid and semisolid product or liquid medium containing the solid foods, it creates the mechanical and cavitation effect in liquid medium by the sonication process. These effects lead to the various microbial and enzyme inactivation principles, that is, disruption or permeation of cell wall membrane, DNA of cell walls, denaturation of protein, etc.

The cavitation effect of liquid molecules generates the chemical species (free radicals) which is responsible for enzyme and microbes inactivation. Inactivation efficiency of microbes and enzyme also depends on various parameters, that is, ultrasonic wave amplitude, treatment time, ultrasonic power, temperature, etc., out of which the ultrasonic intensity and temperature are highly influencing parameters.[47] Its application is continuously increasing for the preservation of foods because it can inactivate the several microorganism and enzymes at room temperature. To achieve a higher inactivation of microorganisms, it can also be used in combination with high or medium temperature (thermosonication), pressure (manosonication), and combination of both temperature and pressure (manothermosonication).[12]

Ultrasound treatment of food material can effectively destroy the various harmful pathogens such as *Staphylococcus aureus*, *Salmonella enteritidis*, *Escherichia coli*, and *Listeria innocua*, etc., and also inactivates enzymes like polyphenol oxidase, lipoxygenase, and peroxidase.[13,47] Ultrasonic bath treatment with mild heat creates an effective lethality to inactivate *Salmonella typhimurium* microbes inoculated in the liquid whole egg and skim milk. Survivability of *S. typhimurium* is lowest at 50°C temperature and 30 min of treatment time.

Based on *S. typhimurium* reduction, it suggests that cavitation induced by ultrasonic waves can kill the Gram-negative bacteria in milk.[66] It efficiently inactivates the various microorganism such as *E. coli* and *Listeria monocytogenes* within 10 min. But this process is ineffective for reducing the activity of alkaline phosphatase and lactoperoxidase enzymes during the ultrasonic pasteurization of milk.[8] *Alicyclobacillus acidoterrestris* spore is highly resistant to the ultrasonic wave when it was treated with the ultrasound of 600 W power and 20 kHz for 30 min. For the inactivation of microbes present in the apple juice, the temperature of the process can be 30 ± 1 and 44 ± 1°C.[20] So, ultrasound processing alone is not too effective in the dairy and food industry for the inactivation of microbes and enzymes. For increasing the effectiveness, it can be used in combination of some other process as heat and pressure,[2] natural antimicrobial,[19] pulse light,[20] etc.

1.7.2 FREEZING

Freezing is one of the oldest and effective method for processing and preservation that maintains the quality and sensory characteristics of foods. It slows the biochemical and physiochemical processes that are responsible for the deterioration of foods. The freezing process converts the water present in food into ice crystal through the process of crystallization. The quality of frozen foods depends on the size of the developed ice crystal, freezing time, and rate of freezing. Slower freezing of foods develops larger ice crystal of needle shape in inside and outside of cells that damages the tissue of cells. To overcome this effect, ultrasonic freezing is an efficient process that influences the initiation of an effective size crystal nucleation and controls the crystal growth rate.[51]

Ultrasonic waves produce cavitational effect, which helps to form the primary nuclei of ice crystal. The collapse of gas bubbles generates the strong force that breaks the dendritic ice crystal into small size ice crystal which act as new nuclei.[57] These nuclei then with the help of cavitation promote the secondary nucleation formation through microjet generation.[31] Due to this effect of microjet produced by cavitation of gas bubbles, the heat transfer efficiency for freezing solution is improved. Hence it improves the freezing rate, quality of frozen foods, and reduces the freezing time between crystallization to complete ice formation.[14] It helps to produce small and uniform ice crystals. Ultrasonic waves minimize the microstructure changes of potato during the freezing process that maintains the better cellular structure of potato tissue. Ultrasound power of 15.84 W prevents the intercellular voids and cell disruption from freezing.[62]

1.7.3 DRYING

Drying is one of most effective method for reducing the water activity of foods and for increasing the shelf life of food products. Sun drying and hot air drying are most common method for reducing the moisture content of foods.[33,34] Several other technologies like vacuum, infrared, microwave, and radiofrequency are in current trends for drying of different food materials. In addition of these processes, ultrasound technology has gained the attention of researchers for drying process. It can be applied in three ways: air-borne sonication, direct contact of food, and as a pretreatment process.[45] Cavitation phenomena is beneficial for release of moisture from the food product.

High intensity of ultrasound wave in air creates the high microstreaming effect at the interface of product and air, which lead to reduced diffusion boundary layer during the mass transfer process. It then results in the enhancement of water diffusion process during the drying.[30] Ultrasonication by the acoustic effect of ultrasound waves also reduces the external and internal resistance during the process which is the most important hurdle for drying process.[45] Ultrasound treatment is used to reduce the required drying time and also preserves the quality of dried food material. Drying rate of food material in ultrasound technology depends on food material and emitters contact condition (direct and indirect). Drying rate is better in the case of direct contacting type ultrasound as compare to air-borne ultrasound.[30] Air-borne ultrasound in combination with air drying reduces the drying time by 63% (200 W and 70°C) for raspberries[35] and 44% for strawberries.[25] At constant temperature and air velocity conditions, drying time can also be significantly decreased by 50% for 45 W and 74% for 90 W.[54]

1.7.4 EXTRACTION OF COMPOUNDS

Extraction of compound from the fruit material is used in food processing industries. It is one of the prominent technologies to address the issues of cost, treatment, time, energy, and quality of extracted products. Due to the bubble cavitation effect of ultrasound waves, the disruption and permeability of cell wall membrane increases that help to release out the bound material from different food cell matrices. It increases the diffusion rate of compounds for extraction through the cell wall of the sample.

There are various processing parameters, that is, ultrasound frequency, ultrasound intensity, solvent to product ratio, treatment time and temperature which effects the extraction yield and quality of extracted compounds. Most commonly, the probe-type ultrasound is used for extraction in the laboratory. But the prolong use of probe-type ultrasound affects the volatile compounds present in food samples. Sometimes indirect contact type sonicator (bath type) is also used for extraction purpose. Ultrasound extraction process is a green and environment-friendly technology because it reduces the utilization of solvent for extraction. This technology is widely in use for cash crop as it improves the quality and purity of extracted bioactive compounds.[65] It extracts the compound in less extraction time.[42] Extraction yield of polyphenol compounds from *Nephelium lappaceum* L. fruit peel,[42] grape seed,[53] and orange peel[32] is higher in case of ultrasound-assisted extraction.

1.7.5 CUTTING

Cutting is used in food industries to cut the product like confectionary products, bakery products, candy bars, cheese, etc. In conventional method of cutting, the product losses are more, obtained final product can be crushed, distorted, and crumbled, and also the underweight or overweight cutting can be there.[44] Ultrasonic cutting of food produces the high quality and accurate final product. This technique uses the high-frequency vibratory knives. This vibration is applied to tool used for cutting creates the low friction cutting surfaces that reduce the resistance and hence can slice the food very thin and clean.[41] The edges of cutters used in ultrasonic cutter are less sharp than the mechanical knives, as the vibratory motion is main factor for the good cutting effect not the sharpness of the cutting tool.[38] The foods cut by ultrasound have reduced smearing, less deformation, and excellent cut faces. It can be used to cut the hard and soft cheeses, sandwiches, wraps, pizzas, breads or cake products, and other frozen snacks.[1,59]

1.7.6 FILTRATION

It is the process to separate the two products on the basis of size. Conventionally, membranes are used to filter the product which is available in wide range. After sometime, the cake is formed in front of the filter surface which acts as resistance to transfer the material effectively through the filter. So, in conventional filtration, clogging of membranes is one of main factor responsible for low efficiency.[28] To overcome this problem, ultrasonic filtration is in trend in the food industries. On applying power ultrasonics to slurry or liquid the phenomenon like cavitation, collapse of bubbles, and microjets are generated. This intense force modifies the flow rates and fluid dynamics.[36] The ultrasound also generates the vibrational energy which maintains the suspension of particles and allow particle to continuously move. This leaves the filter surfaces frictionless and free of congestion and facilitates the passage for the solvent to pass. Thus, it improves the flow and reduces the process time. The filter life is also increases by the use of ultrasound due to removed clogging and fouling materials.[37,61] Ultrasonic filtration is considerably more efficient than conventional filtration.

1.7.7 MIXING AND HOMOGENIZATION

Mixing is a unit operation in which the components that are in a partially mixed or unmixed state are treated to make the each component in contact with each other as much near as possible. So, mixing can achieve homogenization of mixture, emulsification of one component into another, and also suspends the solid in the liquid. Efficient mixing is dependent on two factors, that is, mixing time and residence time distribution.[26]

Homogenization is also the most required unit operation that reduces the particle size to distribute the particles evenly, small, and uniformly. Ultrasound is applied to liquid–liquid and solid–liquid suspensions to enhance the homogenization process.[15,18] The bubble formation, collapse of bubbles, high pressure and high-speed liquid jets, and cavity formation are common phenomenon of ultrasound which is responsible for homogenization. Ultrasound generates high hydraulic shear forces which break droplets, cell tissues, and agglomerates into small fragments which produces uniform and even-size product.[9] Low-power ultrasound level is not effective for homogenization, so adequate exposure time and power are needed to produce extremely small-sized fat globules. It can be concluded that for effective ultrasound homogenization increased power level is to be used but also consider the energy utilization in the process.

Ultrasonic homogenization can also be used very efficiently to reduce size of soft or hard particle. These homogenizers have a few number of moving and wetted part which reduces cleaning time and frictional wears. It also has the exact control on operational and other parameters, that is, amplitude and pressure which influence the cavitation and allows very gentle or even very destructive product processing.

1.7.8 EMULSIFICATION

Mixing of two immiscible phase like water and oil into an emulsion or homogeneous dispersion by using surface-active or emulsifying agent is termed as emulsification. Size of droplets in emulsions is between 10 and 1 mm. Energy input is required for both mechanical or ultrasound means of emulsification to form a small-sized droplet. The ultrasound at a frequency from 20 to 100 kHz is used to create emulsions.[4] The emulsification by ultrasound consists of two stages, first is instability of oil–water interface by ultrasound waves which erupts the droplets of dispersed phase into continuous phase and mixes of two phases. The second stage is physical effect after generation of cavitation bubbles.

The cavitation created by ultrasound releases microjets at high energy near the interface which facilitates the emulsification process.[16,40,58] This cavitation only occurs at certain minimum pressure amplitude of sound source and this is termed as cavitation threshold. Emulsification initiates in system of oil/water at cavitation threshold condition. Emulsion can also be obtained in the absence of emulsifying agent. The threshold cavitation pressure depends on the viscosity of liquid; it increases with increase in viscosity, so it is easier for less viscous liquid to undergo cavitation and to become continuous phase in emulsion.[23]

In comparison to mechanical agitation method, the ultrasonic emulsion requires less amount of emulsifying agent or surfactants and also produce more stable and small droplets.[16,40,58] It is also environment friendly and nontoxic method of emulsification. The factors which affects emulsification by ultrasound are: sonication time, power, frequency, and solution temperature.[10]

1.7.9 CLEANING

Cleaning is the important unit operation to maintain the hygienic condition and food safety of product. The pipelines and other equipment like heat exchangers can carry off the contaminating substances. The residue transfer from surface to product must be avoided.[48] To remove adherent residues and inactivate microorganisms, the traditional methods are based on chemical, mechanical, or thermal principles but all these methods consumes a good amount of time and energy. The ultrasonic cleaning separates adherents from the surface by using an aqueous medium for the transmission of sound. The detergent solution is used to remove the adherents. Similar to other application mechanisms, cavitation is responsible for cleaning or to remove the insoluble impurities. Cavitation occurs above the strong enough threshold power of ultrasound for breakage of molecular bonds between the impurities and surface.[48] Application of ultrasonics in cleaning can effectively reduce the cost, labor, and cleaning time. It can clean the surfaces of contact areas, glass bottles, and conveyor belts. The efficiency of ultrasonic cleaning depends on material to be cleaned; it is high for hard materials like stainless steels and lower for soft surface like plastic material. It can clean the areas that are difficult to clean manually and can directly inactivate microorganisms.

1.7.10 MODIFICATION OF FOOD PROPERTIES

The physical properties of food like consistency and texture affect the sensory quality and palatability of food.[21,46] Food structure is one of the main factor that affects stability, glycemic index, flavor release, and digestibility and bioavailability of nutrients of food products.[22] Ultrasound processing is used to modify the structure. There is physicochemical effect of ultrasound, due to the bubble cavitation process.[63] The collapse of cavitated bubbles extremely increases the pressure and temperature up to 50 MPa and 5000 K, respectively. The generated shear force and water molecules cleavage form hydrogen and hydroxyl free radicals. These free radicals are highly reactive, so responsible for cross-linkage of polysaccharides and toughen the tissues which results in chemical modification and structural changes of macromolecules.[6]

The stable cavitating bubble interacts with acoustic field, which creates the shock waves that result in strong shear stress and microstreaming and alters permeability of membrane of cells. All these changes thus facilitate biochemical reaction which modifies the structure of food.[55] The physical effects by cavitation cause change in structure due to disintegration of particles and macromolecules and also due to breakdown of interaction force of intermolecules that are weak, like Vander Waal interaction and hydrogen bonds. Parameters like power, intensity, frequency, and geometry influence the mechanism of modification. It can also be utilized for assessment of the composition of poultry, fish, meat products, protein analysis, and to detect adulteration in honey.

1.8 IMPORTANCE OF ULTRASOUND PROCESSING IN FOODS

Ultrasound is a green and emerging technology. It reduces the energy consumption and maximizes production. It can be used in any food type like flours, cocoa, meat, dairy, sugar, oils, fruits, vegetables, etc., and can be applied in modification in food properties, filtration, cutting, cleaning, emulsification, homogenization, mixing, extraction, drying, freezing, etc. All ultrasound-assisted processes are efficient enough in comparison to conventional methods. Low-power ultrasounds are attractive nonthermal method. It overcomes the problem of heat treatment like nutritional and organoleptic loss. It also reduces cost, labor, and energy with maximum productivity, value, and safety of the product. It is a relatively cheaper technology.

1.9 SUMMARY

Ultrasound processing has been used in wide range for food processing with enhanced final product properties and good energy and cost efficiency as compare to conventional methods. To aid the food processing, it has already been used in laboratory scale from many years but now, the food processing industry is also incorporating ultrasound in its process for better efficiency of the process. The present chapter indicates the successful use of ultrasound in the food processing, such as drying, freezing, extraction, inactivation of microorganisms, cleaning, cutting, emulsification, etc. The targeted use depends on frequency and intensity to create acoustic cavitation and the medium to generate the ultrasound. It results in less compromise for the nutritional and overall acceptance of final food product. Low-energy ultrasound is nondestructive technology, so can provide information of physicochemical property of food and high-energy ultrasound is useful for processing by physical disruptions and accelerated chemical reaction.

KEYWORDS

- **food preservation**
- **food processing**
- **food safety**
- **novel technology**
- **ultrasound**

REFERENCES

1. Arnold, G.; Leiteritz, L.; Zahn, S.; Rohm, H. Ultrasonic Cutting of Cheese: Composition Affects Cutting Work Reduction and Energy Demand. *Int. Dairy J.* **2009,** *19* (5), 314–320.
2. Arroyo, C.; Cebrián, G.; Pagán, R.; Condón, S. Synergistic Combination of Heat and Ultrasonic Waves under Pressure for *Cronobacter sakazakii* Inactivation in Apple Juice. *Food Control* **2012,** *25* (1), 342–348.
3. Arvanitoyannis, I. S.; Kotsanopoulos, K. V.; Savva, A. G. Use of Ultrasound in the Food Industry - Methods and Effects on Quality, Safety, and Organoleptic Characteristics of Foods: A Review. *Critic. Rev. Food Sci. Nutr.* **2017,** *57* (1), 109–128.
4. Ashokkumar, M.; Bhaskaracharya, R.; Kentish, S.; Lee, J.; Palmer, M.; Zisu, B. The Ultrasonic Processing of Dairy Products - An Overview. *Dairy Sci. Technol.* **2010,** *90* (2–3), 147–168.
5. Awad, T. S.; Moharram, H. A.; Shaltout, O. E.; Asker, D.; Youssef, M. M. Applications of Ultrasound in Analysis, Processing and Quality Control of Food: A Review. *Food Res. Int.* **2012,** *48* (2), 410–427.
6. Barteri, M.; Diociaiuti, M.; Pala, A.; Rotella, S. Low Frequency Ultrasound Induces Aggregation of Porcine Fumarase by Free Radicals Production. *Biophys. Chem.* **2004,** *111* (1), 35–42.
7. Brennan, J. G. (Ed.). *Food Processing Handbook.* Wiley-VCH Verlag GmbH & Co: Weinheim, 2006; p 607.
8. Cameron, M.; Lynn D. Mc.M.; Trevor J. B. Impact of Ultrasound on Dairy Spoilage Microbes and Milk Components. *Dairy Sci. Technol.* **2009,** *89* (1), 83–98.

9. Campoli, S. S.; Rojas, M. L.; do Amaral, J. E. P. G.; Canniatti-Brazaca, S. G.; Augusto, P. E. D. Ultrasound Processing of Guava Juice: Effect on Structure, Physical Properties and Lycopene In Vitro Accessibility. *Food Chem.* **2018,** *268*, 594–601.
10. Canselier, J. P.; Delmas, H.; Wilhelm, A. M.; Abismaïl, B. Ultrasound Emulsification: An Overview. *J. Dispers. Sci. Technol.* **2002,** *23* (1–3), 333–349.
11. Cao, X.; Cai, C.; Wang, Y.; Zheng, X. Effects of Ultrasound Processing on Physicochemical Parameters, Antioxidants, and Color Quality of Bayberry Juice. *J. Food Qual.* **2019,** *2019*, E-article 7917419.
12. Chemat, F.; Zill-E-Huma; Khan, M. K. Applications of Ultrasound in Food Technology: Processing, Preservation and Extraction. *Ultrason. Sonochem.* **2011,** *18* (4), 813–835.
13. Chen, F.; Zhang, M.; Yang, C. H. Application of Ultrasound Technology in Processing of Ready-to-Eat Fresh Food: A Review. *Ultrason. Sonochem.* **2020,** 63, E-article 104953.
14. Cheng, X.; Zhang, M.; Xu, B.; Adhikari, B.; Sun, J. The Principles of Ultrasound and Its Application in Freezing Related Processes of Food Materials: A Review. *Ultrason. Sonochem.* **2015,** *27*, 576–585.
15. da Silva, L. F. R.; Gomes, A. D. S.; Castro, D. R. G.; Souza, F. D. C. D. A.; Mar, J. M.; da Silva, L. S.; Sanches, E. A.; Bezerra, J. de A.; Bakry, A. M.; Campelo, P. H. Ultrasound-Assisted Homogenization and Gum Arabic Combined to Physicochemical Quality of Cupuaçu Juice. *J. Food Process. Preserv.* **2019,** *43* (9), E-article 14072.
16. Dapčević Hadnadev, T.; Dokić, P.; Krstonošić, V.; Hadnadev, M. Influence of Oil Phase Concentration on Droplet Size Distribution and Stability of Oil-in-Water Emulsions. *Eur. J. Lipid Sci. Technol.* **2013,** *115* (3), 313–321.
17. Dolatowski, Z. J.; Stadnik, J.; Stasiak, D. Applications of Ultrasound in Food Technology. *Acta Scient. Polon. Technol. Aliment.* **2007,** *63* (6), 89–99.
18. Fasaki, I.; Siamos, K.; Arin, M.; Lommens, P.; Van Driessche, I.; Hopkins, S. C.; Glowacki, B. A.; Arabatzis, I. Ultrasound Assisted Preparation of Stable Water-Based Nanocrystalline TiO_2 Suspensions for Photocatalytic Applications of Inkjet-Printed Films. *Appl. Catal. A* **2012,** *411–412*, 60–69.
19. Ferrante, S.; Guerrero, S.; Alzamora, S. M. Combined Use of Ultrasound and Natural Antimicrobials to Inactivate *Listeria Monocytogenes* in Orange Juice. *J. Food Protect.* **2007,** *70* (8), 1850–1856.
20. Ferrario, M.; Alzamora, S. M.; Guerrero, S. Study of the Inactivation of Spoilage Microorganisms in Apple Juice by Pulsed Light and Ultrasound. *Food Microbiol.* **2015,** *46*, 635–642.
21. Foegeding, E. A. Rheology and Sensory Texture of Biopolymer Gels. *Currt. Opin. Colloid Interface Sci.* **2007,** *12* (4–5), 242–250.
22. Fundo, J. F.; Quintas, M. A. C.; Silva, C. L. M. Molecular Dynamics and Structure in Physical Properties and Stability of Food Systems. *Food Eng. Rev.* **2015,** *7* (4), 384–392.
23. Gaikwad, S. G.; Pandit, A. B. Ultrasound Emulsification: Effect of Ultrasonic and Physicochemical Properties on Dispersed Phase Volume and Droplet Size. *Ultrason. Sonochem.* **2008,** *15* (4), 554–563.
24. Gallo, M.; Ferrara, L.; Naviglio, D. Application of Ultrasound in Food Science and Technology: A Perspective. *Foods* **2018,** *7* (10), 164–170.
25. Gamboa-Santos, J.; Montilla, A.; Cárcel, J. A.; Villamiel, M.; Garcia-Perez, J. V. Air-Borne Ultrasound Application in the Convective Drying of Strawberry. *J. Food Eng.* **2014,** *128*, 132–139.

26. Ghanem, A.; Lemenand, T.; Della Valle, D.; Peerhossaini, H. Static Mixers: Mechanisms, Applications, and Characterization Methods - A Review. *Chem. Eng. Res. Des.* **2014,** *92* (2), 205–228.
27. Gogate, P. R.; Kabadi, A. M. A Review of Applications of Cavitation in Biochemical Engineering/Biotechnology. *Biochem. Eng. J.* **2009,** *44* (1), 60–72.
28. Grossner, M. T.; Belovich, J. M.; Feke, D. L. Transport Analysis and Model for the Performance of an Ultrasonically Enhanced Filtration Process. *Chem. Eng. Sci.* **2005,** *60* (12), 3233–3238.
29. Guerrero, S. N.; Ferrario, M.; Schenk, M.; Carrillo, M. G. Hurdle Technology Using Ultrasound for Food Preservation. Chapter 3; In: *Ultrasound: Advances for Food Processing and Preservation*; Bermudez-Aguirre, D. (Ed.); Elsevier Science: London – United Kingdom, 2017; pp 39–99.
30. Huang, D.; Men, K.; Li, D.; Wen, T.; Gong, Z.; Sunden, B.; Wu, Z. Application of Ultrasound Technology in the Drying of Food Products. *Ultrason. Sonochem.* **2020,** *63*, 104950.
31. James, C.; Purnell, G.; James, S. J. A Review of Novel and Innovative Food Freezing Technologies. *Food Bioproc. Technol.* **2015,** *8* (8), 1616–1634.
32. Khan, M. K.; Abert-Vian, M.; Fabiano-Tixier, A. S.; Dangles, O.; Chemat, F. Ultrasound-Assisted Extraction of Polyphenols (Flavanone Glycosides) from Orange (*Citrus Sinensis*) Peel. *Food Chem.* **2010,** *119* (2), 851–858.
33. Kohli, D.; Shahi, N. C.; Kumar, A. Drying Kinetics and Activation Energy of Asparagus Root (*Asparagus Racemosus* Wild.) for Different Methods of Drying.*Curr. Res. Nutr. Food Sci.* **2018,** *6* (1), 191–202.
34. Kohli, D.; Shahi, N. C.; Pandey, J. P.; Singh A. Drying of Asparagus Roots in Solar and Fluidised Bed Dryer. *Int. J. Agric. Sci.* **2017,** *9* (13), 4072–4076.
35. Kowalski, S. J.; Pawłowski, A.; Szadzińska, J.; Łechtańska, J.; Stasiak, M. High Power Airborne Ultrasound Assist in Combined Drying of Raspberries. *Innov. Food Sci. Emerg. Technol.* **2016,** *34*, 225–233.
36. Kyllönen, H. M.; Pirkonen, P.; Nyström, M. Membrane Filtration Enhanced by Ultrasound: A Review. *Desalination* **2005,** *181* (1–3), 319–335.
37. Li, W.; Gamlath, C. J.; Pathak, R.; Martin, G. J. O.; Ashokkumar, M. Ultrasound – The Physical and Chemical Effects Integral to Food Processing. Chapter 1.21; In: *Innovative Food Processing Technologies*; Knoerzer, K. and Muthukumarappan, K. (Eds.); Elsevier: London, 2021; Vol. 1; pp 329–358.
38. Lin, S. Y.; Chung, C. T.; Cheng, Y. Y. Combination of Ultrasonic Vibration and Cryogenic Cooling for Cutting Performance Improvement of Inconel 718 Turning. *AIP Conf. Proc.* **2011,** *1315*, 1163–1168.
39. Liu, D.; Vorobiev, E.; Savoire, R.; Lanoisellé, J. L. Comparative Study of Ultrasound-Assisted and Conventional Stirred Dead-End Microfiltration of Grape Pomace Extracts. *Ultrason. Sonochem.* **2013,** *20* (2), 708–714.
40. Maali, A.; Mosavian, M. T. H. Preparation and Application of Nanoemulsions in the Last Decade (2000-2010). *J. Dispers. Sci. Technol.* **2013,** *34* (1), 92–105.
41. MacBeath, A. Ultrasonic Bone Cutting. PhD Thesis, University of Glasgow: UK, 2006; pp 170.
42. Maran, J. P.; Manikandan, S.; Nivetha, C. V.; Dinesh, R. Ultrasound Assisted Extraction of Bioactive Compounds from *Nephelium lappaceum* Fruit Peel Using Central Composite Face Centered Response Surface Design. *Arab. J. Chem.* **2017,** *10*, S1145–S1157.

43. Mason, T. J. Large Scale Sonochemical Processing: Aspiration and Actuality. *Ultrason. Sonochem.* **2000,** *7* (4), 145–149.
44. Mason, T. J. Power Ultrasound in Food Processing–the Way Forward. Chapter 6; In: *Ultrasounds in Food Processing*; Povey, M. J. W. and Mason, T. J. (Eds.); Blackie Academic & Professional: London – UK, 1998; pp 105–126.
45. Miano, A. C.; Rojas, M. L.; Augusto, P. E. D. Other Mass Transfer Unit Operations Enhanced by Ultrasound. Chapter 15; In: *Ultrasound: Advances in Food Processing and Preservation*; Bermudez-Aguirre, D. (Ed.); Academic Press: London–UK, 2017; pp 369–389.
46. Moelants, K. R. N.; Cardinaels, R.; Van Buggenhout, S.; Van Loey, A. M.; Moldenaers, P.; Hendrickx, M. E. Review on the Relationships between Processing, Food Structure, and Rheological Properties of Plant-Tissue-Based Food Suspensions. *Comprehens. Rev. Food Sci. Food Saf.* **2014,** *13* (3), 241–260.
47. O'Donnell, C. P.; Tiwari, B. K.; Bourke, P.; Cullen, P. J. Effect of Ultrasonic Processing on Food Enzymes of Industrial Importance. *Trends Food Sci. Technol.* **2010,** *21* (7), 358–367.
48. Otto, C.; Zahn, S.; Rost, F.; Zahn, P.; Jaros, D.; Rohm, H. Physical Methods for Cleaning and Disinfection of Surfaces. *Food Eng. Rev.* **2011,** *3* (3–4), 171–188.
49. Paliwal, S.; Mitragotri, S. Ultrasound-Induced Cavitation: Applications in Drug and Gene Delivery. *Expert Opin. Drug Deliv.* **2006,** *3* (6), 713–726.
50. Paniwnyk, L. Applications of Ultrasound in Processing of Liquid Foods: A Review. *Ultrason. Sonochem.* **2017,** *38*, 794–806.
51. Petzold, G.; Aguilera, J. M. Ice Morphology: Fundamentals and Technological Applications in Foods. *Food Biophys.* **2009,** *4* (4), 378–396.
52. Piyasena, P.; Mohareb, E.; McKellar, R. C. Inactivation of Microbes Using Ultrasound: A Review. *Int. J. Food Microbiol.* **2003,** *87* (3), 207–216.
53. Porto, C. D.; Porretto, E.; Decorti, D. Comparison of Ultrasound-Assisted Extraction with Conventional Extraction Methods of Oil and Polyphenols from Grape (*Vitis Vinifera* L.) Seeds. *Ultrason. Sonochem.* **2013,** *20* (4), 1076–1080.
54. Puig, A.; Perez-Munuera, I.; Carcel, J. A.; Hernando, I.; Garcia-Perez, J. V. Moisture Loss Kinetics and Microstructural Changes in Eggplant (*Solanum Melongena*) during Conventional and Ultrasonically Assisted Convective Drying. *Food Bioprod. Process.* **2012,** *90* (4), 624–632.
55. Raviyan, P.; Zhang, Z.; Feng, H. Ultrasonication for Tomato Pectinmethylesterase Inactivation: Effect of Cavitation Intensity and Temperature on Inactivation. *J. Food Eng.* **2005,** *70* (2), 189–196.
56. Richards, W. T.; Loomis, A. L. The Chemical Effects of High Frequency Sound Waves I. A Preliminary Survey. *J. Am. Chemic. Soc.* **1927,** *49* (12), 3086–3100.
57. Saclier, M.; Peczalski, R.; Andrieu, J. A Theoretical Model for Ice Primary Nucleation Induced by Acoustic Cavitation. *Ultrason. Sonochem.* **2010,** *17* (1), 98–105.
58. Santana, R. C.; Perrechil, F. A.; Cunha, R. L. High- and Low-Energy Emulsifications for Food Applications: A Focus on Process Parameters. *Food Eng. Rev.* **2013,** *5* (2), 107–122.
59. Schneider, Y.; Zahn, S.; Linke, L. Qualitative Process Evaluation for Ultrasonic Cutting of Food. *Eng. Life Sci.* **2002,** *2* (6), 153–157.
60. Show, K. Y.; Mao, T.; Lee, D. J. Optimisation of Sludge Disruption by Sonication. *Water Res.* **2007,** *41* (20), 4741–4747.

61. Sousa, L. dos S.; Cabral, B. V.; Madrona, G. S.; Cardoso, V. L.; Reis, M. H. M. Purification of Polyphenols from Green Tea Leaves by Ultrasound Assisted Ultrafiltration Process. *Seper. Purif. Technol.* **2016,** *168*, 188–198.
62. Sun, D.; Li, B. Microstructural Change of Potato Tissues Frozen by Ultrasound-Assisted Immersion Freezing. *J. Food Eng.* **2003,** *57* (4), 337–345.
63. Terefe, N. S.; Sikes, A. L.; Juliano, P. Ultrasound for Structural Modification of Food Products. Chapter 8; In: *Innovative Food Processing Technologies: Extraction Separation, Component Modification and Process Intensification*; Knoerzer, K.; Juliano, P. and Smithers, G. (Eds.); Woodhead Publishing: Oxford – UK, 2016, pp 209–230.
64. Tiwari, B. K.; Mason, T. J. Ultrasound Processing of Fluid Foods. Chapter 6; In: *Novel Thermal and Non-Thermal Technologies for Fluid Foods*. Cullen, P. J.; Tiwari. B. K. and Valdramidis, V. P. (Eds.); Academic Press: London – UK, 2012; pp 135–165.
65. Wen, C.; Zhang, J.; Zhang, H.; Dzah, C. S.; Zandile, M.; Duan, Y.; Ma, H.; Luo, X. Advances in Ultrasound Assisted Extraction of Bioactive Compounds from Cash Crops – A Review. *Ultrason. Sonochem.* **2018,** *48*, 538–549.
66. Wrigley, D. M.; Llorca, N. G. Decrease of Salmonella Typhlmurium in Skim Milk and Egg by Heat and Ultrasonic Wave Treatment. *J. Food Protect.* **1992,** *55* (9), 678–680.

CHAPTER 2

MICROWAVE HEATING OF FOOD PRODUCTS

GAZIA NASIR, SADAF ZAIDI, and FARHANA M. ALLAI

ABSTRACT

This chapter focuses on microwave heating and its mechanism including, the influence of microwaves on the chemical constituents of food material and their sensory characteristics. It sheds light on the microwave dielectric properties and the factors affecting the same. The mechanism of microwave drying along with the applications of microwaves in freeze drying, vacuum drying, osmotic dehydration, and infrared drying are elaborated. The mechanisms of pasteurization and sterilization assisted by microwaves are stated. This is followed by a discussion on microwave-assisted extraction, with a section detailing the factors that affect the performance of MAE.

2.1 INTRODUCTION

In food processing, microwave heating has numerous applications since decades like baking of food items, pasteurization, drying, tempering, and thawing. There are several beneficial features of microwave heating such as uniform heating, decrease in time of cooking, less maintenance, safe, as well as secure handling and operation. As analogous to heating by convention, microwave heating is efficient in retaining the nutritional and synergistic value of food to a larger extent while reheating and cooking.[1] The frequency of microwaves ranges from 300 MHz to 300 GHz. The operational frequency of microwave systems in industries ranges from 915 MHz to 2.45 GHz,

Novel and Alternative Methods in Food Processing: Biotechnological, Physicochemical, and Mathematical Approaches. N. Veena, Megh R. Goyal, & Ritesh B. Watharkar (Eds.)

whereas in domestic use, the operational frequency of microwave appliances is 2.45 GHz.[17]

In food processing, uninterrupted growth in innovation is necessary to fulfill the demand of consumers in order to have foods free from chemical additive, minimally processed, and are microbiologically safe. A volumetric heating method such as microwave heating has the advantage of delivering safe foods with increased sensory and nutritional properties over the conventional heating process as these methods have less rate of transfer of heat and have effects such as hot wall with cold center. At the global level in domestic, industrial, and commercial food processes, microwave heating has enormous applications.[5]

The benefits provided by microwave heating in food processes are a major function of its faster processing rates that are a result of effects such as volumetric heating that is linked with microwave power absorption. The interaction between dielectrics and electromagnetic fields gives rise to the absorption of microwaves by food substances. Hence, microwave processing of food substances not only relies upon dielectric properties but also depends upon sample thickness. At some specific thickness of the sample, the absorbed power gains resonance which leads to increased processing rates as suggested by earlier researchers.[8] During the processing of food, microwave heating is broadly employed in efficient and rapid heating of food material but it has the disadvantage of nonuniform heating that obstructs its other applications. The mode stirrer, variable frequency, and rotating turntable are the practical applications to increase the uniformity as suggested by certain engineering advances.

Mathematical modeling using computer simulation is commonly used to unfold the different changes that take place in various physical processes. It also assists the designing and optimizing of products by adopting microwave energy which is cost as well as time effective in order to enhance the uniformity during heating by microwaves. Sample's dielectric properties, position of the waveguide, sample size, cavity connection, shape, and speed of the conveyor are some of the many components that affect the heating uniformity in the case of microwave heating.[72]

In this chapter, microwave heating has been highlighted and discussed in detail as an appropriate method for processing of foods.

2.2 MICROWAVE HEATING

Microwave heating is widely studied and accepted as an emerging and latest technology because of its numerous benefits, such as less maintenance and plant space requirement, cleaner working environment, easy operation, and decreased processing time because of higher heating rates.[61]

2.2.1 MECHANISM OF MICROWAVE HEATING

The transformation of energy by electromagnetic field into thermal energy due to the polarization effect of the electromagnetic radiation is the mechanism of heating by microwaves. During microwave heating, electric fields having high frequency are used to convert the energy into the heat which in turn alters the properties of the food material involving dielectric properties. Energy generation and its conversion to heat occur when the energy absorption takes place by the dipolar molecules in the ionic constituents during heating by microwaves. The interaction of electric field constituents of electromagnetic radiation with ionic constituents of food material for heat generation depends on the dielectric characteristics. The food composition, temperature of operation, frequency of microwaves, and the moisture content are the key factors on which dielectric properties of food material are dependent.[42]

2.2.2 INFLUENCE OF MICROWAVES ON CHEMICAL CONSTITUENTS OF FOOD MATERIALS

Microwave heating affects the various chemical characteristics of food material like bioactive substances, loss of cooking, antioxidant capacity, antinutritional factors involving phytates, trypsin inhibitors, saponins, tannins and haemagglutinin activity, etc.[23] Several researchers have worked on the preservation of antioxidant activity and bioactive substances in various vegetables by using microwave heating. Akdaş and Bakkalbaşı investigated that treatment by microwaves without making use of water was found satisfactory for Kale cooking as the total ascorbic acid, total chlorophyll, and total carotenoid of Kale were retained at 89.4%, 44.7%, and 99.8%, respectively. However, water was added in a significant amount to decrease the level of nutrients to a greater extent.

Dolinsky et al.[21] studied that cooking by microwave using water was not an adequate method for some vegetables as it reduces the soluble and hydrolysable polyphenols in the case of green beans (reduction of 22.9%), kale and tomatoes with decrease of 23.4% and 21.9%, respectively. Steaming method was found to be very effective as osmotic exchange takes place and also helps in regulating and increasing the polyphenol concentration.

2.2.3 INFLUENCE OF MICROWAVES ON SENSORY CHARACTERISTICS OF FOOD MATERIALS

The alterations linked with the sensory characteristics of food substances can be seen in color and textural attributes of food substances in heating by microwaves. Throughout thermal processing, the changes in color of food substances can be affected by pigment degradation, browning both enzymatic and nonenzymatic and ascorbic acid oxidation.[34] Połtorak et al.[53] investigated that shrinkage rate value and the shear force of bovine gluteus medium muscle roasted using microwaves were considerably greater than roasted by conventional convective heating.

Additionally, Jouquand et al.[27] reported that beef burgundy cooked using microwaves was harder in textural attributes as compared to cooking using convective heating. Although Choi et al.[15] observed in their study that microwave heating had lowered the value of hardness in chicken steak than when prepared by grilling and boiling. Pellegrini et al.[51] observed that as analogous to steaming and boiling, heating by microwave was found to be the best method for retention of the color of frozen and fresh brassica vegetables such as cauliflower, Brussel sprouts and broccoli.

2.3 MICROWAVE DIELECTRIC PROPERTIES

The response of the matter when exposed to microwave field or radio frequency (RF) for the purpose of processing, heating, or drying relate to the dielectric properties of matter to a greater extent. Most beneficial quantities required in the thermal processes design in microwave heating are best described by the depiction of dielectric properties that are important for understanding the response of the matter to microwaves. Permeability (considering magnetic losses) and permittivity (considering electric losses) of the matter under investigation, constrained by the frequency of microwave used, are taken into consideration during the quantitative and macroscopic description of microwave–matter interactions. An appropriate illustration of the permittivity has been stated for a large number of systems involving heterogeneous solids, pure liquids, emulsions, crystals and mixture of liquids, etc. For these systems, the temperature and frequency reliance of the permittivity are known and the results are associated with the theoretical predictions.[25]

The reorientation phenomena and electric field-induced polarization are the main reasons for most of the matter and microwaves interaction that is chemically relevant. Practically, the extent of transformation of electromagnetic energy into heat depends upon the permittivity, γ^*, a complex number that has real and imaginary parts, as shown in Eq. (2.1):

$$\gamma^* = \gamma' + i\gamma'' \tag{2.1}$$

where γ' is the dielectric constant that denotes the capacity of the matter to be polarized by the application of an external electric field and therefore it is a relative measure of microwave energy density. It is indicated relative to the permittivity of free space γ'' by Eq. (2.2)

$$\gamma'' = \gamma_r \gamma_0 \tag{2.2}$$

where γ'' represents the loss factor that quantifies the efficiency of electromagnetic energy that gets transformed to heat.[37]

The loss tangent (tan δ) gives the ratio of the dissipative and capacitive behavior of matter; the dissipative behavior involves electrical conductivity. In some cases, this parameter along with the linear association of loss factor and dielectric constant is taken for losses. According to the simplified equation, the loss tangent tan δ is as mentioned below[64]:

$$\text{Tan}\ \delta = \frac{-\gamma''}{\gamma'} \tag{2.3}$$

The capacity of the materials that are heated can be easily correlated with tan δ; the higher the value the higher will be the capacity. The huge portion of energy imparted by microwaves will be proportionate to the material's magnetic and dielectric characteristics and also to the magnetic and electric field intensities.[65] But it needs to be taken into consideration that there are some disturbances like the alteration of dielectric characteristics along with chemical changes and temperature of the process.[67]

In addition to the moderation of the dielectric characteristics that prevail, the deepness at which the dissipation of power occurs, that is, the thickness of the heated layer also influences the microwave heating. Microwave irradiation can be categorized as localized, selective, and volumetric due to the reliance on the magnetic and dielectric characteristics of the material being considered. The presence of varying values of permittivity will cause a differential power density in the biphasic system or a compounded material and thus will result in an increase in temperature, for a specified electromagnetic field intensity of the material.[31]

2.3.1 FACTORS AFFECTING DIELECTRIC PROPERTIES KEEPING THE FREQUENCY CONSTANT

2.3.1.1 TEMPERATURE AND MOISTURE CONTENT

In most of the materials, the dielectric characteristics change with several varying factors. The quantity of water present inside the material is usually an influential factor in the case of hygroscopic materials like agri-foods. The components which influence the dielectric properties are the material temperature, structure, density, composition, and the applied alternating electric field frequency. The other constituent that affects the dielectric characteristics is the bulk density of air-particle mixture in the case of granulated materials. Certainly, the dielectric characteristics of the materials are based upon the chemical composition and in particular on the dipole moments that are combined with water and other molecules that makeup the concerned matter.

In most of the materials, the dielectric properties change significantly with the applied electric field frequency except in those substances that do not take up energy essentially from microwave fields and RF. An essential phenomenon promoting the dependence of frequency on the dielectric properties is the polarization that arises due to the orientation of molecules having permanent dipole moments with the applied electric field. Many researchers have reviewed that dielectric properties worked at dissimilar frequencies in the case of food products and agricultural practices. The dielectric constant may rise or fall with temperature depending upon the material. The presence of water in foods influences the microwave heating to a larger extent. The greater the moisture present, the greater will be the heating by microwave since water absorbs microwave energy in foods efficiently. The purest form of water is a classic example of a polar dielectric material.[66]

2.3.1.2 EFFECT OF COMPOSITION

The chemical composition of the food materials also determines the dielectric properties. The process by which salt (or ash) and water content are confined or obstructed in their motion by other components of food deduces their impact on dielectric properties. This makes the prediction more complex, and the data for dielectric characteristics of blend are deriving from data for particular ingredients. In comparison with water or aqueous ionic fluids with the organic components of food that are dielectrically inert (Ɣ'<3 and Ɣ'<0.1), the organic components may be determined explicit to energy.[39] The constituents having low specific heat become key factors in heating when the residual amount of water is confined and remains unaltered by readily changing microwave fields at a very less value of moisture.

2.3.1.3 PHYSICAL STRUCTURE OF THE OBJECT MATERIAL

The physical characteristics (such as homogeneity, bulk density, and particle size) are responsible for varying dielectric properties. Nelson[41] studied the correlation of dielectric properties of powdered and whole grains with varying frequency, bulk densities, and moisture content. It has been identified that bulk density is a key parameter of difference for Ɣ' and Ɣ".[40] The dependency of density on the dielectric properties of materials in elaborating functions should be considered for moisture content determination in grains.[38] In addition, to control the continuous on-line grain processing, this relation could be used.

2.3.1.4 DEPENDENCE ON DENSITY

Density has an influence on the dielectric characteristics as the effect of a dielectric substance depends on its interacting mass with the electromagnetic fields.[41] Particulate dielectrics like granular or pulverized materials are especially notable with this. The relationship of the dielectric characteristics of solid substances and the mixtures of air constituents, for example, pulverized or granular samples of such solids are essential to understand the nature of dependence of density on the dielectric characteristics of particulates. The dielectric characteristics of the solids may be required when only accessible form of the materials is particulate samples, in some cases. This was in agreement with cereal grains, in which kernels were very much smaller for the dielectric sample holders for estimation.[40]

2.4 MICROWAVE-ASSISTED TECHNOLOGIES

In food processing, microwaves are utilized to improve and enhance the traditional as well as nontraditional methods. Microwave-assisted technologies also improve the product quality development which is not so easily rendered in case of conventional methods. The addition of microwave energy leads to the development of good quality product, effective operational cost, and time and energy conservation that proves a boon for both technologies.[13]

2.5 MICROWAVE DRYING

2.5.1 MECHANISM OF MICROWAVE DRYING

Microwave drying is a complicated technique that involves the transfer of both heat and mass. Examples include microwave-assisted convective drying, vacuum drying, freeze drying (FD), far infrared (IR) combination drying, and hot-air microwave drying, based upon volumetric heating.[54] Inside a food material, vapors are produced and with the help of an internal pressure gradient it spreads. Due to the strong penetration of microwaves, both inner and outer part of the food is heated at similar time and the temperature increase takes place at the same time. Drying by microwaves transforms electromagnetic energy of higher frequencies into heat thus the moisture in the liquid form gets evaporated and is carried toward the surface of the food substance.[33] During microwave drying, two consecutive stages should be taken into consideration namely, liquid evaporation,[4] and drying; the latter consisting of the stages of heating, constant rate drying period, and falling rate drying period.[6]

2.5.2 MIGRATION AND DISTRIBUTION OF MOISTURE DURING DRYING

The phenomenon of effective moisture diffusion exhibits a mass transport property of water in the food substance throughout constant and falling rate periods of drying, involving vapor and liquid diffusion, vaporization–condensation, hydrodynamic flow, and other possible mass transfer processes. The effective moisture coefficient diffusion involves several factors that influence its composition, porosity, temperature, and moisture content of food substances as explained by Fick's law of diffusion which describes the technique of transfer of moisture content to the surface.[23]

2.5.3 EFFECT OF MICROWAVES ON QUALITY CHARACTERISTICS OF FOOD DURING DRYING

In microwave drying, the quality of food items changes to some extent. These changes involve optical characteristics, namely, appearance and color, sensory evaluation namely flavor, taste, and odor, structural features like density, rehydration properties, namely, rehydration capacity and rehydration rate, specific volume, porosity, textural, and nutritional characteristics namely, proteins and vitamins.[23]

2.6 MICROWAVE-ASSISTED DRYING

2.6.1 MICROWAVE-ASSISTED FREEZE DRYING (MAFD)

FD is a popular technique for dehydration of foods that are heat sensitive, (e.g., raspberries or tomatoes), hindering chemical composition, and stimulating an easier rehydration process. FD is a costly process and takes a longer time for drying. It involves higher energy costs and results in lower production.[22] Microwave radiation can easily be tuned with FD, known as microwave-assisted freeze drying (MAFD) in order to alleviate the shortcomings of FD. As analogous to conventional FD, the key features of MAFD are retention of volatiles, effective drying in the falling rate period, significantly reduced time of processing, high energy saving, and faster energy dissipation all around the material.[22]

2.6.2 MICROWAVE-ASSISTED VACUUM DRYING (MAVD)

Over the decades, MAVD has gained attention because of its application of reduced pressure and heating absorption volumetrically. MAVD employs microwaves in conjunction with vacuum drying. The major benefits of MAVD comprise of rapid dehydration that reduces the physical and chemical changes in products. MAVD has found its application in a range of products that include fruits and vegetables as compared to other products.[71] In a study, MAVD with hot air convective drying was used to investigate the quality attributes of thawed blueberries after drying. In comparison to the other methods used, MAVD exhibited higher DPPH

(2,2-diphenyl-1-picrylhydrazyl) radical scavenging activity, and better retention of anthocyanins was reported.[73]

2.6.3 MICROWAVE-ASSISTED OSMOTIC DEHYDRATION

The process of removal of water exploring the difference in concentration by the use of a semipermeable membrane that results in the condition of equilibrium on either ends is called as osmotic dehydration (OD).[1] It is a time-consuming process that takes 24–48 h in some mountings accompanied with changed water activity (a_w) values.[1] The foods containing higher levels of sugar and salt leads to a reduced level of a_w (below 0.90) resulting in unavailability of moisture in osmotically dewatered material. The advantages of OD include no need of latent heat of vaporization, moderate temperature needed for operation, and reduced time for the subsequent finish drying of the product thus leading to lowered quality damage. Microwaves can effectively be synergized with OD to furnish a faster process with increased even heating, resulting in a decreased time of dehydration, and increased uptake of solute by altering the dielectric features. In a study, Patel and Sutar[50] osmotically dried elephant foot yam in a vacuum pulsed microwave oven, thus resulting in good quality reduced moisture yam slices.

2.6.4 MICROWAVE-ASSISTED OHMIC HEATING

Ohmic heating (OH) is a process that involves the transformation of electrical energy into thermal energy by the implementation of an alternating current along a food product. This technique is 100% energy-efficient process.[58] In single-phase foods, OH has contributed significantly in the improvement of uniform heating of food products.[44] In particulate or multiphase foods, it is challenging to achieve uniform OH due to the consideration of electrical conductivity.

The association of microwaves with OH method allows the probability of microwave heating the particles of food without considering electrical conductivity. Apart from this, the electric current heats the liquid which increases the evenness of heating. Choi et al.[14] investigated the lags in temperature in particulate food by the combined application of OH and microwave. The experimental food taken was cubes of carrot in 0.1% NaCl solution. When the microwave heating and OH were applied separately, thermal lag took place between the liquid and the cubes of carrot at a temperature of 18.9°C and 11.2°C, respectively. The interaction between the OH and microwave heating showed that carrot cubes initially exhibited lead in temperature on the application of microwave irradiation, then the pattern was transposed during the second phase comprising OH. In the end of the process, the difference in temperature across the liquid and the carrot cubes ultimately dropped to less than 2°C.

2.6.5 *MICROWAVE-ASSISTED INFRARED (IR) DRYING*

Dehydration using IR has gained a lot of attention as a great alternate for dehydrating large number of agro-products. IR heating has the limitation of low penetration; therefore, the combined effect of microwave energy and IR radiation can overcome this limitation. Öztürk et al.[47] carried out a study involving the drying properties of banana and kiwifruit to examine the correlation between microwave–IR heating and dielectric properties. It was found that the synergistic effect of microwave and IR heating had a rapid rate of drying (saving roughly 98% time) with low range of ultimate moisture content (0.011–0.15 kg water/ kg dry matter) for either of the fruits in comparison with the traditional drying.

2.6.6 *MICROWAVE-ASSISTED ULTRASONICATION*

Currently, ultrasound finds numerous applications in food processes because of its valuable effects like aiding thermal treatments, improvement in food preserving methods, and beneficial effects on food quality parameters. Moreover, it decreases the cost of processing by eradicating few consequent purification steps, etc. Hence, the applications of ultrasound in food processes are diverse, namely, extraction, drying, freezing, and sterilization.[63] Although, it must be taken into consideration that physicochemical influence of ultrasonication might be deterioration of quality, in the form of depletion of components, off flavors, and alteration of physical parameters.[11]

2.7 MICROWAVE PASTEURIZATION

The purpose of doing pasteurization and sterilization is to kill or inactivate microorganisms in order to enhance the shelf life of food items.[45] The food items are kept up at a specific temperature for a specified time period so as to verify that pathogenic microorganisms are destroyed. In pasteurization process, bacteria and other pathogenic microorganisms are killed by the thermal action. Also, pasteurization inactivates undesirable enzymes that lead to the loss of cloud in some juices. By novel thermal (OH and RF) and nonthermal (high-intensity ultrasound, oscillating magnetic field, high hydrostatic pressure, ionizing radiation, UV treatment, and pulsed electric field) techniques, pasteurization can be achieved without influencing the nutritive value, color, or flavor of food items.[52]

2.7.1 MECHANISM OF MICROWAVE PASTEURIZATION

The theories that describe the deterioration of microorganisms or enzymes by RF or microwaves at sublethal temperatures are electroporation, magnetic field coupling, selective heating, and cell membrane rupture.[29] The theory of electroporation suggests that the electrical potential over the cell membrane forms pores, which leads to the leakage of cellular matter. The theory of magnetic field coupling suggests that because of the coupling of electromagnetic energy with critical molecules like DNA or proteins, the internal constituents of the cell are distorted. According to the theory of selective heating, microbes are preferentially heated because of microwaving and attain a higher temperature than the encompassing fluid which results in faster destruction of microbes. The cell membrane rupture theory suggests that due to the application of a voltage across the cellular membrane, the cell membrane gets disrupted.[29]

However, several theories showed that in the absence of pH and temperature or other nonthermal techniques, the energy is not sufficient to inactivate microbes. Yet a significant effect of thermal processes might have involved the use of microwave energy.[30]

2.7.2 PASTEURIZATION BY MICROWAVE OF FLUID STATE FOOD SUBSTANCES

The implementation of pasteurization by microwave technique is widely employed to fluid foods like milk and fresh juices. Pasteurization by microwaves of fruit juice and sterilized milk was studied by Salazar-Gonzalez et al.[56] Inactivation of enzymes and microbes of the variety of liquid foods, namely, apple cider, sweet potato puree, apple juice, coconut water, grape water juice, and milk were investigated. It was observed that the lethality of the process and the adequate conditions for the implementation of microwave energy can be attained according to the need with the help of information regarding dielectric characteristics.[56]

2.7.3 PASTEURIZATION BY MICROWAVE OF SOLID STATE FOOD SUBSTANCES

Microwaves can help to achieve in-shell egg pasteurization.[20] It was seen that there was no notable difference in the rate of heating in yolk and albumin during microwave heating despite this being known that the albumin had a higher value of dielectric properties than the yolk. The collaboration of egg size, geometry, and dielectric characteristics of the egg might cause boost in interior heating. Without compromising the eggshell, microwave pasteurization can be achieved as shown by Dev et al.[20] *Salmonella typhimurium* inactivation in the yolk of shell eggs was achieved through microwave pasteurization according to a different study.[57] Using microwave irradiation for 15 s, a 22% decrease in microbial load was obtained while using moist heat treatment a total decrease of 36% was obtained for 15 min.[57]

2.8 MICROWAVE STERILIZATION TECHNIQUE

People around the globe are more health conscious and are aware of the various food-borne diseases that are caused due to bacteria and other harmful microorganisms. Food-borne diseases have reportedly resulted in about 48 million illness, hospitalizations of 28,000, and death of 3000 persons on annual basis, excluding the unrevealed victims in the USA.[12] Enterococci, Salmonella, *Escherichia coli*, Shigella, *Listeria monocytogenes*, *Staphylococcus aureus*, and Campylobacter are the most severe concerns.[43] The temperature attained at different parts of the food, the frequency applied, and constituents of the food are the major causes for the production of heat in microwave-based food sterilization.

2.8.1 MECHANISM OF MICROWAVE STERILIZATION

Extending the storage life and improving the safety are the primary purpose of sterilization. Magnetic field coupling, selective heating, rupture of cell membrane, and electroporation are the mechanisms that explain microwave pasteurization or sterilization.[29] For instance, microbial bodies can attain a greater temperature as compared to the surrounding fluid, due to microwave selective heating which leads to the rapid destruction of microbes. According to the mechanism of electroporation, pores are generated due to the application of electric potential along the cell membrane that leads to the leaked cellular material. As far as the mechanism of magnetic field coupling is considered, essential components of the cell like the DNA or proteins having association with the electromagnetic field can be eliminated.

2.8.2 MICROWAVE STERILIZATION

The potential microbes found in food can be made ineffective using sterilization by microwaves and radiation by microwaves comprising water-assisted heating by microwaves. The colony count of food microorganisms can be decreased using microwave sterilization. The effectiveness of microwave sterilization could be improved by raising the sterilization temperature and sterilization power and prolonging the sterilization time.[65] The microbes in food can be substantially decreased using microwave sterilization. In general, various kinds of foods inoculate with various types and microbial level, which shows further the need for various microwave sterilizations. The deactivation of microbes in certain foods has been shown in Table 2.1.

2.8.3 MICROWAVE-ASSISTED THERMAL STERILIZATION

Microwave-assisted thermal sterilization (MATS) is an evolving technique that is an effective union of microwave and thermal energies. It is used to disinfect packaged food items thus inactivating microorganisms equivalent to thermal sterilization in addition to reducing the detrimental effect of heat on the food constituents.[61] In this system, temperature of the packaged food items is raised to 120°C using a hot water bath under pressure together with 915 MHz of microwave energy followed by rapidly cooling. Food and Drug Administration (FDA) of the USA has approved the use of MATS at the commercialized level for both homogenous and non-homogeneous food products.[62] In the years 2009, 2010, and 2012, FDA adopted this technique for mashed potatoes, salmon filet, and items of meat, respectively.[7,61] This technique has been evaluated by Tang et al.[62] and patented by Washington State University.

TABLE 2.1 Deactivation of Microorganisms in Various Foods or Microbiological Media by Microwave Pasteurization and Sterilization.

Food material	Pathogen	Method used	Power	Time	Temperature	Pathogen deactivation (Log reduction)	References
Frankfurters	*Listeria monocytogens*	Microwave pasteurization	550 W	2–19 min	65°C	0.41/min	27
Grape and tomatoes	*Salmonella enterica*	Microwave sterilization	700 W	50 s	–	>2.05	36
Jalapeno pepper	*Salmonella typhimuriam*	Water-assisted microwave sterilization	950 W	25 s	63°C	5.12	20
Milk	*Staphylococcus aureus*	Microwave Pasteurization	600 W	85–100 s	82–93°C	5	29

2.8.4 BENEFITS OF MICROWAVE-ASSISTED THERMAL STERILIZATION (MATS) OVER CONVENTIONAL RETORTING

The major difference between MATS and traditional microwave technology is because of its particular characteristic of utilizing water in relation to heat (for a marginal extent as analogous to retorting) following the procedure of microwaving.[7,61] The association of these technologies and the water used as a means would decrease the problem of uneven heating caused by microwave and also decrease the exposure time of the food to the lethal temperature analogous to traditional retorting. As analogous to retort processing, MATS has the capability of delivering a cook value 58 at the time of processing 9 min at 121.1°C while in the case of retort processing at similar temperature it takes 33 min, whereas at the same lethality it would take the cook value 39 as desirable at a cold spot as studied by Tang.[61] The MATS system is thus fashioned in order to attain uniform heating and processing time that provides less sensory losses, reduced exposure to heat, more retention of nutrients, and prolonged shelf life.

As analogous to conventional methods, microwave sterilization is an innovative technology which is recently popular among the researchers, although further analysis is required for exploring its potential and practical inferences in food industry. Certain key research areas in MATS are flexible packaging usage, clean label credentials, attaining fresh like quality, and the potential to treat food products higher in heat sensitivity. In addition to this, mathematical models are used to solve the issues of validations instead of using complex laboratory assessments. The attainment of fresh like quality is a major accomplishment of MATS due to lesser time of exposure to heat as compared to conventional retorting system.[60]

2.9 MICROWAVE-ASSISTED EXTRACTION (MAE)

Extraction depicts the principal step in obtaining crude extract from plant sources. The identification of active compounds and further analyses have to be done in the obtained extract. The conventional techniques such as water percolation, soaking, maceration, and Soxhlet extraction require prolonged extraction time that leads to the thermal destruction of thermolabile active components.[9] Since the last decades, different novel extraction techniques have been established and studied and it has been found that a great number of them have claimed to be better in efficiency, consumption of solvent, and time of extraction.[25,46,69]

Generally, MAE system can be subdivided into focused mode and multimode systems. The focused mode system is also known as mono-mode system. The focused (mono-mode) system permits only focused radiation in a confined area of cavity, whereas in a multimode system, microwave radiations disperse randomly in the cavity using mode stirrer. The focused mode system operates under atmospheric pressure or high pressure while the multimode system is usually linked to elevated pressure operation. In order to avoid uncertainty, MAE can be classified as "closed system" and "open system" that signifies operation above and under atmospheric pressure, respectively.[18,34]

The closed MAE system consists of a sealed container in which extractions are performed with varying the microwave intensity. Under uniform heating, the extraction normally occurs. Rapid and effective extraction takes place under high temperature and pressure. The temperature of the system is maintained higher than the boiling point of the solvent used in extraction while the pressure in the container should be regulated such that it does not overreach the working pressure of the container.[70] Although the closed system renders rapid and effective extraction with low-level consumption of solvent, labile components are vulnerable to losses due to the restricted throughput of the sample.

Due to the high temperature and pressure operating conditions, this system has safety issues. To overcome them the open system has been designed. It operates at milder conditions that have increased sample throughput and facilitates the addition of solvent in large amounts.[34]

2.9.1 FACTORS AFFECTING THE PERFORMANCE OF MAE

The effectiveness of MAE depends majorly upon the parameters and the conditions of operation influencing the extraction yield and mechanisms. Solvent to feed ratio, temperature, effect of stirring, nature of solvent, time of extraction, sample characteristics, and microwave power are the key factors that influence the performance of MAE. Thus, it is necessary to perceive the effects and functioning of these components on the process of MAE.

2.9.1.1 NATURE OF SOLVENT AND SOLVENT TO FEED RATIO

The selection of an appropriate solvent in the process of MAE relies largely on its dielectric constant, interaction with sample matrix, solvability of the target analyte, and solvent's transfixation. Water enhances the transfixation of solvent into the sample casting, which in turn improves the heating efficiency. Therefore, the aqueous solutions of some organic solvents are desirable for some extractions.[3] Certain other organic solvents are effective in extraction namely, acetone, methanol, and ethanol.

2.9.1.2 TEMPERATURE OF EXTRACTION AND POWER BY MICROWAVE

During extraction from plants, the increased power of microwaves might lead to inferior yield of extraction because of the destruction of thermally sensitive components. Usually, with the escalation of power of microwaves, the yield of the extraction increases in proportion up to a certain limit until the increase turns negligible.[36] Since the microwave power produces localized heating, therefore prompts MAE to break the plant matrix in order to diffuse out the analyte and disperse it in the solvent. Generally, the yield of extraction could be increased by increasing the power which in turn results in shorter time of extraction.

Alfaro et al.[3] investigated a new term called as energy density, which is the ratio of power and mass per unit of time, to study the influence of the power of microwaves on MAE, since power level solely does not provide adequate information with respect to absorption of microwave energy in to the system. They found that on destruction of the plant matrix by microwave radiation, the active components get discharged. Also, a higher level of power does not provide any significant benefaction between the solvent for extraction and the microwaves with the specimen.

2.9.1.3 EXTRACTION TIME AND CYCLE

Besides the collaborative influence of temperature, the impact of power of microwave could be expanded to the time of extraction. It was observed that the extraction yield decreased due to exposure even at less power of operation and less temperature. This was attributed to the destruction of active components.[24,68] Generally, the time of extraction in MAE is kept up to a maximum of half an hour to prevent oxidation and thermal degradation. In MAE without solvent, the time of extraction is prolonged to 1 h to necessitate the complete extraction of essential oil. Moreover, extended time of extraction in supercritical fluid microwave-assisted extraction is detrimental to the extracted hydrocarbons as they are under a grave risk of being hydrolyzed and oxidized.[69] In case of the need of longer time of extraction, the risk of thermal deterioration can be decreased via extraction cycle, which operates by repeating the extraction steps and by resorting to adding fresh solvent to the residue to make sure that the extraction process is complete.

2.9.1.4 CHARACTERISTICS OF PLANT MATRIX

The properties of the plant specimen also have an effect on the performance of MAE. For an optimal yield of extract, the extracted plant sample is generally dried and made in to very fine powder before the extraction. Due to the very fine size of particles, it is difficult to separate the residue from the extract. Hence additional steps for cleaning are applied. Further, the sample is treated with solvent for 90 min to improve the heating efficiency of MAE, enhance the movement of desired components to the solvent, and stimulate the diffusion process.[48] In some cases, it was found that prolonged time for pretreatment did not enhance the yield of extract because the active components might have leaked out from the sample casting prior to extraction.[49] The sample casting that is pretreated using water prior to drying aids in localized heating of microwave. Further, the extraction process is enhanced by heating up of moisture found in the sample, which gets evaporated and produces the internal stress within the cells which in turn erupts the cell wall and thus release of active components takes place that further increases the yield of the extract.[69]

2.9.1.5 STIRRING EFFECT

Stirring influences the mass transfer process significantly. By the introduction of stirring in MAE, the harmful effects of the ratio of lower solvent to feed on yield of extraction can be minimized. Further, the obstruction created in mass transfer by the concentration of the desired components in a specific region because of the insufficiency of the solvent can also be reduced that results in a higher yield of extract. It could be said that agitation speeds up the extraction process by speeding up the dissolution and desorption of active components adhering to the sample matrix.[55]

2.10 SUMMARY

Microwave heating finds profitable application in a number of food operations, namely, pasteurization, sterilization, drying, and the extraction of food items. In this chapter, illustrative studies on different food processing techniques assisted with microwaves have been highlighted. The proficiency in dielectric characteristics of heating by microwaves of food items is very advantageous. In microwave heating, shape, temperature, and uniformity are greatly influenced by position, shape, and size of food substances. Usually, the greater number of hot spots is generated in the center region of a food item as compared to the other regions.

The pasteurization by microwaves was found to be more efficient in the killing of pathogens or enzyme inactivation because of remarkable intensification of thermal effects. The combination of different drying methods with microwaves such as air, IR, freeze, or vacuum drying yielded better attributes of drying in comparison to their respective methods of drying or drying by microwaves alone. Drying by microwaves resulted in less energy consumption and yielded products with improved sensory characteristics. The elevated power of hot air heating by microwaves can effectively lead to reduced ultimate water content and recovery rate, thus enhancing the rate of drying, and shrinkage and rehydration of the samples consequently.

The expedient extraction technique, MAE has led to improves extraction in synergy with other modern technologies. The wide use of microwaving in food processing, necessitates more in depth studies in certain fields like the unevenness of microwave fields in order to attain even processing of foods, which result in hot and cold spots in food items to reduce the harmful effects of this technology for enhanced food safety and quality and to make it cost-effective.

KEYWORDS

- **dielectric properties**
- **drying**
- **extraction**
- **microwave mechanism**
- **pasteurization**
- **sterilization**

REFERENCES

1. Ahmed, I.; Qazi, I. M.; Jamal, S. Developments in Osmotic Dehydration Technique for the Preservation of Fruits and Vegetables. *Innov. Food Sci. Emerg. Technol.* **2016,** *34*, 29–43.
2. Akdaş, Z. Z.; Bakkalbaşı, E. Influence of Different Cooking Methods on Color, Bioactive Compounds, and Antioxidant Activity of Kale. *Int. J. Food Propert.* **2016,** *20* (4), 877–887.
3. Alfaro, M. J.; Bélanger, J. M. R.; Padilla, F. C.; Paré, J. R. J. Influence of Solvent, Matrix Dielectric Properties, and Applied Power on the Liquid-Phase Microwave-Assisted Extraction of Ginger (*Zingiber officinale*). *Food Res. Int.* **2003,** *36* (5), 499–504.
4. Arballo, J. R.; Campãnone, L. A.; Mascheroni, R. H. Modeling of Microwave Drying of Fruits. *Dry. Technol.* **2010,** *28* (10), 1178–1184.
5. Atuonwu, J. C.; Tassou, S. A. Quality Assurance in Microwave Food Processing and the Enabling Potentials of Solid-State Power Generators: A Review. *J. Food Eng.* **2018,** *234*, 1–15.
6. Bal, L. M.; Kar, A.; Satya, S.; Naik, S. N. Drying Kinetics and Effective Moisture Diffusivity of Bamboo Shoot Slices Undergoing Microwave Drying. *Int. J. Food Sci. Technol.* **2010,** *45* (11), 2321–2328.
7. Barbosa-Cánovas, G. V.; Medina-Meza, I.; Candoğan, K.; Bermúdez-Aguirre, D. Advanced Retorting, Microwave Assisted Thermal Sterilization (MATS), and Pressure Assisted Thermal Sterilization (PATS) to Process Meat Products. *Meat Sci.* **2014,** *98*, 420–434.
8. Bhattacharya, M.; Basak, T. Analysis of Microwave Power and Heating Characteristics for Food Processing: Asymptotes and Resonances. *Food Res. Int.* **2006,** *39* (10), 1046–1057.
9. Castro, M. D. L.; Garcia Ayuso, L. E. Soxhlet Extraction of Solid Materials: An Outdated Technique with a Promising Innovative Future. *Anal. Chim. Acta* **1998,** *369* (1–2), 1–10.
10. Chandrasekaran, S.; Ramanathan, S.; Basak, T. Microwave Food Processing-A Review. *Food Res. Int.* **2013,** *52* (1), 243–261.
11. Chemat, F.; Zille, H.; Khan, M. K. Applications of Ultrasound in Food Technology: Processing, Preservation and Extraction. *Ultrason. Sonochem.* **2011,** *18* (4), 813–835.
12. Chiller, T. M. Salmonella/ Foodborne Outbreaks in USA. *Pathology* **2019,** *51*, S59–S60.
13. Chizoba Ekezie, F. G.; Sun, D. W.; Han, Z.; Cheng, J. H. Microwave-Assisted Food Processing Technologies for Enhancing Product Quality and Process Efficiency: A Review of Recent Developments. *Trends Food Sci. Technol.* **2017,** *67*, 58–69.
14. Choi, W.; Nguyen, L. T.; Lee, S. H.; Jun, S. A Microwave and Ohmic Combination Heater for Uniform Heating of Liquid-Particle Food Mixtures. *J. Food Sci.* **2011,** *76* (9), 576–585.
15. Choi, Y. S.; Hwang, K. E.; Jeong, T. J.; Kim, Y. B.; Jeon, K. H.; Kim, E. M.; Sung, J. M.; Kim, H. W.; Kim, C. J. Comparative Study on the Effects of Boiling, Steaming, Grilling, Microwaving and Superheated Steaming on Quality Characteristics of Marinated Chicken Steak. *Korean J. Food Sci. Anim. Resour.* **2016,** *36* (1), 1–7.
16. Clark, D. E.; Sutton, W. H. Microwave Processing of Materials. *Annu. Rev. Mater. Sci.* **1996,** *26*, 299–331.
17. Datta, A. K.; Davidson, P. M. Microwave and Radio Frequency Processing. *J. Food Sci.* **2000,** *65* (8), 32–41.
18. Dean J. R.; Xiong, G. Extraction of Organic Pollutants from Environmental Matrices: Selection of Extraction Technique. *Trends Anal. Chem.* **2000,** *19* (9), 553–564.

19. De La Vega-Miranda, B.; Santisteban-Lopez, N. A.; Lopez-Malo, A.; Sosa-Morales, M. E. Inactivation of *Salmonella typhimurium* in Fresh Vegetables using Water-Assisted Microwave Heating. *Food Control* **2012,** *26* (1), 19–22.
20. Dev, S. R. S.; Raghavan, G. S. V.; Gariepy, Y. Dielectric Properties of Egg Components and Microwave Heating for In-shell Pasteurization of Eggs. *J. Food Eng.* **2008,** *86* (2), 207–214.
21. Dolinsky, M.; Agostinho, C.; Ribeiro, D.; Rocha, G. D. S.; Barroso, S. G.; Ferreira, D.; Polinati, R.; Ciarelli, G.; Fialho, E. Effect of Different Cooking Methods on the Polyphenol Concentration and Antioxidant Capacity of Selected Vegetables. *J. Culin. Sci. Technol.* **2015,** *14* (1), 1–12.
22. Duan, X.; Liu, W. C.; Ren, G. Y.; Liu, L. L.; Liu, Y. H. Browning Behavior of Button Mushrooms during Microwave Freeze-Drying. *Dry. Technol.* **2016,** *34* (11), 1373–1379.
23. Guo, Q.; Sun, D. W.; Cheng, J. H.; Han, Z. Microwave Processing Techniques and their Recent Applications in the Food Industry. *Trends Food Sci. Technol.* **2017,** *67*, 236–247.
24. Hao, J. Y.; Han, W.; Huang, S. D.; Xue, B. Y.; Deng, X. Microwave Assisted Extraction of Artemisinin from *Artemisia annua* L. *Separat. Purif. Technol.* **2002,** *28* (3), 191–196.
25. Herrero, M.; Mendiola, J. A.; Cifuentes, A.; Ibáñez, E. Supercritical Fluid Extraction: Recent Advances and Applications. *J. Chromatogr. A* **2010,** *1217* (16), 2495–2511.
26. Huang, L. H.; Sites, J. Automatic Control of a Microwave Heating Process for in Package Pasteurization of Beef Frankfurters. *J. Food Eng.* **2007,** *80* (1), 226–233.
27. Jouquand, C.; Tessier, F. J.; Bernard, J.; Marier, D.; Woodward, K.; Jacolot, P. Optimization of Microwave Cooking of Beef Burgundy in Terms of Nutritional and Organoleptic Properties. *LWT - Food Sci. Technol.* **2015,** *60* (1), 271–276.
28. Kindle, G.; Busse, A.; Kampa, D.; Meyer-König, U.; Daschner, F. D. Killing Activity of Microwaves in Milk. *J. Hosp. Infect.* **1996,** *33* (4), 273–278.
29. Kozempel, M. F.; Annous, B. A.; Cook, R. D.; Scullen, O. J.; Whiting, R. C. Inactivation of Microorganisms with Microwaves at Reduced Temperatures. *J. Food Protect.* **1998,** *61* (5), 582–585.
30. Kozempel, M. F.; Cook, R. D.; Scullen, O. J.; Annous, B. A. Development of a Process for Detecting Nonthermal Effects of Microwave Energy on Microorganisms at Low Temperature. *J. Food Process. Preserv.* **2000,** *24* (4), 287–301.
31. Leonelli, C.; Mason, T. J. Microwave and Ultrasonic Processing: Now a Realistic Option for Industry. *Chem. Eng. Process.* **2010,** *49* (9), 885–900.
32. Ling, B.; Tang, J.; Kong, F.; Mitcham, E. J.; Wang, S. Kinetics of Food Quality Changes during Thermal Processing: A Review. *Food Bioprocess Technol.* **2014,** *8* (2), 343–358.
33. Li, Z. Y.; Wang, R. F.; Kudra, T. Uniformity Issues in Microwave Drying. *Dry. Technol.* **2011,** *29* (6), 652–660.
34. Luque-García, J. L.; Luque de Castro, M. D. Where Is Microwave-based Analytical Equipment for Solid Sample Pre-treatment Going? *Trends Anal. Chem.* **2003,** *22* (2), 90–98.
35. Lu, Y.; Turley, A.; Dong, X.; Wu, C. Reduction of *Salmonella enterica* on Grape Tomatoes using Microwave Heating. *Int. J. Food Microbiol.* **2011,** *145* (1), 349–352.
36. Mandal, V.; Mandal, S. C. Design and Performance Evaluation of a Microwave based Low Carbon Yielding Extraction Technique for Natural Bioactive Triterpenoid: Oleanolic Acid. *Biochem. Eng. J.* **2010,** *50* (1–2), 63–70.
37. Metaxas, A. C. (Ed.) *Foundations of Electro Heat: A Unified Approach.* John Wiley & Sons: New York, 1996; p 530.

38. Meyer, W.; Schilz, W. A Microwave Method for Density Independent Determination of the Moisture Content of Solids. *J. Phys. D* **1980,** *13* (10), 1823–1830.
39. Mudgett, R. E. Dielectric Properties of Foods. In: *Microwaves in the Food Processing Industries*; Schweigert, B. S. S.; Decareau, R. V. and Mudgett, R. E. (Eds.); Elsevier Science: New York, 1985; pp 15–37.
40. Nelson, S. O.; You, T. S. Microwave Dielectric Properties of Corn and Wheat and Soybean. *Trans. ASAE* **1989,** *32* (1), 242–249.
41. Nelson, S. O. Estimation of Permittivity of Solids from Measurements on Pulverized or Granular Materials. Chapter 6; In: *Dielectric Properties of Heterogeneous Materials*; Priou, A. (Ed.); Elsevier Science Ltd.: New York – USA, 1992; Vol. 6; pp 231–267.
42. Nelson, S. O. (Ed.) *Dielectric Properties of Agricultural Materials and their Applications*; Elsevier: London – UK, 2015; p 271.
43. Newell, D. G.; Koopmans, M.; Verhoef, L.; Duizer, E.; Aidara-Kane, A.; Sprong, H.; Opsteegh, M.; Langelaar, M.; Threfall, J.; Scheutz, F.; van der Giessen, J.; Kruse, H. Food-Borne Diseases–The Challenges 20 Years Ago Still Persist While New Ones Continue to Emerge. *Int. J. Food Microbiol.* **2010,** *139* (Suppl. 1), S3–S15.
44. Nguyen, L. T.; Choi, W.; Lee, S. H.; Jun, S. Exploring the Heating Patterns of Multiphase Foods in a Continuous Flow, Simultaneous Microwave and Ohmic Combination Heater. *J. Food Eng.*, **2013,** *116* (1), 65–71.
45. Nott, K. P.; Bows, J. R.; Patrick, M. L. Three-Dimensional MRI Mapping of Microwave Induced Heating Patterns. *Int. J. Food Sci. Technol.* **1999,** *34* (4), 305–315.
46. Ondruschka, B.; Asghari, J. Microwave-Assisted Extraction–A State-of-the-Art Overview of Varieties. *CHIMIA Int. J. Chem.* **2006,** *60 (6),* 321–325.
47. Öztürk, S.; Şakıyan, Ö.; Özlem Alifakı, Y. Dielectric Properties and Microwave and Infrared-Microwave Combination Drying Characteristics of Banana and Kiwifruit. *J. Food Process Eng.* **2017,** *40* (3), E-article 12502.
48. Pan, X.; Niu, G.; Liu, H. Microwave-Assisted Extraction of Tea Polyphenols and Tea Caffeine from Green Tea Leaves. *Chem. Eng. Process.* **2003,** *42* (2), 129–133.
49. Pan X.; Niu, G.; Liu, H. Microwave-Assisted Extraction of Tanshinones from *Salvia miltiorrhiza bunge* with Analysis by High Performance Liquid Chromatography (HPLC). *J. Chromatogr. A* **2001,** *922* (1–2), 371–375.
50. Patel, J. H.; Sutar, P. P. Acceleration of Mass Transfer Rates in Osmotic Dehydration of Elephant Foot Yam (*Amorphophallus paeoniifolius*) Applying Pulsed-Microwave-Vacuum. *Innov. Food Sci. Emerg. Technol.*, **2016,** *36*, 201–211.
51. Pellegrini, N.; Chiavaro, E.; Gardana, C.; Mazzeo, T.; Contino, D.; Gallo, M. Effect of Different Cooking Methods on Color, Phytochemical Concentration, and Antioxidant Capacity of Raw and Frozen *Brassica* Vegetables. *J. Agric. Food Chem.* **2010,** *58* (7), 4310–4321.
52. Pereira, R. N.; Vincente, A. A. Environmental Impact of Novel Thermal and Non-Thermal Technologies in Food Processing. *Food Res. Int.* **2010,** *43* (7), 1936–1943.
53. Połtorak, A.; Wyrwisz, J.; Moczkowska, M.; Marcinkowska-Lesiak, M.; Stelmasiak, A.; Rafalska, U.; Wierzbicka, A; Sun, D. W. Microwave versus Convection Heating of Bovine *Gluteus Medius* Muscle: Impact on Selected Physical Properties of Final Product and Cooking Yield. *Int. J. Food Sci. Technol.* **2015,** *50* (4), 958–965.
54. Pu, Y. Y.; Sun, D. W. Combined Hot-Air and Microwave-Vacuum Drying for Improving Drying Uniformity of Mango Slices Based on Hyperspectral Imaging Visualization of Moisture Content Distribution. *Biosyst. Eng.* **2017,** *156*, 108–119.

55. Ruan, G. H.; Li, G. K. The Study on the Chromatographic Fingerprint of *Fructus xanthii* by Microwave Assisted Extraction Coupled with GC-MS. *J. Chromatogr. B* **2007,** *850* (1–2), 241–248.
56. Salazar-González, C.; Martín-González, F. S.; López-Malo, A.; Sosa-Morales, M. E. Recent Studies Related to Microwave Processing of Fluid Foods. *Food Bioprocess Technol.* **2012,** *5*, 31–46.
57. Shenga, E.; Singh, R. P.; Yadav, A. S. Effect of Pasteurization of Shell Egg on its Quality Characteristics under Ambient Storage. *J. Food Sci. Technol.* **2010,** *47* (4), 420–425.
58. Shim, J.; Lee, S. H.; Jun, S. Modeling of Ohmic Heating Patterns of Multiphase Food Products using Computational Fluid Dynamics Codes. *J. Food Eng.* **2010,** *99* (2), 136–141.
59. Spencer, P. L. Method of Treating Foodstuffs. U.S. Patent No. 2,495,429, 1950; p 3.
60. Soni, A.; Smith, J.; Thompson, A.; Brightwell, G. Microwave-Induced Thermal Sterilization: Review on History, Technical Progress, Advantages and Challenges as Compared to the Conventional Methods. *Trends Food Sci. Technol.* **2020,** *97*, 433–442.
61. Tang, J. Unlocking Potentials of Microwaves for Food Safety and Quality. *J. Food Sci.* **2015,** *80* (8), 1776–1793.
62. Tang, J.; Liu, F.; Pathak, S. K.; Eves, E. L. Apparatus and Method for Heating Objects with Microwaves. Patent No. WO2005023013A2, 2005; p 64.
63. Tao, Y.; Sun, D. W. Enhancement of Food Processes by Ultrasound: A Review. *Critic. Rev. Food Sci. Nutr.* **2015,** *55* (4), 570–594.
64. Ulaby, F. T. *Fundamentals of Applied Electromagnetics*. Prentice Hall: Upper Saddle River – NJ, 2004; p 456.
65. Valero, A.; Cejudo, M.; García-Gimeno, R. M. Inactivation Kinetics for *Salmonella enteritidis* in Potato using Microwave Heating Treatments. *Food Control* **2014,** *43*, 175–182.
66. Venkatesh, M. S.; Raghavan, G. S. V. An Overview of Microwave Processing and Dielectric Properties of Agri-Food Materials. *Biosyst. Eng.* **2004,** *88* (1), 1–18.
67. Veronesi, P.; Siligardi, C.; Franchini, M.; Leonelli, C. Microwave Industrial Applications in the Ceramic Field. *Int. Ceram. J.* **2001,** *2001*, 57–62.
68. Wang, J.; Zhang, J., Wang, X., Zhao, B.; Wu, Y.; Yao, J. A Comparison Study on Microwave-Assisted Extraction of *Artemisia Sphaerocephala* Polysaccharides with Conventional Method: Molecule Structure and Antioxidant Activities Evaluation. *Int. J. Biol. Macromol.* **2009,** *45* (5), 483–492.
69. Wang, L.; Weller, C. L. Recent Advances in Extraction of Nutraceuticals from Plants. *Trends Food Sci. Technol.* **2006,** *17* (6), 300–312.
70. Wang, Y.; You, J.; Yu, Y.; Qu, C.; Zhang, H; Ding, L.; Zhang, H; Li, X. Analysis of Ginsenosides in Panax Ginseng in High Pressure Microwave-Assisted Extraction. *Food Chem.* **2008,** *110* (1), 161–167.
71. Wray, D.; Ramaswamy, H. S. Novel Concepts in Microwave Drying of Foods. *Dry. Technol.* **2015,** *33* (7), 769–783.
72. Zhou, J.; Yang, X.; Chu, Y.; Li, X.; Yuan, J. A Novel Algorithm Approach for Rapid Simulated Microwave Heating of Food Moving on a Conveyor Belt. *J. Food Eng.* **2020,** *282*, E-article 110029.
73. Zielinska, M.; Michalska, A. Microwave-Assisted Drying of Blueberry (*Vaccinium corymbosum* L.) Fruits: Drying Kinetics, Polyphenols, Anthocyanins, Antioxidant Capacity, Color and Texture. *Food Chem.* **2016,** *212*, 671–680.

CHAPTER 3

POTENTIAL OF NONTHERMAL TECHNOLOGIES IN THE FOOD INDUSTRY

R. KAAVYA, R. PANDISELVAM,
SERGIO I. MARTINEZ-MONTEAGUDO, ANJINEYULU KOTHAKOTA,
N. U. SRUTHI, and KALI RAMESH

ABSTRACT

In the current decade, consumers demand for more quality and safe food thereby maintaining sensorial and nutritional characteristics with shelf life extension. Such preferences in consumers have led to the exploration of nonthermal food processing applications in the food business. This chapter presents the emerging innovative technologies (such as microfiltration (MF), ultra-high voltage electric fields, high-pressure processing (HPP), pulsed electric field, and irradiation). These provide insights into operation of the process, inactivation mechanism, process optimization, and recent developments in the application of this technique. An elucidation of the variety of foods utilized in each technique and process conditions will pave way for upscaling these technologies and for already existing technical knowledge in the food process industry, such as MF, HPP, and irradiation modeling.

3.1 INTRODUCTION

Food preservation has been the greatest concern in the food processing industries, microorganism plays a vital task in the spoilage of foods and

Novel and Alternative Methods in Food Processing: Biotechnological, Physicochemical, and Mathematical Approaches. N. Veena, Megh R. Goyal, & Ritesh B. Watharkar (Eds.)

making it hazardous for consumption. The throwaway technique of food safeguarding and substitute of conventional food preservation techniques such as extreme heat up, salting, acidification, exposure to hot air, and chemical preservation is replaced by the latest techniques of preservation. There is an increased demand for foods with highly nutritious, tasty, natural, and safe to handle food products forced the researchers to develop innovative preservation techniques to safeguard the foods. The conventional food processing technologies, such as pasteurization, elevated temperature sterilization, and hot air exposure can cause a reduction in the microbial contamination; however, this processing technique can destroy some valuable food nutrients. Foods are processed at elevated temperatures and prolonged processing time also produce some caustic components that bring threat to human.[39]

Nowadays, growing population increases the consumer demand for superior-quality food products with fresh characteristics, the introduction of nonthermal technologies with potential benefits such shorter time with lower temperature brings foods to no change in color, flavor, and crucial nutrients.[15,63]

The purposeful role of this chapter is to discuss the advancements in applications of various nonthermal technologies for food preservation.

3.2 MICROFILTRATION (MF)

Membrane technology, a filtration technique for clarification, concentration, stabilization, fractionation, desalting, as well as purifying of a wide range of liquid food products is an extensively used technique in recent times. Its operation is based on the pressure difference, with reduced energy consumption serves an alternate technique to thermal treatment. Membrane processing outweighs in both nutritional and sensorial characteristics. It has the ability to separate fine particles, microbes, and other impurities from liquid foods and used for concentrating liquid foods like milk, juices, etc. This methodology also results in increased yield and avoids the use of fining agents resulting in minimization of cost and disposal problems.[22]

3.2.1 APPLICATION OF MF IN FOOD INDUSTRY

Pressure-driven membrane processes are gaining attention and popularity in both processing and separation methods. MF along with other membrane technologies can be used for the separation of fat globules, removal of microorganisms, fractionation and standardization of milk proteins, treatment of whey, etc. The dairy industry is also a largest contributor in wastewater and membrane processes also find application in treatment of the discharged volume of wastewater.[27] The use of MF in clarification and sterilization of vinegar is also gaining popularity.[24] The traditional removal of yeast after the beer-making process by centrifugation can be replaced by a perfect combination of membrane technologies. MF eliminates the use of filter aids in the recovery of sediment from the tank bottom. It also eliminates the addition of finings in the clarification of beer from chill conditioning tanks.[3] The traditional method of evaporation for concentrating juices can be replaced by MF at relatively lower and safer temperatures, thus preserving the heat-sensitive compounds.[14]

3.2.2 ADVANTAGES OF MF

Improved product quality, lesser or no thermal degradation, lower energy consumption, and lesser operating cost are the major advantages of MF over other conventional techniques. It also optimizes the exploitation of raw materials with a lessened environmental impact during processing. It also aids in recovering the components that are valuable and goes in diluted effluents and by-products bringing more benefits to the industries. Furthermore, it eliminates the use of chemicals and additives thus making it a greener approach in food processing. Cross-flow MF could also replace the conventional filtration technique and reduces the possibility of fouling of the membrane. The process also eliminates any mechanical damage to the dissolved constituents and valuable compounds like which happens during conventional separation like centrifugation.

3.2.3 MF CHARACTERISTICS

MF removes particles or biological entities in the size range of 0.025–10.0 μm through a membrane filter. These membranes are efficient in separating large suspended solids such as bacteria, colloidal particulates, and fat; however, permitting sugars, salts, proteins, and low molecular weight molecules to cross through.[45] The membrane consists of an asymmetric pore structure, with firmer surface pores to manage rejection, and more exposed macrovoids in the membrane cross-section, so as to optimize flux. The membrane structure includes filters depth which traps particles at constrictions inside the membranes and screen filters to collect retained substances on the surface. Cross-flow filtration and dead-end filtration are the two configurations in which the membrane generally act.

In a cross-flow filtration, there is a tangential flow toward the membrane, while the fluid is allowed to pass perpendicularly in a dead-end filtration where all the particles larger than the pore size are stopped at the membrane surface.

For the dead-end flow process, mostly plate-and-frame configurations are used. This arrangement is compact and possesses a robust design with reduced capital; but still the cost of operation is high.

These modules are most appropriate for small-scale laboratory applications to filter dilute solutions. The particular module used for a cross-flow filtration is a spiral wound module. This module is favored when heavily concentrated solutions are to be handled and in situations of high temperatures and extreme pH. This configuration of MF is mostly used in large-scale industrial process.[35] Applications of MF are presented in Table 3.1.

TABLE 3.1 Application of MF.

Product (juice and emulsion)	Processing conditions				References
	Temperature (°C)	Transmembrane pressure (bar)	Pore size (μm)	Cross-flow velocity (m/s)	
Acai	25	1	0.2–0.8	3	53
Bottle gourd	–	0.14–1.6	0.2	–	16
Cactus pear	25	2.2	0.2	–	21
Cashew apple	30	0.8–2.3	0.1–1.4	7	1
Citrus	40	3	0.2	6	64
Elderberry	30	3.5	0.8	–	34
Oil–water emulsion	–	2.1	3.06–2.16	1.5	32
Passion fruit	20	1	0.4	–	28
Red plum	20–30	0.5–2.5	0.22	0.2–0.5	57
Red raspberry	22	1	0.2	0.07	76
Sugarcane	25–35	2–4	0.4	0.49–0.87	67
Umbu	35	0.5–1.1	0.2	4–6	75
Watermelon	50–60	0.5–2	0.2–1.4	6	23

3.3 ULTRA-HIGH VOLTAGE ELECTRIC FIELDS (UHVEF)

The use of electric field (EF) in food processing has indicated their potential for microbial destruction, enzyme inactivation, freezing, drying, extraction, and wine making. However, electric fields can be applied in two different ways: pulsed electric field (PEF) and UHVEF. While PEF applications are at least vaguely established in the industry level, UHVEF developments are at a primitive stage of research and their industrial applications are yet to be explored widely. The principle of UHVEF can be explained by the phenomenon of electric breakdown in water.[17] On applying UHVEF (30–40 kV) into a liquid medium, sudden release of energy occurs, which is also supplemented to the plasma channels. This consequently leads to secondary effects such as emission of high-intensity UV light, shock waves, hydroxyl radicals, and cavitation which are predominant as compared to the EF effects.[65] These effects lead to high-intensity cellular damage can also accelerate the intracellular compounds extraction from the cytoplasm of cell.

3.3.1 APPLICATIONS OF UHVEF

The major area which utilizes UHVEF in food industry is to enhancement of mass transfer processes, in particular, to extract the intracellular blends from by-products and waste stream from food industries. UHVEF has been stated as more efficient as compared to PEF in terms of extraction yield, rate, energy, and cost. Food preservation by drying is an oldest techniques, tremendously lowers water activity, and inhibits the bacterial growth. On application of UHVEF as a pretreatment in food drying, the periphery layer linking the food surface and drying air gets disturbed for improving the evaporation rate.

Bai et al.[7] compared the electrohydrodynamic (EHD) drying, which is used to dry the sea cucumber at 18°C with oven drying (80°C) and exposure to air (18°C) in attributes like drying rate, quality and energy consumption. The results depicted that EHD drying resulted in obtaining a quality product with higher drying rate and consuming lesser energy compared to the other two methods. UHVEF treatment significantly improved the thawing rates and was also found to inhibit microbial growth during post-thawing storage.[38] The mechanism of microbial inactivation by UHVEF can be mainly attributed on two factors: the hydraulic shock waves generated by discharging of high voltage electric current in the aqueous solution and the free radicals that serve lethal to specific intracellular components essential for cellular metabolism.

The process efficiency mainly depends on the components present in food and electrical parameters like energy input, strength of electric field, and duration of treatment. At threshold EF strength, the potential difference causes conformational modifications in the membrane structure, bring about notable loss of its semipermeable properties. Above the critical value, the fraction of surviving cells declines with a further increase in EF. Similar is the trend, when treatment time exceeds the critical value. Table 3.2 presents various treatment conditions of UHVEF on agricultural produce.

TABLE 3.2 Summary on the Processing Conditions of UHVEF Treated Foods.

Food matrix	Optimized treatment conditions				References
	Electric voltage (kV)	Frequency (Hz)	Time (min)	Temperature (°C)	
Blackberry	40	0.5	300	50	12
Grape Stems	40	0.5	–	50	18
Grape pomace	40	0.5	–	30	12
Mango peels	40	–	50	20	61
Olive kernel	40	0.5	20	20	68
Papaya peel	40	–	272	50	60
Pomegranate peel	20	0.5	30	70	80
Sesame cake	40	0.5	20	60	69
Stevia	40	0.5	–	50	20
Vine shots	40	0.5	–	50	66

UHVEF is recently evolved as an emerging technique and more experimental studies on this technology from the perspective of industrial usage in conditions of cost-effectiveness and energy utilization has to be evaluated. On assessing PEF and high voltage electric discharge (HVED) for the pretreatment of solid food materials, results with HVED have been observed to be the cheapest technique in energy utilization and the most efficient method for extracting a particular compound when carried out at an equal value of specific energy. However, more advanced research is required in this aspect considering some spotted limitations during the treatment.

3.4 PULSED ELECTRIC FIELD (ELECTROPORATION)

The studies on electric current usage in preserving food have been sight seen from the period that electrical energy was made available commercially. PEF deals with treating food with electric pulses of 100–300 V/cm to 20–80 kV/cm for very little time in order to destroy food-borne microbes and to check troublesome microorganisms that cause spoilage.[49] As food contains several ions and dipolar molecules, it is capable of transferring electricity. On applying electric field, the current flows through the liquid food and gets transmitted to all direction. This charges the cell membrane which increases the transmembrane potential around the cytoplasmic membrane. When a threshold value of possible transmembrane has been achieved (0.2–1.0 V), small metastable hydrophilic pores are created which increase in number and size as the treatment proceeds. This is accompanied by the leakage of intracellular components and the entrance of extracellular substances.[69]

At higher electric fields (>20 kV/cm), PEF outweighs conventional thermal processing due to the given reasons. Pathogenic microorganisms and deleterious enzymes are inactivated by retaining or negligibly altering sensorial, nutritional, as well as health-promoting aspects of food product. On the other hand, at lower electric fields, the pulses cause the biological membrane to get electrically pierced which temporarily or permanently destroys its semipermeable nature thus allowing selective recovery of value-added compounds from various mediums.[11,13,26]

3.4.1 PULSE CHARACTERISTICS: SHAPE, POLARIZATION, AND FREQUENCY

The efficacy of PEF treatment depends on many critical factors, of which the pulse characteristics are of much importance. High voltage electric pulses are of different shapes or waveforms, namely, square, exponential, oscillatory, and bipolar, of which the exponential and square waveforms are of practical relevance in food processing. The functional electric field in the case of the square pulse is maintained for the pulse duration, making it superior to the exponential pulse, in which, the field intensity changes from a peak electric field to cycles of sublethal electric field intensities till it arrives at a ground level. The efficacy of the treatment using exponentially decaying pulses declines linearly with maximum efficiency at peak electric field and minimum critical value has been attained.[5] The pulse frequency regulates the extent of electrical energy supplied per unit of time, while in turn causing the rise of product temperature owing to Joule effect. The capacity of a continuous PEF processing system can be raised by intensifying the pulse frequency.

3.4.2 APPLICATIONS OF PEF IN MICROBIAL INACTIVATION

Preservation of liquid foods and beverages by PEF treatment has been successfully reported by many researchers where a wide variety of enzymes and microorganisms were inactivated. [41,56,72,85] PEF has the ability to lower the microbial load by log 5 units given as mandatory by the regulation of FDA and the rates of microbial inactivation had an increment when the number of pulses increased.

Electric field, when applied in a suspension of cells, triggers a potential difference around the membrane which induces a sharp surge in the membrane conductivity and permeability and when the potential surpasses the critical value (1.0 V), membrane destruction occurs. The inactivation rate depends on some critical factors together with the type of microorganism, its culture, treatment medium, time of treatment, temperature, and pulse shape. The rate when inactivation is found to increase with increment with the electric current intensity provided the applied field is above the significant transmembrane potential.[57] The combined role of the number of pulses and pulse width constitute the treatment time or the actual time during which a particular suspension is subjected to the applied field strength.

Microbial inactivation is found to increase when an increase in any one of the above said parameters increase. The size, shape, species, and the growth stage of the target pathogens are the other determinant factors affecting the PEF treatment. During PEF treatment, large cells permitted greater induction of electric field into the membrane than smaller ones. The conductivity, resistivity, ionic strength, dielectric properties, composition, and pH of the treated media are the other intrinsic factors that influence PEF application significantly. Table 3.3 summarizes different PEF-assisted treatments applied to food products.

PEF is widely accepted as a processing tool because of the phenomenon of disruption through electroporation and increasing the membrane permeability. It has also provided more energy-efficient possibilities of processing in a variety of food matrices. PEF is known to not produce any hazardous or toxic by-product, which makes it a greener approach. However, in future research studies, the PEF should be focused, such as in meat industry could be given a thrust, as the tenderization effect of PEF could find a commercial relevance in future.

TABLE 3.3 Summary on the Application of PEF in Foods.

Food matrix	Optimized treatment conditions					References
	Electric field strength (kV/cm)	Number of pulses	Pulse width (μs)	Total treatment time (μs)	Temperature (°C)	
Apple tissue	5	10	10	100	–	77
Beer (7% alcohol)	45	46.3	1.5	70	53	56
Carrot puree	1.1	150	20	3000	5	51
Grape juice	27	15	3.0	45	48.8	41
Milk formula (Infant)	35	280	2.5	700	28	62
Potato tissue	1.5	45	120	5400	–	79
Red pepper	2.5	–	30	4E + 6	28	78
Spinach leaves	–	25	100	2500	–	29
Whole milk	28	0.85	20	17	55	72
Wine	5	100	10	1000	–	31

Similarly, in spite of the advantages stated for the PEF application in wine industry, the treatment influences of PEF on the composition of wine have not been identified so far in detail. To add to that, the bioavailability or bioaccessibility of the extracted or retained compounds using PEF treatment is yet to be given appropriate research as adequate information in this regard is deficient. The application of PEF treatment for wastewater treatment is another area, where there is scope for future studies.

3.5 HIGH-PRESSURE PROCESSING (HPP)

3.5.1 PROCESS DESCRIPTION

A batch-type HPP operation involves a prepacked food placed at specific pressure and time. Though there are both batch and continuous process in HPP, the industrial implementation remains a batch process. The important process design of high-pressure equipment is comprising of combined top and bottom closures pressure vessel, yoke system of mechanism to protect pressure vessel during operation, pressure generation pump and intensifier, food loading and unloading holders, and controllers for various process monitoring and recordings.[54]

A typical HPP process is similar to that of thermal retort or canning. In general, the food which has to treated mostly in vacuum package and then loaded into cylindrical carrier basket and further into stainless steel vessel. The closed end pressure vessel packed with water as a transmitting fluid for pressure. The combined action of the pressure pump and intensifier on pressure transmitting fluid ultimately achieves the targeted pressure. Upon pressurization, foods containing high moisture experience a physical compression of about 15% of its volume. After decompression, processed food products get back on close to the their initial volume. Once the predetermined pressure is attained, the product is kept for holding to complete the process time. After completing the process duration, instantaneously the vessel is depressurized and unloads the processed products. This process cycle is about 10 min.[8]

3.5.2 APPLICATIONS OF HPP

- **Pasteurization**—The greater effective inactivation capacity of HPP on various pathogenic and spoilage microorganisms, yeast and mold, viruses and vegetative cells. The inactivation of pathogenic microbe is depending on the level of pressure which is used (400–600 MPa), at the temperature of 25–50°C, on time of 3–10 min, food characteristic such as composition, pH, water activity (a_w), and redox potential.[54] The efficiency of HPP for pasteurization of solid and liquid foods like meats, seafoods, salads, vegetable products, and fruit juices has been inspected by different researchers.

- **Sterilization**—The high pressure applied on a preheated sample has shown to inactivate bacterial spores and therefore achieving commercial sterilization. This pressure-assisted thermal processing (PATP) technology involves preheating of food material to about 75–90°C with the subsequent high-pressure application up to 500–600 MPa.[55] A typical target temperature for PATP is between 100–120°C for 3–10 min. The uniqueness and beneficial part of PATP is the rapid increase in the temperature due to compression and decompression upon consequent cooling.[10] The complete process results in lesser treatment time as compared to the traditional canning, which produces imperceptible heat transfer rate. Although PATP has not been marketed yet, this processing method has the possible delivery platform for different processed foods, namely, baby foods, vegetables, milk-based egg products, RTE foods, gravies, desserts, soups, and sauces.[74]
- **Extraction**—HPP enhances the biocomponent extraction with valuable compounds from different biological sources. The main processing variables responsible for bioextraction are pressure, type of solvent used, and treatment time. A mixture of solvents (organic) has experimented with pressure-assisted antioxidants extraction and other bioactive substances.[46] Further decompression changes the chemical bonds present in the extracting matrix which disturbed by pressure and making available for extracting compounds for easier extraction.
- **Pressure-Ohmic Thermal Processing (POTP)**—This improved newer processing technology involved in ordered and real-time applications under elevated pressure and ohmic heating. This technique helps in preserving food products with low-acid content for extending the shelf life. The synergistically combined technology in compression of heat effect at elevated pressures along with ohmic heating to reduces thermal exposure on the product quality.[9] The temperature rate increases inside the pressure vessel during this process can be tuned by controlling the electrical heating in pressure, POTP drawing

as well ignores the requirement of preheating of pressure vessel by this means reducing energy need and thermal stress on the pressure equipment.

3.6 IRRADIATION

γ-irradiation is a nonthermal technology which is an effectual means of preserving the food by decontaminating key microorganisms. The use of ionizing radiation, such as γ-rays, electron beams, or X-ray, is the basic principle for the radiation process. Irradiation affects the physiological metabolism of cells, which inhibits their growth.[30] When the microbes are exposed to those radiations, energy transfer takes place within the body, by breaking the bonds on the DNA structure, the protein and enzyme denatures resulting in reduced normal physiological activities. It also loses its ability to replicate and division of chromosomes leading to aging and cell death.[52]

The intensity level of the treatment is the energy (J) to which the mass (kg) of the product is exposed. The dose is measured in terms of amount of radiation the product is exposed. The dose is defined as the energy per unit mass of the substance.[50] Irradiation doses can be categorized into three types (1) low treatment dose (<1 kGy), (2) intermediate treatment dose (1–10 kGy), and (3) high dose treatment (10–60 kGy) according to the nature of the product to which is exposed. However, the irradiated food can be influenced by the various parameters such as irradiation dose, food composition, and microbial species.

3.6.1 RECENT RESEARCH DEVELOPMENTS IN FOOD IRRADIATION

3.6.1.1 FRUITS AND JUICES

In recent times, the consumers demand for fresh and pathogen-free produce with all nutritional and better organoleptic properties. Microorganisms such as yeast and mold and the bacteria which are mostly acid tolerant which causes spoilage to the fruit juices and they have higher value of D_{10} than other bacterial pathogens. In the case of yeast and mold, the D_{10} values are in the range of 1–3 kGy[52] and 0.3–0.7 kGy, respectively.[2] Various research carried out for different fruits and fruit juices, focused mainly on the inhibition of the microbial growth, extension of shelf life by retaining nutritional properties such as the bioactive components, maintain the quality standards, and making it available to the consumers. The anthocyanin stability of up to 80% was retained in the pomegranate juice when it was exposed to 0.5 and 10 kGy.[2]

Study on the fresh pomegranate by Shahbaz et al.[71] revealed that fruits treated at 0.4 kGy and 1 kGy showed a better juice taste, consistency and appearance, the total phenolic content, and the antioxidant assay were not affected at 0.4 kGy, being a lower dose. In the research by Arjeh et al.[4] on the sour cherry juice, doses with ≥3 kGy, the physicochemical properties, reduced microbial load, and antioxidant activity were also maintained as per the standards. In cherry tomatoes, a reduction of food-borne pathogens of 5–11 log was obtained when it was exposed to a irradiation dose of 3.2 kGy resulting in minor changes in the physiochemical characteristics.[36]

Kaavya et al.[47] explained sugarcane juice was preserved for a month when the dosage of 5 kGy applied and storing at 10°C. There was no change in phenolics and flavonoids contents during the storage temperature of 10°C. The physicochemical characteristics on Mango juice from eight different cultivars were studied by Naresh et al.[58] who revealed that TSS and pH was unaffected with exposing 1 kGy. In the case of titratable acidity, there was no change until 0.5 kGy but at 1 and 3 kGy doses there was a slight decrease. A combined treatment of γ-irradiation 1.2 kGy with carboxymethyl cellulose prevented mold growth in the peach fruit.[43]

3.6.1.2 FRESH VEGETABLES AND SPICES

Intermediate dose of gamma irradiation (1.0–3.0 kGy) is most commonly used for delaying the maturity or senescence in most vegetables and also to decontaminate the microbes.[52] Improvements in the physicochemical properties of the eggplant by combined treatment with 1.0 kGy and ascorbic acid can inhibited the polyphenol oxidase action, maintained the quality attributes, and the creamy white layer.[42] Studies on the fresh cut lettuce conducted by Zhang et al.[83] reported that total bacterial counts and coliform count reduced to 2.35 log CFU/g and 30 MPN (most probable number)/100 g, respectively, and also no loss was observed in the vitamin C content while irradiated at 1.0 kGy.

Similar studies on the minimally processed carrot and lettuce were carried out by Frimpong et al.[33] who revealed that the pathogenic organisms such as *Staphylococcus aureus, Cronobacter sakazakii, Bacillus cereus,* and *Klebsiella* spp. was inactivated at 2 kGy thereby ensuring the food safety. The bioactive phytochemical contents and antioxidant activities were enhanced in spinach and fenugreek by treating them at dose of 0.25–1.5 kGy and also the visual characteristics and color were well retained. Spices were irradiated for surface decontamination and to retain the quality attributes. A γ-irradiation dose of 5 kGy decreased the microbial count without causing any color change.[73] Research by Zafar et al.[82] reported the red chilies treated with dose of 6 kGy minimized the fungal growth and toxicity level by 5 and 2 logs, respectively.

3.6.1.3 MEAT PRODUCTS

Meat and sea foods come under raw foods and the application of γ-irradiation was done to maintain the freshness and taste by eliminating the microbial contamination. Low dose of 2.5 kGy combined with edible coating of chitosan (2%) and grape seed extract (0.1%) was applied in the chicken meat.[37] It was observed that addition of bioactive coating reduced the lipid oxidation in meat products. Therefore, a significant effect on the reduction of bacterial growth and also on the chemical changes like pH, this is due to the acidic properties of the chitosan that inhibited the microbial growth in the chicken meat. Studies show the reduction in the residual nitrate and improved lipid oxidation stability in chicken sausages when γ-radiation combinedly treated with mugwort extract and ascorbic acid.[44] Bu et al.[19] studied the quality effect of yellow fin fillets when treating with electron irradiation and bayberry polyphenols. In total, 5 kGy dose accelerated the degradation of adenosine triphosphate and lipid oxidation. Incorporation of bayberry had a significant effect in inhibiting the growth of bacteria, loss of freshness, discoloration of the fillets, and histamine formation. However, 1–3 kGy was optimum to expand the shelf life of the product up to 7 days at 4°C. In smoked salmon fish, 3 kGy irradiation dose inactivated 6.05 and 6.59 log CFU/g of *Vibrio parahaemolyticus* and *Listeria monocytogenes* without any effect on the sensory and quality attributes.[6]

Research studies by Hocaoğlu et al.[40] explained that high doses of 5 kGy and low temperature (−18°C) treatment enhanced the storage shelf life of the shrimp to 90 days by reducing their microbial populations. This might be due to the lower concentrations of total volatile base-nitrogen and trimethylamine-N oxide during refrigeration.

3.6.1.4 CEREALS AND PULSES

The presence of various essential nutrients and bioactive compounds offers possible health benefits. The quality and price of the soybean have been devasted by the spoilage caused by fungal pathogenic organisms such as *Alternaria*, *Phomopsis*, and *Fusarium*. Therefore, exposing to 10 kGy eliminated the fungi and reduced the level of aflatoxin content in the soybean at about 62.20% and 76.38% when incubated for 30 and 60 days, respectively.[84] Electron beam irradiation (EBI) generates high energy in a short span of time; therefore, when EBI is applied to high fat and protein content foods, there are very less unexpected flavors and tastes.

Xue et al.[81] studied the physiochemical properties of the corn flour which depicted that EBI (0–5.4 kGy) obtained smooth surfaces, uniform size, and spherical structures. The pasting peak viscosity of the corn flour was decreased dramatically, which indicates the inhibition of corn flour gelatinization. The rich source of hazelnut includes Vitamin E, fats, and proteins are often subjected to storage loses due to the presence of some pests, which positively create a failure on the sensory and chemical properties during their long-time storage. Koc et al.[48] stated that 0.5 kGy was the optimum dosage for the hazelnut to preserve the sensory qualities, low free fatty acids, peroxide value, and vitamin E content remains unchanged. However, during the long-term storage of about 18 months, changes were observed in the physicochemical characters.

3.7 SUMMARY

Current research and development endeavors have successfully demonstrated that nonthermal technologies are a promising aid in the spectrum of food processing and safeguarding operations. In comparison with thermal treatments, the nonthermal technologies offer various advantages such as shorter processing time, less energy consumption, and higher levels of safety. In fact, it signifies a quick, effective, and trustworthy substitute to develop the sensory and nutritional characteristic of food products and extended shelf life. Further industrial implementation of the nonthermal technologies requires more deeper studies on the mechanisms, misconceptions of each technology, approval of standard regulations, and most importantly the change in consumer mindset about these technologies.

KEYWORDS

- **applications**
- **microbial inactivation**
- **nonthermal processing**
- **processing conditions**

REFERENCES

1. Abreu, F.; Perez, A. M.; Dornier, M.; Reynes, M. Potentialities of Tangential Microfiltration on Mineral Membranes for the Clarification of Cashew Apple Juice. *Fruits* **2005,** *60* (1), 33–40.
2. Alighourchi, H.; Barzegar, M.; Abbasi, S. Effect of Gamma Irradiation on the Stability of Anthocyanins and Shelf-life of Various Pomegranate Juices. *Food Chem.* **2008,** *110* (4), 1036–1040.
3. Ambrosi, A.; Cardozo, N. S. M.; Tessaro, I. C. Membrane Separation Processes for the Beer Industry: A Review and State-of-the-Art. *Food Bioprocess Technol.* **2014,** *7* (4), 921–936.
4. Arjeh, E.; Barzegar, M.; Sahari, M. A. Effects of Gamma Irradiation on Physicochemical Properties, Antioxidant and Microbial Activities of Sour Cherry Juice. *Radiat. Phys. Chem.* **2015,** *114* (1), 18–24.
5. Arvanitoyannis, I. S.; Ladas, D.; Mavromatis, A. Novel Food Processing Technologies. *Int. J. Food Sci. Technol.* **2006,** *41* (5), 470–474.
6. Badr, H. M. Control of Potential Health Hazards of Smoked Fish by Gamma Irradiation. *Int. J. Food Microbiol.* **2012,** *154* (3), 177–186.
7. Bai, Y.; Qu, M.; Luan, Z. Electro Hydrodynamic Drying of Sea Cucumber (*Stichopus japonicus*). *LWT-Food Sci. Technol.* **2013,** *54* (2), 570–576.
8. Balasubramaniam, V. M.; Martinez-Monteagudo, S. I.; Gupta, R. Principles and Application of High Pressure–Based Technologies in the Food Industry. *Ann. Rev. Food Sci. Technol.* **2015,** *6* (1), 435–462.
9. Balasubramaniam, V. M.; Park, S.; Sastry, S. K. High Pressure Ohmic Thermal Sterilization Methods and Apparatus. *US Provision. Patent Appl.* **2012,** 733.
10. Balasubramaniam, V. M.; Ting, E.; Stewart, C. M.; Robbins, J. A. Recommended Laboratory Practices for Conducting High-Pressure Microbial Experiments. *Innov. Food Sci. Emerg. Technol.* **2004,** *5* (3), 299–306.
11. Barba, F. J.; Brianceau, S.; Turk, M.; Boussetta, N.; Vorobiev, E. Effect of Alternative Physical Treatments (Ultrasounds, Pulsed Electric Fields, and High-Voltage Electrical Discharges) on Selective Recovery of Bio-compounds from Fermented Grape Pomace. *Food Bioprocess Technol.* **2015,** *8*, 1139–1148.
12. Barba, F. J.; Galanakis, C. M.; Esteve, M. J.; Frigola, A.; Vorobiev, E. Potential Use of Pulsed Electric Technologies and Ultrasounds to Improve the Recovery of High-added Value Compounds from Blackberries. *J. Food Eng.* **2015,** *167* (part A), 38–44.

13. Barba, F. J.; Parniakov, O.; Pereira, S. A.; Wiktor, A.; Grimi, N.; Boussetta, N.; Saraiva, J. A.; Raso, J.; Martin-Belloso, O.; Witrowa-Rajchert, D.; Lebovka, N.; Vorobiev, E. Current Applications and New Opportunities for the Use of Pulsed Electric Fields in Food Science and Industry. *Food Res. Int.* **2015,** *77* (part 4), 773–798.
14. Bhattacharjee, C.; Saxena, V. K.; Dutta, S. Fruit Juice Processing using Membrane Technology: A Review. *Innov. Food Sci. Emerg. Technol.* **2017,** *43*, 136–153.
15. Birmpa, A.; Sfika,V.; Vantarakis, A. Ultraviolet Light and Ultrasound as Non-thermal Treatments for the Inactivation of Microorganisms in Fresh Ready-to-Eat Foods. *Int. J. Food Microbiol.* **2013,** *167* (1), 96–102.
16. Biswas, P. P.; Mondal, M.; De, S. Comparison Between Centrifugation and Microfiltration as Primary Clarification of Bottle Gourd (*Lagenariasiceraria*) Juice. *J. Food Process. Preserv.* **2016,** *40* (2), 226–238.
17. Boussetta, N. J.; Lanoiselle, L.; Bedel-Cloutour, C.; Vorobiev, E. Extraction of Soluble Matter from Grape Pomace by High Voltage Electrical Discharges for Polyphenol Recovery: Effect of Sulphur Dioxide and Thermal Treatments. *J. Food Eng.* **2009,** *95*, 192–198.
18. Brianceau, S.; Vitrac X.; Turk, M.; Vorobiev, E. High Voltage Electric Discharges Assisted Extraction of Stilbenes from Grape Stems. In *First World Congress on Electroporation and Pulsed Electric Fields in Biology, Medicine and Food & Environmental Technologies*, 2016; pp 113–116.
19. Bu, T.; Jin, Y.; Li, X.; Zhang, J.; Xu, D.; Yang, W.; Lou, Q. Effect of Electron Irradiation and Bayberry Polyphenols on the Quality Change of Yellowfin Tuna Fillets during Refrigerated Storage. *Radiat. Phys. Chem.* **2017,** *138* (1), 67–71.
20. Carbonell-Capella, J. M.; Žlabur, J. S.; Brnčić, S. R.; Barba, F. J.; Grimi, N.; Koubaa, M.; Brnčić, M.; Vorobiev, E. Electrotechnologies, Microwaves, and Ultrasounds Combined with Binary Mixtures of Ethanol and Water to Extract Steviol Glycosides and Antioxidant Compounds from *Stevia rebaudiana* Leaves. *J. Food Process. Preserv.* **2017,** *41* (5), 131–179.
21. Cassano, A.; Conidi, C. Integration of Membrane Technologies into Conventional Existing Systems in the Food Industry. Chapter 13. In *Bioenergy Systems for the Future: Prospects for Biofuels and Biohydrogen*; Dalena, F., Basile, A., Rossi, C., Eds.; Woodhead Publishing: Duxford –UK, 2017; pp 451–479.
22. Cassano, A.; Conidi, C.; Drioli, E. Physicochemical Parameters of Cactus Pear (*Opuntiaficus indica*) Juice Clarified by Microfiltration and Ultrafiltration Processes. *Desalination* **2010,** *250* (3), 1101–1104.
23. Chaparro, L.; Dhuique-Mayer, C.; Castillo, S.; Vaillant, F.; Servant, A.; Dornier, M.; Concentration and Purification of Lycopene from Watermelon Juice by Integrated Microfiltration-Based Processes. *Innov. Food Sci. Emerg. Technol.* **2016,** *37*, 153–160.
24. Chen, X.; Lu, L.; Yang, H. Effects of Membrane Filtration on Vinegar Ingredients. *China Brew.* **2012,** *2*, 50–55.
25. Dalvi-Isfahan, M.; Hamdami,N.; Le-Bail, A. Effect of Freezing Under Electrostatic Field on the Quality of Lamb Meat. *Innov. Food Sci. Emerg. Technol.* **2016,** *37* (part A), 68–73.
26. Deng, Q.; Zinoviadou, K. G.; Galanakis, C. M. The Effects of Conventional and Non-Conventional Processing on Glucosinolates and its Derived Forms, Isothiocyanates: Extraction, Degradation, and Applications. *Food Eng. Rev.* **2015,** *7* (3), 357–381.
27. Dhineshkumar, V.; Ramasamy, D. Review on Membrane Technology Applications in Food and Dairy Processing. *J. Appl. Biotechnol. Bioeng.* **2017,** *3* (5), 399–407.

28. Domingues, R. C. C.; Ramos, A. A.; Cardoso, V. L.; Reis, M. H. M. Microfiltration of Passion Fruit Juice using Hollow Fibre Membranes and Evaluation of Fouling Mechanisms. *J. Food Eng.* **2014,** *121* (1), 73–79.
29. Dymek, K.; Dejmek, P.; Galindo, F. G.; Wisniewski, M. Influence of Vacuum Impregnation and Pulsed Electric Field on the Freezing Temperature and Ice Propagation Rates of Spinach Leaves. *LWT-Food Sci. Technol.* **2015,** *64* (1), 497–502.
30. Ekezie, F. G. C.; Cheng, J. H.; Sun, D. W. Effects of Nonthermal Food Processing Technologies on Food Allergens : A Review of Recent Research Advances. *Trends Food Sci. Technol.* **2018,** *74*, 12–25.
31. El Darra, N.; Turk, M. F.; Ducasse, M. A.; Grimi, N.; Maroun, R. G.; Louka, N.; Vorobiev, E. Changes in Polyphenol Profiles and Color Composition of Freshly Fermented Model Wine due to Pulsed Electric Field, enzymes and Thermovinification Pretreatments. *Food Chem.* **2016,** *194*, 944–950.
32. Emani, S.; Uppaluri, R.; Purkait, M. K. Cross Flow Microfiltration of Oil–Water Emulsions using Kaolin based Low Cost Ceramic Membranes. *Desalination* **2014,** *341*, 61–71.
33. Frimpong, G. K.; Kottoh, I. D.; Ofosu, D. O.; Larbi, D. Effect of Gamma Irradiation on Microbial Quality of Minimally Processed Carrot and Lettuce: A Case Study in Greater Accra region of Ghana. *Radiat. Phys. Chem.* **2015,** *110* (1), 12–16.
34. Furulyas, D.; Nyeki, F.; Steger-Mate, M.; Stefanovits-Banyai, E.; Banvolgyi, S. Effects of Pectolytic Enzyme Treatment and Microfiltration on Antioxidant Components of Elderberry Juice. *Acta Uniers. Sapien.* **2017,** *10* (1), 116–126.
35. Geankoplis, C. J., Ed. *Transport Processes and Separation Process Principles (Includes Unit Operations)*, 4th ed.; Pearson: London, UK, 2003; p 1040.
36. Guerreiro, D.; Madureira, J.; Silva, T.; Melo, R..; Santos, P. M. P.; Ferreira, A.; Trigo, M. J.; Falcão, A. N.; Margaça, F. M. A.; Cabo Verde, S. Postharvest Treatment of Cherry Tomatoes by Gamma Radiation: Microbial and Physicochemical Parameters Evaluation. *Innov. Food Sci. Emerg. Technol.* **2016,** *36*, 1–9.
37. Hassanzadeh, P.; Tajik, H.; Rohani, S. M. R.; Moradi, M.; Hashemi, M.; Aliakbarlu, J. Effect of Functional Chitosan Coating and Gamma Irradiation on the Shelf-life of Chicken Meat During Refrigerated Storage. *Radiat. Phys. Chem.* **2017,** *141*, 103–109.
38. He, X.; Liu, R.; Nirasawa, S.; Zheng, D.; Liu, H. Effect of High Voltage Electrostatic Field Treatment on Thawing Characteristics and Post-Thawing Quality of Frozen Pork Tenderloin Meat. *J. Food Eng.* **2013,** *115* (2), 245–250.
39. Hellwig, M.; Henle, T. Baking, Ageing, Diabetes: A Short History of the Maillard Reaction. *Angew. Chem.* **2014,** *53* (39), 10316–10329.
40. Hocaoğlu, A.; Demirci, A. Ş.; Gümüs, T.; Demirci, M. Effects of Gamma Irradiation on Chemical, Microbial Quality and Shelf Life of Shrimp. *Radiat. Phys. Chem.* **2012,** *81* (12), 1923–1929.
41. Huang, K.; Yu, L.; Wang, W. Comparing the Pulsed Electric Field Resistance of the Microorganisms in Grape Juice: Application of the Weibull model. *Food Control* **2014,** *35* (1), 241–251.
42. Hussain, P. R.; Omeera, A.; Suradkar, P. P.; Dar, M. A. Effect of Combination Treatment of Gamma Irradiation and Ascorbic Acid on Physicochemical and Microbial Quality of Minimally Processed Eggplant (*Solanum melongena* L.). *Radiat. Phys. Chem.* **2014,** *103*, 131–141.

43. Hussain, P. R.; Suradkar, P. P.; Wani, A. M.; Dar, M. A. Potential of Carboxymethyl Cellulose and Irradiation to Maintain Quality and Control Disease of Peach Fruit. *Int. J. Biol. Macromol.* **2016,** *82*, 114–126.
44. Hwang, K. E.; Kim, H. W.; Song, D. H.; Kim, Y. J.; Ham, Y. K.; Lee, J. W.; Choi, Y. S.; Kim, C. J. Effects of Antioxidant Combinations on Shelf Stability of Irradiated Chicken Sausage during Storage. *Radiat. Phys. Chem.* **2015,** *106* (1), 315–319.
45. Ilame, S. A.; Satyavir, V. S. Application of Membrane Separation in Fruit and Vegetable Juice Processing: A Review. *Critic. Rev. Food Sci. Nutr.* **2015,** *55* (7), 964–987.
46. Jung, S. Applications and Opportunities for Pressure-Assisted Extraction. In *High Pressure Processing of Foods*; Balasubramaniam, V., Barbosa-Cánovas, G., Lelieveld, H., Eds.; Springer: New York, 2016; pp. 173–191.
47. Kaavya, R.; Pandiselvam, R.; Kothakota, A.; Banuu-Priya, E. P.; Arun-Prasath, V. Sugarcane Juice Preservation: A Critical Review of the State-of-the-Art and Way Forward. *Sugar Tech.* **2019,** *21*, 9–19.
48. KoçGüler, S.; Boston, S. Z.; Çon, A. H. Effects of Gamma Irradiation on Chemical and Sensory Characteristics of Natural Hazelnut Kernels. *Postharv. Biol. Technol.* **2017,** *123* (1), 12–21.
49. Koubaa, M.; Rosello-Soto, E.; ŠicŽlabur, J.; Jambrak, A. R.; Brnčić, M.; Grimi, N.; Boussetta, N.; Barba, F. J. Current and New Insights in the Sustainable and Green Recovery of Nutritionally Valuable Compounds from *Stevia rebaudiana* Bertoni. *J. Agric. Food Chem.* **2015,** *63* (31), 6835–6846.
50. Kuntz, F.; Strasser, A. The Specifics of Dosimetry for Food Irradiation Applications. *Radiat. Phys. Chem.* **2016,** *129*, 46–49.
51. Leong, S. Y.; Oey, I.; Clapperton, D.; Aganovic, K.; Toepfl, S. Chilling Prior to Low Intensity Pulsed Electric Field Processing Improved Vitamin C Stability of Carrot Purée (*Daucus carota*). *Int. J. Food Sci. Technol.* **2015,** *50* (8), 1757–1763.
52. Lung, H. M.; Cheng, Y. C.; Chang, Y. H.; Huang, H. W.; Yang, B. B.; Wang, C. Y. Microbial Decontamination of Food by Electron Beam Irradiation. *Trends Food Sci. Technol.* **2015,** *44* (1), 66–78.
53. Machado, R. M. D.; Haneda R. N.; Trevisan, B. P.; Fontes, S. R. Effect of Enzymatic Treatment on the Cross-Flow Microfiltration of Açaí Pulp: Analysis of the Fouling and Recovery of Phytochemicals. *J. Food Eng.* **2012,** *113* (3), 442–452.
54. Martinez-Monteagudo, S. I.; Balasubramaniam. V. M. Fundamentals and Applications of High-Pressure Processing Technology. In *High Pressure Processing of Foods*; Balasubramaniam, V., Barbosa-Cánovas, G., Lelieveld, H., Eds.; Springer: New York, USA, 2016; pp 3–17.
55. Martinez-Monteagudo, S. I.; Saldaña, M. D. A. Chemical Reactions in Food Systems at High Hydrostatic Pressure. *Food Eng. Rev.* **2014,** *6*, 105–127.
56. Milani, E. A.; Alkhafaji, S.; Silva, F. V. M. Pulsed Electric Field Continuous Pasteurization of Different Types of Beers. *Food Control* **2015,** *50*, 223–229.
57. Mohamed, M. E. A.; AmerEissa, A. H. Pulsed Electric Fields for Food Processing Technology. Chapter 11. In *Structure and Function of Food Engineering*; Eissa, A. A., Ed.; IntechOpen online: London, United Kingdom, 2012; pp. 275–306.
58. Naresh, K.; Varakumar, S.; Shekhar, P.; Sharma, A. Effect of γ -Irradiation on Physico-Chemical and Microbiological Properties of Mango (*Mangifera indica* L.) Juice from Eight Indian Cultivars. *Food Biosci.* **2015,** *12*, 1–9.

59. Nourbakhsh, H.; Alemi, A.; Emam-Djomeh, Z.; Mirsaeedghazi, H. Effect of Processing Parameters on Fouling Resistances during Microfiltration of Red Plum and Watermelon Juices: A Comparative Study. *J. Food Sci. Technol.* **2014,** *51* (1), 168–172.
60. Parniakov, O.; Barba, F. J.; Grimi, N.; Lebovka, N.; Vorobiev, E. Impact of Pulsed Electric Fields and High Voltage Electrical Discharges on Extraction of High-Added Value Compounds from Papaya Peels. *Food Res. Int.* **2014,** *65* (part C), 337–343.
61. Parniakov, O.; Barba, F. J.; Grimi, N.; Lebovka, N.; Vorobiev, E. Extraction Assisted by Pulsed Electric Energy as a Potential Tool for Green and Sustainable Recovery of Nutritionally Valuable Compounds from Mango Peels. *Food Chem.* **2016,** *192*, 842–848.
62. Pina-Pérez, M. C.; Martínez-López, A.; Rodrigo, D. Cocoa Powder as a Natural Ingredient Revealing an Enhancing Effect to Inactivate *Cronobactersakazakii* Cells Treated by Pulsed Electric Fields in Infant Milk Formula. *Food Control* **2013,** *32* (1), 87–92.
63. Pina-Pérez, M. C.; Rodrigo, D.; Martínez, A. Non-Thermal Inactivation of *Cronobacter-sakazakii* in Infant Formula Milk: A Review. *Critic. Rev. Food Sci. Nutr.* **2014,** *56* (10), 1620–1629.
64. Polidori, J., Dhuique-Mayer, C.; Dornier, M. Cross Flow Microfiltration Coupled with Diafiltration to Concentrate and Purify Carotenoids and Flavonoids from Citrus Juices. *Innov. Food Sci. Emerg. Technol.* **2018,** *45*, 320–329.
65. Puértolas, E.; Barba, F. J. Electrotechnologies Applied to Valorization of By-products from Food Industry: Main Findings, Energy and Economic Cost of their Industrialization. *Food Bioprod. Process.* **2016,** *100* (part A), 172–184.
66. Rajha, H. N.; Boussetta, N.; Louka, N.; Maroun, R. G.; Vorobiev, E. Electrical, Mechanical, and Chemical Effects of High-Voltage Electrical Discharges on the Polyphenol Extraction from Vine Shoots. *Innov. Food Sci. Emerg. Technol.* **2015,** *31*, 60–66.
67. Rezzadori, K.; Serpa, L.; Penha, F. M.; Petrus, R. R.; Petrus, J. C. C. Cross Flow Micro-filtration of Sugarcane Juice: Effects of Processing Conditions and Juice Quality. *Food Sci. Technol. Int.* **2014,** *34* (1), 210–217.
68. Rosello-Soto, E.; Barba, F. J.; Parniakov, O.; Galanakis, C. M.; Lebovka, N.; Grimi, N.; Vorobiev, E. High Voltage Electrical Discharges, Pulsed Electric Field, and Ultrasound Assisted Extraction of Protein and Phenolic Compounds from Olive Kernels. *Food Bioprocess Technol.* **2015,** *8*, 885–894.
69. Sarkis, J. R.; Boussetta, N.; Tessaro, I. C.; Marczak, L. D. F.; Vorobiev, E. Application of Pulsed Electric Fields and High Voltage Electrical Discharges for Oil Extraction from Sesame Seeds. *J. Food Eng.* **2015,** *153*, 20–27.
70. Saulis, G. Electroporation of Cell Membranes: the Fundamental Effects of Pulsed Electric Fields in Food Processing. *Food Eng. Rev.* **2010,** *2* (2), 52–73.
71. Shahbaz, H. M.; Ahn, J. J.; Akram, K.; Kim, H. Y.; Park, E. J.; Kwon, J. H. Chemical and Sensory Quality of Fresh Pomegranate Fruits Exposed to Gamma Radiation as Quarantine Treatment. *Food Chem.* **2014,** *145*, 312–318.
72. Sharma, P.; Bremer, P.; Oey, I.; Everett, D. W. Bacterial Inactivation in Whole Milk Using Pulsed Electric Field Processing. *Int. Dairy J.* **2014,** *35* (1), 49–56.
73. Song, W. J.; Sung, H. J.; Kim, S. Y.; Kim, K. P.; Ryu, S.; Kang, D. H. Inactivation of *Escherichia coli* O157 : H7 and *Salmonella typhimurium* in Black Pepper and Red Pepper by Gamma Irradiation. *Int. J. Food Microbiol.* **2014,** *172*, 125–129.
74. Tonello, C. Case of Studies on High-Pressure Processing of Foods. Chapter 4. In *Nonthermal Processing Technologies for Foods*; Zhang, H. Q., Barbosa-Cánovas, G. V., Balasubramaniam,V. M., Dunne, C. P., Farkas, D. F., Yuan, J. T. C., Eds.; Wiley-Blackwell: Oxford, UK, 2011; pp 36–50.

75. Ushikubo, F. Y.; Watanabe, A. P.; Viotto, L. A. Microfiltration of Umbu (*Spondias tuberosa*) Juice. *J. Membr. Sci.* **2007,** *288* (1–2), 61–66.
76. Vladisavljević, G. T.; Vukosavljević, P.; Veljović, M. S. Clarification of Red Raspberry Juice using Microfiltration with Gas Backwashing: A Viable Strategy to Maximize Permeate Flux and Minimize a Loss of Anthocyanins. *Food Bioprod. Process.* **2013,** *91* (4), 473–480.
77. Wiktor, A.; Śledz, M.; Nowack, M.; Chudoba, T.; Witrowa-Rajchert, D.; Pulsed Electric Field Pretreatment for Osmotic Dehydration of Apple Tissue: Experimental and Mathematical Modeling Studies. *Dry. Technol.* **2014,** *32* (4), 408–417.
78. Won, Y. C.; Min, S. C.; Lee, D. U. Accelerated Drying and Improved Color Properties of Red Pepper by Pretreatment of Pulsed Electric Fields. *Dry. Technol.* **2015,** *33* (8), 926–932.
79. Wu, Y.; Zhang, D. Effect of Pulsed Electric Field on Freeze-Drying of Potato Tissue. *International J. Food Eng.* **2014,** *10* (4), 857–862.
80. Xi, J.; He, L.;Yan, L.Continuous Extraction of Phenolic Compounds from Pomegranate Peel using High Voltage Electrical Discharge. *Food Chem.* **2017,** *230*, 354–361.
81. Xue, P.; Zhao, Y.; Wen, C.; Cheng, S.; Lin, S. Effects of Electron Beam Irradiation on Physicochemical Properties of Corn Flour and Improvement of the Gelatinization Inhibition. *Food Chem.* **2017,** *233*, 467–475.
82. Zafar, S.; Ahmad, I.; Rafique, M.; Zuber, M.; Shahid, M.; Parveen, I. Effect of Gamma Irradiation on Fungal Load and Aflatoxins Reduction in Red Chillies. *Radiat. Phys. Chem.* **2013,** *82*, 80–84.
83. Zhang, L.; Lu, Z.; Lu, F.; Bie, X. Effect of Gamma Irradiation on Quality-Maintaining of Fresh-Cut Lettuce. *Food Control* **2006,** *17* (3), 225–228.
84. Zhang, Z. S.; Xie, Q. F.; Che, L. M. Effects of Gamma Irradiation on Aflatoxin B_1 Levels in Soybean and on the Properties of Soybean and Soybean Oil. *Appl. Radiat. Isotop.* **2018,** *139* (9), 224–230.
85. Zhao, W.; Yang, R.; Shen, X. Zhang, S.; Chen, X. Lethal and Sublethal Injury and Kinetics of *Escherichia coli*, *Listeria monocytogenes* and *Staphylococcus aureus* in Milk by Pulsed Electric Fields. *Food Control* **2013,** *32* (1), 6–12.

CHAPTER 4

THERMOPHYSICAL AND PHYSICOCHEMICAL TECHNOLOGIES FOR DRYING OF FOOD PRODUCTS

NAMRATA A. GIRI, T. KRISHNAKUMAR, G. JEEVARATHINAM, and M. S. SAJEEV

ABSTRACT

Drying is a thermophysical and physicochemical food process in which the excess moisture from a food product is removed. Dehydration is a process in which products are dried to very low-temperature levels. To expedite the process of drying, it is very important to understand the nature of air and the composition of products. During the drying process, the moisture present in the solids gets evaporated and diffuse in a dilute environment. This chapter summarizes the various pretreatments such as blanching, salting, acid and alkali treatment, sulfating and freezing treatment, and advanced drying technologies such as hybrid drying, infrared drying, refractance window drying, radio frequency drying, heat pump drying, and superheated steam drying.

4.1 INTRODUCTION

Drying is the most effective unit operation for the protection of food grains and safe storage of agricultural products. It removes the available moisture present in products to a safe moisture level so that it will be stored without any quality loss to increase the shelf-life (e.g., cereals). Dehydration means complete removal of moisture into bone dry conditions of the produce/

Novel and Alternative Methods in Food Processing: Biotechnological, Physicochemical, and Mathematical Approaches. N. Veena, Megh R. Goyal, & Ritesh B. Watharkar (Eds.)

product where the water activity in the material is well below the safe level (e.g., fruits and vegetables). Drying helps to minimize postharvest losses, stability in the price of the produce, and extend the availability of particular farm produce beyond the season. During drying, the total heat content of the grain is increased, and thereby vapor pressure inside the grain is increased. This causes the transfer of moisture from the internal part of the grain or seeds to the outside atmosphere. In most cases, the movement of moisture takes place because of diffusion, cell contraction, and vapor pressure gradient.

The drying or heat capacity of the given drying agent (air) mainly depends on air temperature, grain type, grain maturity, grain moisture, and the relationship between moisture content and relative humidity. Depending upon the purpose, the drying air temperature should be selected. Excessive drying air temperature leads to quick drying but induces chemical changes or fissures within the produce, which results in increased brokenness in milling. During drying, food materials are subjected to various quality changes through physical, chemical, and biological means that affect sensory characteristics (aroma, flavor, and texture) and nutritional value.

Thus, the overall objective of drying and dehydration is to make food grains and products acceptable for safe storage and conserve them against biotic environmental factors such as insects, molds, microorganisms, etc., during storage. The classification of dryers is mainly based on the three major principal factors namely; a method in which heat is passed into the material, air temperature (low, medium, and high), and pressure of operation (atmospheric or vacuum condition), and a method over the food material is controlled within the dryer. There are two major important critical process factors to be considered in the unit operation of drying: heat transfer for providing the necessary latent heat of vaporization; and water vapor movement across the food produce and dissociation of water vapor from food produce.

This chapter describes the components and working principles of some of the advanced drying technologies such as hybrid drying, infrared (IR) drying, refractance window drying, radio frequency drying, heat pump drying, and superheated steam drying in detail with the required pictorial representation.

4.2 PRETREATMENTS OR PREDRYING TREATMENTS

To obtain a good quality dried product, the harvested raw produce must be handled properly throughout the value chain. In general, pretreatments reduce drying time thereby reducing the postharvest processing cost through higher drying efficiency. Biological materials are subjected to particular pretreatments before drying in regards to enzyme inactivation and reduce possible unfavorable changes during drying and followed by further storage. Peroxide enzyme is one of the major key enzymes accountable for off-flavor development. Drying of fruits without suitable pretreatments causes browning of fruits even after the drying process.[31]

Various physicochemical changes take place when drying is combined with pretreatments that eventually improve the movement of water vapor transfer and quality of products.[13] The major quality criterion of dehydration products is good rehydration characteristics, color, and storage stabilities. As pretreatments influence, the selective physicochemical properties of dried products, a selection of appropriate pretreatments should be made to have superior quality dried products. Thus, food produce must be dried in combination with selecting suitable pretreatments such as blanching, salting, alkaline, acid, sulfuring, and freezing treatment. Different pretreatments that shall affect product quality before drying are discussed in this section.

4.2.1 BLANCHING

Blanching is an important heat process done using hot water or steam in the preparation of fruits and vegetables before dehydration. Saturated steam is used for steam blanching. The loss of water-soluble material is less in steam blanching and it is easier to stabilize the blancher than water blanchers. The hot water blanching is thermostatically controlled at a temperature of 70–96 °C and a duration of 1–10 min. It is intended for fruits and vegetables in advance of drying to inactivate enzymes namely, polyphenol oxidase in fruits, peroxidase in vegetables, and catalase in cabbage. The other beneficial effects are: (a) using enzyme inactivation arrest the growth of objectionable odors and flavors during storage; (b) reduce drying time; (c) removes inter-molecular void space from the tissues; (d) causes softening of texture; (e) retains carotene and ascorbic acid during storage; and (f) removes pungency (onions) and reduces the loss of soluble solids.

In fruits (such as apples, apricots, and peaches), blanching provides the desirable semitransparent characteristics to the dehydrated product. There are two important critical factors such as time and temperature to be considered to attain the desirable character of the dried products.[25] Thompson seedless grapes treated with high-humidity hot air impingement blanching of different time-temperature combinations found that the polyphenol oxidase activity got reduced with raise in time and temperature of blanching. Moreover, the color of the treated grapes was green–yellow as desirable.[5] Thompson seedless grapes treated with hot water blanching treatment alone, showed a quick reduction in the drying time, whereas the final product's quality was not desirable.[49]

The rate of drying in bell pepper subjected to blanching before drying was high due to rupturing of the membrane that facilitates the faster removal of moisture.[61] The peroxidase enzyme present in the chilies was inactivated by hot water blanching (90°C, 3 min). The physical characteristics of the dried chilies were examined to be best in the treatment of gum acacia solution (0.2%) for 15 min at an atmospheric temperature.[20] Limited or reduced blanching temperature and increased blanching time have been recommended nowadays for better retention of texture and nutrients.[36]

4.2.2 SALTING

Salting (known as curing) a method of osmotic dehydration that is a liquid–solid operation, is considered a suitable pretreatment, for food preservation due to energy saving.[52] Osmotic dehydration as a pretreatment for water removal and enhancing taste has been confirmed in various studies.[7] It is mostly applied to maintain color, improve drying characteristics, and quality of products such as fish, meat, and cheese products. The commonly used salt, which consists of sodium chloride, commercially sold as solar salt, evaporated salt, and rock salt, is used as a curing agent in the salting process. Salting is a cheap and old method of food preservation and acts as a pretreatment that is carried out before drying or smoking. There are different salting processes such as dry salting, wet salting, pickle salting, kench salting, mona curing, pit curing, and Colombo curing that are adopted in fish preservation. Salting assists in the development of characteristic flavor and aroma qualities in the dried products in combination with drying.[59]

4.2.3 ALKALINE TREATMENT

This treatment is widely used for drying whole fruits especially apricot, berries, grapes, and plum. Chilies pretreated with lye solution consisting of 0.5% KOH in hot water at 80°C followed by drying in a tray dryer at 65.55°C results in a variation of physicochemical qualities and dehydration ratio of chili cultivars.[62] The rate of drying of grapes (var. sultans) increased for an increase in K_2CO_3 concentration from 4% to 7%.[33] The drying time required for Thompson seedless grapes was comparatively lower in NaOH treatment as compared with other chemical treatments.[49]

4.2.4 ACID TREATMENT

The use of an acid predip before sulfuring provides better color stability to a product. The use of ascorbic acid plays the unique role of a reducing agent and lowers the pH of the product. The enzymatic browning in peaches during the drying process can be arrested by dipping the peaches in 1% ascorbic acid along with 0.25% malic acid solution. Peaches treated with 1.0% ascorbic acid for 3 min followed by sun-drying has better quality than nontreated fruits. Steeping of mushrooms in 0.5–1.0% citric acid for 16 h resulted in better quality dried products in terms of color, appearance, rehydration properties, and keeping quality. Chili pretreated with citric acid (0–5%), dried in microwave-assisted vacuum dryer (1120 W power, 600 mm Hg vacuum) for 60 min yielded dried chili color alike fresh chili.[69]

Sweet potato flour was made from sweet potato chips that were pretreated with 1% ascorbic acid. The carotenoid level of the sweet potato flour was improved during the first month of storage, but these levels were not maintained for over 4–6 months of storage.[6] Fried sweet potato chips that were pretreated with citric acid and predried for 50 min had the overall liking and best sensory attributes based on the organoleptic evaluation.[47]

Mango fruit chips pretreated with citric acid at the rate of 1% showed better color retention and rehydration properties after drying as compared with blanching.[44] A higher rate of moisture reduction in mango slices, which pretreated with lemon juice followed by a salt solution, hot water blanching compared with control samples has been noticed.[12]

4.2.5 SULFURING/SULFITING/SULFUR DIOXIDE TREATMENT

Sulfur dioxide is traditionally used as a preservative in the fruit and vegetable processing industries. It inhibits the enzyme-assisted oxidative changes, preserves better color and aroma, and inhibits the growth of spoilage microorganisms. Fruits are generally given sulfur treatment by placing them in a vented compartment containing sulfur gas, whereas vegetables are given sulfite solution treatment for controlled absorption. Sulfur dioxide, an antioxidant has been used for preservation besides the protection of carotene and ascorbic acid content of dried fruits. Onion slices dipped in 0.2, 0.3, and 0.4%

potassium metabisulfite for 5 min, showed an increase in energy absorption due to shorter drying time and rise in product temperature.[50] Sulfuring pretreatment has been proven to improve the storage stability of dried fruits. Soak treatment of hydrogen peroxide (30%) for 30 min improves the whiteness of dehydrated *Agaricus bisporus* but adversely affects the texture, taste, and flavor.

4.2.6 FREEZING TREATMENT

Generally freezing of foods before drying has a positive effect on the drying process. There is only limited research that has been conducted so far on the effect of freezing on tissue structure before drying. It is evident that moderate duration of freezing and temperature results in better rehydration of dried products. The faster drying rate of the banana fruit was noticed in the freezing treatment than blanching and chiling, whereas texture and flavor of the banana samples were reduced significantly.[10]

4.3 DRYING CHARACTERISTICS AND KINETICS

Grain drying involves exposing the agricultural produce to the heating agent (dry air), which ultimately evaporates and take away the moisture from the agricultural produce. As the process of drying greatly influence the quality of produce, it is vital to apprehend the basic principles of grain drying. The moisture removal predominantly depends on atmospheric conditions rather than the type of product. In general, agricultural products are dried under different drying rates. The duration of constant rate drying is very short for agricultural products and continues till attaining the critical moisture content. Constant rate drying is mainly depending on the relative humidity of drying air, the velocity of drying air, and water vapor pressure of grains. Food grains are subjected to falling rate drying once the constant rate drying completes. The critical moisture content of food grains occurs between constant rate and falling rate period and it mostly depends on the nature of products and their thickness.

The drying rate is directly proportional to the moisture content of the product. It largely depends on different properties of the product (moisture content, geometry, surface temperature, water activity, moisture loss) and air (temperature, relative humidity, velocity, heat transfer coefficient). The rate of drying is usually expressed in percent per hour. To design the dryer system for agricultural produce, the information on the physical characteristic parameters of the produce must be determined. The drying constant of a drying process provides information about the rate of drying of a product. The kinetic equation in the drying process is mostly used to express the moisture ratio as a function of time. This is an appropriate technique to express the moisture content reduction in food products during the drying process.

4.4 NEED FOR NEW DEVELOPMENTS IN DRYING TECHNOLOGY

Industrial products are generally dried using different mechanical dryers. The basic three heating methods are conduction, convection, and radiation. Excessive drying air temperature causes quick drying but induces chemical changes or fissures or flow or strain within the food grains, which results in increased brokenness in milling. In the case of malted tea, high temperatures impart desired chemical changes within the material so that the final product is suitable for preparing drinks.

Drying and storage are part of the processing activity. These operations should be done very carefully in a "system approach." For example, drying paddy grains to a moisture level below 10% may help for long storage but during milling, breakage will be more. Hence these operations should be done to the optimum level. The dehydration/drying resulted in physical, chemical, and biological changes. Many of these changes are undesirable. Therefore, while dehydrating the fruits and vegetables the processor must consider some important parameters to get a product of superior quality. These parameters are retention of volatile flavors and aroma, retention of color, texture, and biological purity, retention of nutritional quality, and change of shape-shrinkage. In the case of designing a new drier, the following major design considerations should be taken into account:

- Heating medium.
- Mechanical aid if any, to improve drying.
- Method of circulating the air and its influence if any, on the measurement of the moist material.
- Method of heating.
- Methods of moving the moist material through the drier.
- Nature of the wet material and the manner of its introduction into the drier.
- Temperature and pressure level in the drier.

- Type of food material to be dried.
- Ways in which the material is supported.

Since there are many limitations in the existing drying technologies, it is the need of the hour to develop innovative new drying technologies by considering the above design considerations.

The latest drying technologies are coming under the category of fourth-generation that includes IR, microwave, fluidized bed, refractance window (RW™), heat pump, radiofrequency, and other hybrid drying systems. They are versatile due to their handling of different raw materials and their efficiency in holding quality parameters of the final product.[8] Presently, drying methods like IR, hybrid, and refractance window are widely used as nonthermal methods for drying food products.

4.5 HYBRID DRYING

Bioproducts are produced in higher quantities in most of the agricultural countries that are dried to enhance the postharvest life, reduction in the cost of packaging, reducing shipping loads, improving appearance, original flavor encapsulation, and maintaining nutritional value. Food drying's essential objective is extending the shelf-life from spoilage, hinging on suitable method, the fresh product may turn into a wholesome finished end product with some important difference in quality. Due to these reasons, more priority was given to select a dryer with more suitability to the product and also the strong-supportive correlation among the dryers and bioproduct to be dried. Dryers are more concentrated among all other works. Dire consequences, quantity and quality losses, and sometimes substantial financial losses occur if the drying technology and product are mismatched. The thirst in the field of discovering new hybrid bioproducts, which produced constantly creates a wider spectrum for dried bioproducts.

Nowadays there is a necessity for the development of two combined drying systems becomes more crucial because of the inabilities faced by a single drying system in satisfying precise quality requirements of the new bio-products. Therefore, the concept of developing new hybrid drying technologies focuses mainly on satisfying consumer needs and also to produce the product with suitable moisture content.

For achieving suitable dryness, drying time, quality of the product, and manufacturing throughput, hybrid drying technologies should employ multiple dryers or multiple heat transfer modes as well as more stages of drying. Normally, the term hybrid means joining two systems. Here, two or more drying systems are combined and named hybrid drying technologies. Combination/integration can be achieved by generating an idea for the development of a new process. Examples of hybrid technology have discussed below.

The system based on IR and hot air oven was compared for fish paste roasting and rice crackers baking.[57] Among these, IR heating system was better in energy savings (45–70%) compared with another system. Ozdemir and Devres[48] reported the color development of hazelnuts processed by the IR roasting method. The synergistic effect was considered as more efficient that can be obtained through the combination of both electromagnetic radiation and conventional heating. The application of combined multihybrid heating systems (IR, microwave oven, and hot air) over postharvest foods was reported.[11]

The specific energy and the quality characteristics of barley during combination drying using far-IR and conventional air were studied.[2] Almost 245% of total energy was saved by using this technology. The combination method of drying gained a synergistic effect. The theoretical representation of the hybrid drying mechanism (IR and hot air) was presented in Figure 4.1. When IR radiation passed over the food material, it intrudes on the surface and radiation penetrates the food.

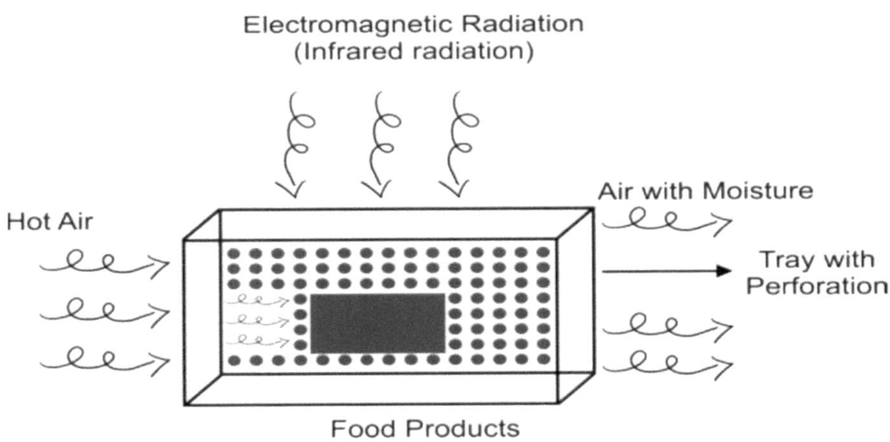

FIGURE 4.1 Conceptual diagram of hybrid drying (IR and convective air).[22]

Sakai and Hanzawa[57] stated that penetration causes more radiation absorption that results in increased vibration of molecules, in turn, generates thermal energy in the surface and inner layers of food material. This quick heating improves the rate of moisture transportation toward the upper layer. The moisture removed from the surface by passing convective air, also reduces the temperature, resulting in an increased mass transfer.

4.5.1 HYBRID DRYING TECHNOLOGIES—MULTIPLE INNOVATIONS

In highly multipart drying systems, the hybridization of two or more systems is involved to obtain a better drying system with more efficiency. Presently, innovations are involved in the development of several combinations of two drying technologies to exploit the benefits of various methods that produce enhanced quality food material, lowering the time took for drying and improving the rate of drying rate.[23] Employing new methods of combined drying technologies such as vacuum, microwave, hot air, and freeze-drying retains the shelf-life, value, and health properties of berries to a larger range were stated.[34]

4.6 IR DRYING

A nonthermal drying method that uses electromagnetic radiation with specified wavelengths is also called IR drying. IR radiation supplies heat that shortens the time of drying making an effective dehydration technique. IR heating is particularly well suited to dry food materials having more surface area exposed to radiation and mainly thin layer materials. The efficiency of heating is between 80% and 90%, the wavelength range of the radiation discharged is in thin form and it is reduced.[29,58]

Heating elements are used only to transfer the energy from IR radiation to the outer layer of the product and it does not transfer to the surrounding air. Due to the air transparency, the drying was carried out at normal air temperature (30 ± 2°C). The radiation intrudes over the material exposed and it penetrates and converts into sensible heat. The material with higher moisture is most preferable for drying. Drying kinetics can be affected when IR energy absorption by moisture is considered an essential variable. Mostly, absorption of IR radiation by solid materials occurs in a thin surface layer. On the other hand, radiation penetrates the wet permeable materials to a particular distance and transmissivity has been chosen by the moisture content.[30] At the time of drying, the reduction in moisture also changes the material's radiation properties. This results in increase in reflectivity and decrease in absorptivity. The absorptivity, transmissivity, and depth of skin are dependent on irradiated material's properties, density, and wavelength of IR heating.

At a short wavelength, water is used to transmit IR radiation, and for a long wavelength, surface absorption of radiation takes place.[57] Therefore, at far-IR radiation (range 25–100 μm), thin layers can be suitable for drying to get more efficiency, for near-IR radiation (range 0.75–3.00 μm) drying thicker bodies/layers had better results.

The benefits of IR drying over conventional drying as the energy transmission happens directly to the surface exposed and the rapid rise of suitable heat levels, which activates the fundamental drying mechanisms (shortens dehydrating time), more energy efficiency, enhanced quality of end food, even heating distribution in the food during drying, and reduction in airflow requirement across the product.[53,60] The temperature difference in the food material decreases shortly due to the intense heating. Hence, the consumption of energy is moderately lesser in the IR drying process. High heat and mass transfer coefficients, less time, and high savings in cost. Equipment and technologies can be compact and automation can be done with the process. It is very much essential due to the fast and the risk of over-dried material.[57] Tunnel-type IR dryers are used to dry vegetables, seaweeds, pasta, and fish flakes. Besides, the application of this type of dryer was found in food analysis, mainly to measure the amount of moisture in food materials.[3]

Because of its more advantages, it is likely to combine with convection or vacuum to develop hybrid technologies. Drying kinetics depends on the wavelength, density, air velocity, space between the IR source released, and the outer layer irradiated. Normally, if the material temperature is high therefore it causes a decrease in the drying medium's relative humidity increases the drying kinetics. The high temperature beyond the optimum causes thermal reduction of phytochemicals. More surface area or decrease in dimensions of material to be dried also shortens the period of exposure of the material to the temperature.

The heat transfer of IR drying has significantly different from convective heating. Surface layers absorb the radiation and changed into heat. In foods with more moisture content, the maximum heating was achieved beneath the surface layer which is irradiated and also hinges on the constant of extinction.

The lesser the constant of extinction, the higher the depth toward the upper layer of product results in higher energy transfer takes place.[18]

Therefore, the temperature produced can conduct at the center and also to the upper surface layer. Convection transfers the high temperature from the upper layer of the product to the air surrounds the system. However, from the midpoint to the upper layer of the material, the water flux is conveyed every time. In end, the energy and fluxes are counter currents in the material and the surface layer is cocurrent. Both fluxes are cocurrent at the surface and during convective drying; there should be a difference in the amount and heating profiles of the air.

An innovation in the IR energy drying model took place.[21] As per assumption, heat absorption occurs at the upper layer of the material, and three divisions of drying take place. Heating the product and constant drying rate period comes under the first division. It is assumed as resistance toward inner moisture movement is very small and surface moisture evaporation takes place. At surface heating, the water vapor pressure is equivalent to the atmospheric vapor pressure. After some time, the next division that is falling rate occurs. The surface becomes moisture less and shrinks and a decrease in drying rate begins. Continuation in drying leads to dryness in the surface and withdrawal of evaporation and movement of moisture toward the center of the body. The third division is described as the evaporation of water as vapor from the dry layer. Heat absorption can be considered in the energy balance; however, the resistance of outside heat transfer is not included. Ratti and Mujumdar[54] developed material and energy balance. The calculations based on theory concluded that the energy input of IR drying was found to be 10 kW/m^2, which is equal to the conventional method, where the heat transfer coefficients occur to a maximum of 200 W/m-K.

4.7 REFRACTANCE WINDOW (RW™) DRYING

This type of drying is an innovative nonthermal method of drying for converting biomaterials, foods with more moisture into concentrate, bars, and powders with value addition. This system can dry the heat-sensitive purees or juices and slices in very lesser exposure times of 2–6 min additionally higher retention of color, vitamins, and antioxidants. This type of drying method is used in meat powder production.[54] It is simple and inexpensive than other dryers such as freeze dryers. The higher capacity seems to be economical. The main principle is the hot surface transfers the heat energy to a layer of juice or puree, which is spread on a plastic conveyor belt as a thin layer. It operates at ambient conditions and is used for the production of varietal products from all types of food materials.

RW is a self-constraining remarkable drying technology that utilizes radiation energy from IR to expel out the moisture from food. Depending on the ability of water to conduct or absorbs IR and the refractance of light, results in it as the signature technique for prolonging the keeping quality of the valuable supplements and phytonutrients found in entire nourishments. In this procedure, significant tangible characteristics of the new entire food, for example, color, fragrance, taste, and health benefits are held. This is a pointer that the dynamic fragrant and color mixes, which give tactile and healthfully priceless properties have been saved all through the drying procedure. This has been utilized as a common process by MCD Technologies for whey drying and other food materials. It does not require high pressure like other dryers. It is reported that only there is the smaller scope of oxidation and free radical formation occurs.

4.7.1 RW™ DRYER WORKING SYSTEM

The drying of liquid and semiliquid products takes place by spreading as a thin film over the belt conveyor is called as Cast Tape Drying.[14,15] This technology was patented in 1986 by[30] MCD Technologies Inc., which is different from the Cast Tape Drying system.[51]

Working of RW drying depends on the electromagnetic radiation of IR, which is absorbed by the hot circulating water.[32] The innovation includes spreading the wet material as a thin layer on the plastic sheet (polyester film) over the conveyor belt.[1] The film transmission depends on the film thickness, which demonstrates more straightforwardness to IR radiations yet its mechanical quality diminishes when thickness brings down. Consequently, the ideal thickness was resolved as 0.25 mm.[73] The warmed water surface was wrapped by the film, which transfers the heat to the wet product by conduction.[32,41] The radiation power of light was absorbed by the sheet through which the water is heated to 95–97°C.[32] The heat from the water to the film and finally transferred to the food material through the conduction, convection, and radiation mode in less time.[1]

The absorption of the radiation by the film is based on the thickness and dampness content of the film.[39,41,53] When the dampness reaches the sheet with high temperature (from the outside boiling water), the opening for electromagnetic energy shaped permitting section for exchange of electromagnetic energy vitality transferred by sheet to the food to be dried out. Therefore, this method is considered an advanced method of drying. The food is dehydrated, meager dampness remains as the IR window continuously closes for less entry of energy constraining the temperature of the food somewhere in the range of 70°C and 75°C. The drying window closes and refracts again to the moisture if the item lost its dampness.[51] The reflection results in warming the water.

Conduction impacts keep on supporting dampness dissipation until the item arrives at basic dampness content. In the last phases of dehydration, related evaporative cooling impacts (advocating there is a difference in item and water temperature at some time of drying) has happened. With a decrease in the pace of warmth move, the item is shielded from high temperature.[46] The simple detachment of the item from the plastic transport line by a scrubber gadget was empowered. The cooling area helps to decrease the item temperature underneath the crystal progress point for the simple evacuation of the last items.

At the stage of completion, the product is brought to normal temperature by contact with the cold water and it is scraped from the belt.[40] The product temperature is brought below the glass transition temperature thereby avoids the stickiness of the product. The water that is used for heating is recirculated for improving thermal efficiency. The whole process is done at normal atmospheric conditions. RW method is a film-drying technique that involves indirect contact. Avoids cross-contamination of the product during dehydration.[51] The three types of heat transfer take place at a dynamic stage (Figure 4.2).

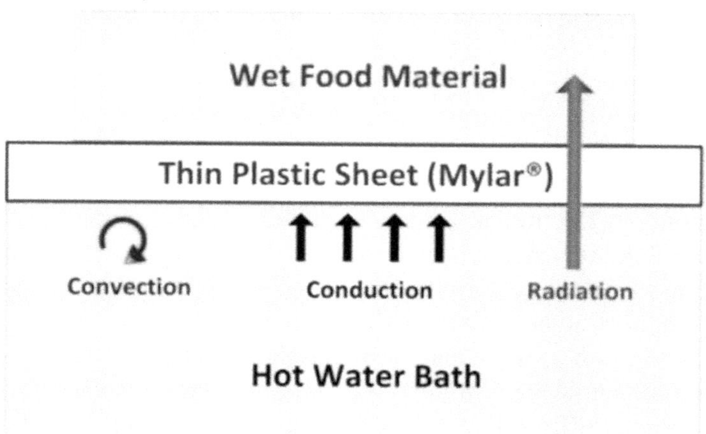

FIGURE 4.2 Heat transfer modes.

4.8 RADIO FREQUENCY DRYING (RFD)

RFD is a nonthermal method of heating, indirect electro heating in which heating of product done by conversion of electric energy to electromagnetic radiation.[30] This novel technique is most demanded in different food processing companies for multiple applications such as cooking, baking, drying, pasteurization, and other applications[45] because its application helps to meet the demand of consumer for tasty, convenient, economical, low-fat, chemical-free, and safe products.

4.8.1 PRINCIPLE AND WORKING OF RFD

A frequency range of 1–300 MHz comes under RF heating, among that 10–50 MHz are frequencies applied for industrial heating.[67] Food components are deprived of electric insulators and the application of electromagnetic fields to food materials leads to scattering electric energy.

A pair of electrode is used and food material is placed in between these electrodes. In RF heating system, an alternate electric field between two electrodes was generated using RF generator.[55] The alternate energy causes polarization of the water molecule present in food material, which is also known as dipolar. These molecules continuously reorient themselves concerning opposite poles of the electrode. These water molecules start oscillation at high frequencies due to following the same electric field with radiation and it resulted to heat the products.

4.8.2 APPLICATIONS OF RFD

RF heating has potential application in the food industry for drying, baking, defrosting, sterilization, pasteurization, and disinfection. Its use is also specific for heat-responsive food materials[8,70,72] and helps to preserve heat-sensitive compounds in nutritious food to retain food attributes than conventional heating. RF heating caused a reduction in processing time and also extends the shelf-life of food products without any chemical treatments or preservatives.[19] The major benefits of RF heating include the ease of targeting the product, not the air surrounding it, which tends to force the wetness from inside out and hence leveling the moisture, considerably decreasing the processing time and prevent extra heating of the product. Its use is also convenient and easy,[45] and can be provided with temperature measurement systems (thermocouple). The examples of applications of RF drying in the food industry are:

Drying and baking: RF drying is most commonly applied in postbaking of biscuits, cookies, crackers, and other snack items.[35] The traditional drying after baking leads to cracking on the surface, discoloration, and flavor damage due to moisture gradient and thermal buildup. These problems could be solved using RF drying due to uniform energy transformation, no moisture gradients, low drying temperature resulted in no discoloration, and flavor loss of baked products. Forced type convective baking oven is used for baking, which requires at least 30 min for proper baking of food materials. The application of RF heating for baking is more prominent for even baking and required less time and reduced to 8–10 min.

Defrosting: Conventional method of slow heating is used for defrosting of food products and its time taking depending on the size of product. The slow heating for defrosting will give chance to microbes for growth and quality deterioration. Slow defrosting also leads to drip loss which is considered an undesirable characteristic of food. The use of RF heating takes short time in minutes instead of hours or days for defrosting, even larger size food products can be defrosted with uniform faster RF heating. The food product inside the packaging material can also be defrosted and also prevent bacterial growth and drip loss.[4] The research work using RF heating mechanism in meat and fish product processing has a considerable consequence on nutrition, quality, quantity, and storage of the produce. Wang et al.[71] stated that the RF heating at 27 MHz and applying an alternating current of 6 kW is applicable for heating of various types of food before their packaging for maintaining the quality better compared to the conventional method of heating.

Sterilization and pasteurization: The application of very high temperatures more than 100°C is employed using saturated heat for sterilization of food material such as retorting whereas temperature less than boiling is used for pasteurization. The use of saturated steam for sterilization and pasteurization is a slow rise in temperature, nonuniform heating, and overheating to outer packaged food resulted in quality loss, burning odor to food, and also loss of heat-sensitive components in food. But RF heating having the ability to heat product eventually and prevent quality deterioration. RF heating is also preferred over conventional due to more economical, efficient, and environmentally friendly as there is no thermal energy loss to the environment. The dry ingredients can also be sterilized by RF heating were more prominent for the growth of microorganisms. RF heating mechanism is reported highly successful for nonperishable ingredients with low moisture content such as cereals, flour, seeds, spices, etc. Tiwari et al.[68] investigated the impact of the RF heating on wheat flour, placed in three cuboidal boxes were subjected to study on its physical properties when used frequency of 27.12 MHz.

Disinfestation: Chemical fumigants (like methyl bromide) are normally applied to kill the pest, but longer use of chemicals may cause environmental damage. The suitable method to control pest is RF heating, which causes fast and consistent heating to disinfect food commodities.

4.9 HEAT PUMP DRYING

It is an advanced technology utilized for drying agricultural perishable commodities having more quantity of free water.[63] It is very simple to use and takes less time for drying as compared with a conventional dryer.[27,56] Heat pump dryer (HPD) is considered as most economical because of the reuse of latent heat for drying purposes.[38] It has been reported that this dryer utilizes very little electricity in comparison with conventional condenser dryers.[16,43] So, it has the advantages of the low operating cost compared to the conventional method. It has the facility to pick up the latent and sensible heat produced from the condensation of moisture from drying air.[26] The energy efficiency is more for HPDs due to the reuse of recover heat for drying purposes. Therefore, it is the most effective and low-temperature technology of drying.[9]

4.9.1 PRINCIPLE AND WORKING OF HPD

Air is entered into the drying system at the entrance of the dryer and reduces the moisture content of produce by pick-up action. It is transferred to the evaporator coil. There are two types of evaporator systems, the use of expansion coil in which the phase of refrigerant directly changed from liquid to vapor to cool and dehumidify the air. Whereas, another related to the use of chilled water system to cool and dehumidify under controlled conditions. Air undergoes dehumidification where air-cooled sensibly to its dew point followed by condensation of water from the air. The refrigerant undergoes boiling when heat is absorbed by the evaporator. The recovery of heat takes place by condenser. When the movement of heat takes place from condenser to dryer, the condenser will absorb the heat to achieve the desired temperature for heating the product. Experimentally, the heat pump is 40–50% more efficient than theoretical Carnot efficiency.[17,66]

4.9.2 *CLASSIFICATION AND APPLICATIONS OF HPD*

HPDs are classified based on the different criteria such as processing modes, the number of drying stages, temperature of the material, etc. It could be effectively utilized with any type of dryer.[24,26] It can be utilized in combining with a convection type of dryer for solving the product deterioration including product quality, energy efficiency, and thermal efficiency.[28] HPD is very much suitable for drying heat-sensitive food materials especially fruits and vegetables. These food materials required low temperature for the prevention of deterioration of quality, loss of heat-sensitive nutrients such as vitamin C, etc. This drying method helps to provide better microbial safety, improved product color with enhanced sensorial quality attributes. The additional benefit of less requirement of energy and maintain the quality of the product is highly noticeable in HPDs.[65]

4.10 SUPERHEATED STEAM DRYING

It is one of the efficient methods of drying in which superheated steam is utilized in place of hot air as a drying medium for the removal of moisture content.[39] A superheated steam dryer (SSD) can be used along with a conventional dryer. These are classified into different types based on the types of dryers used along with superheated steam. The types of dryers that could be associated with superheated steam are flash dryer, fluidized bed dryer, spray dryer, conveyor dryer, impinging jets, agitated bed dryers, packed bed, and through circulation dryers.

4.10.1 *WORKING OF SSD*

The working of superheated steam drying is well explained with the example of drying of Brewer's Spent Grain which is the by-product from beer brewing. A horizontal, rotating drum having a dimension of 3 m in length and 0.37 m in diameter was placed and attached to the grain feeder. The speed of grain-fed was constant into the feed hopper and grain move to the drum with the help of the feed screw and the flow of grain and steam is cocurrent. The action of drying of grain happened in the drum, during the rotating position of the drum along with the contact of grain with superheated steam. Dried gain from drum removed through product screw. The superheated steam present in the drum move to a cyclone, which removes fine particles of grain carried along with steam and the fan is placed to transfer back these particles for drying. A thermocouple is necessary to place before the inlet to the drum. The steam evaporated during drying of grain is detached using shell and tube condenser.[65] There are pros and cons of SSD, such as

Pros:	Cons:
1. More energy efficient 2. Fast drying compared with conventional drying 3. Utilized for various thermal treatment of food products	1. Drying system is much complex, avoid leakage of superheated steam 2. Not feasible for the drying of products where browning reaction is desirable 3. Chances of product undergo glass transitions 4. Cost of the ancillaries is higher than the cost of the dryer

4.11 SUMMARY

Drying is an energy-intensive and mass-transfer industrial process. It is a thermophysical and physicochemical operation by which the excess moisture from a product is removed. Pretreatments performed before the drying process influence various physicochemical properties of the dried products; reduce drying time thereby reducing the postharvest processing cost through higher drying efficiency. This chapter documents the different pretreatments (blanching, salting, acid and alkali, sulfiting, and freezing) and the advanced six drying technologies (HPD, hybrid drying, IR, RF, RW, and SS drying). Presently, hybrid drying and IR drying and RW drying are widely used nonthermal methods for drying food products. If the drying system is efficiently designed, the drying of products will be faster and uniform. Thus, it is the need of the hour to develop cost-effective, energy-efficient drying methods by predominantly considering product quality. In the case of designing a new drier, the major design considerations to be kept in mind are detailed in this chapter. Advanced energy-efficient drying technologies are to be developed with the different combinations of conventional technologies as per the consumer and market requirements.

KEYWORDS

- **dehydration**
- **drying**
- **drying kinetics**
- **hybrid drying**
- **pretreatments**
- **predrying**

REFERENCES

1. Abonyi, B. I.; Tang, J.; Edwards, C. G. Evaluation of Energy Efficiency and Quality Retention for the Refractance Window TM Drying System. Unpublished Research Report, Washington State University: Pullman, WA, 1999; p 38.
2. Afzal, T. M.; Abe, T.; Hikida, Y. Energy and Quality Aspects during Combined FIR-Convective Drying of Barley. *J. Food Eng.* **1999,** *42* (4), 177–182.
3. Anonymous. Determination of Moisture Content in Finnish Honey using an Infrared Dryer. *Food Market. Technol.* **1995,** *9* (1), 40–41.
4. Awuah, G. B.; Ramaswamy, H. S.; Tang, J., Eds. *Radio-Frequency Heating in Food Processing: Principles and Applications.* CRC Press: Boca Raton, FL, 2014; p 373.
5. Bai, J. W.; Sun, D. W.; Xiao, H. W.; Mujumdar, A. S.; Gao, Z. J. Novel High-Humidity Hot Air Impingement Blanching (HHAIB) Pretreatment Enhances Drying Kinetics and Color Attributes of Seedless Grapes. *Innov. Food Sci. Emerg. Technol.* **2013,** *20* (1), 230–237.
6. Bechoff, A.; Westby, A.; Menya, G.; Tomlins, K. I. Effect of Pretreatments for Retaining Total Carotenoids in Dried and Stored Orange-Fleshed-Sweet Potato Chips. *J. Food Qual.* **2011,** *34* (4), 259–267.
7. Changrue, V.; Orsat, V.; Raghavan, G. S. V. Osmotically Dehydrated Microwave-Vacuum Drying of Strawberries. *J. Food Process. Preserv.* **2008,** *32* (5), 798–816.
8. Chou, S. K.; Chua, K. J. New Hybrid Drying Technologies for Heat Sensitive Foodstuffs. *Trends Food Sci. Technol.* **2001,** *12* (10), 359–369.
9. Current Concerns. *Fruit Drying with Heat Pumps*, Current Concerns no. 22, Zürich, Switzerland, 2009.
10. Dandamrongrak, R.; Mason, R.; Young, G. The Effect of Pretreatments on the Drying Rate and Quality of Dried Bananas. *Int. J. Food Sci. Technol.* **2003,** *38* (8), 877–882.
11. Datta, A. K.; Ni, H. Infrared and Hot-Air-Assisted Microwave Heating of Foods for Control of Surface Moisture. *J. Food Eng.* **2002,** *51* (4), 355–364.
12. Dereje, B.; Abera, S. Effect of Pretreatments and Drying Methods on the Quality of Dried Mango (*Mangifera indica*) Slices. *Cogent Food Agric.* **2020,** *6* (1), E-article 1747961.

13. Djendoubi Mrad, N.; Boudhrioua, N.; Kechaou, N.; Courtois, F.; Bonazzi, C. Influence of Air Drying Temperature on Kinetics, Physicochemical Properties, Total Phenolic Content and Ascorbic Acid of Pears. *Food Bioprod. Process.* **2012,** *90* (3), 433–441.
14. Durigon, A.; de Souza, P.G.; Carciofi, B. A. M.; Laurindo, J. B. Cast-Tape Drying of Tomato Juice for the Production of Powdered Tomato. *Food Bioprod. Process.* **2016,** *100* (1), 145–155.
15. Durigon, A.; Parisotto, E. I. B.; Carciofi, B. A. M.; Laurindo, J. B. Heat Transfer and Drying Kinetics of Tomato Pulp Processed by Cast-Tape Drying. *Dry. Technol.* **2018,** *36* (2), 160–168.
16. ESRU. *Heat Pump Background.* A Publication of the Energy Systems Research Unit, University of Strathclyde: Glasgow, UK, 2015; http://www.esru.strath.ac.uk/EandE/Web sites/01-02/heat; (accessed on August 31, 2021)
17. Geeraert, B. Air Drying by Heat Pumps with Special Reference to Timber Drying. In *Heat Pumps and Their Contribution to Energy Conservation*; NATO Advance Study Institute Series, Series E, Applied Science; Noordhoff: Leiden, 1976; pp 219–246.
18. Ginzburg, A. S. Infrared Radiation in Food Industry (in Polish). Wydawnictwo Naukowo-Techniczne: Warszawa, 1969; p 30.
19. Guo, C.; Mujumdar, A. S.; Zhang, M. New Development in Radio Frequency Heating for Fresh Food Processing: A review. *Food Eng. Rev.* **2019,** *11* (1), 29–43.
20. Gupta, P.; Ahmed, J.; Shivhare, U. S.; Raghavan, G. S. V. Drying Characteristics of Red Chilli. *Dry. Technol.* **2002,** *20* (10), 1975–1987.
21. Hasatani, M.; Itaya, Y.; Miura, K. Hybrid Drying of Granular Materials by Combined Radiative and Convective Heating. *Dry. Technol.* **1988,** *6* (1), 43–68.
22. Hebbar, H. U.; Vishwanathan, K. H.; Ramesh, M. N. Development of Combined Infrared and Hot Air Dryer for Vegetables. *J. Food Eng.* **2004,** *65* (4), 557–563.
23. Huang, L.; Zhang, M. Trends in Development of Dried Vegetable Products as Snacks. *Dry. Technol.* **2012,** *30* (5), 448–461.
24. Islam, M. R.; Mujumdar, A. S. Heat Pump-assisted Drying. In *Drying Technologies in Food Processing*; Chen, X. D., Mujumdar, A. S., Eds.; Blackwell Publishing: Oxford, UK, 2008; pp. 190–224.
25. Jackson, J. C.; Bourne, M. C.; Barnard, J. Optimization of Blanching for Crispness of Banana Chips using Response Surface Methodology. *J. Food Sci.* **1996,** *61* (1), 165–166.
26. Jangam, S. V.; Joshi, V. S.; Mujumdar, A. S.; Thorat, B. N. Studies on Dehydration of Sapota (*Achras zapota*). *Dry. Technol.* **2008,** *26* (3), 369–377.
27. Jangam, S. V.; Mujumdar, A. S. Classification and Selection of Dryers for Foods. In *Drying of Foods, Vegetables and Fruits*; Jangam, S. V., Lim, L. C, Mujumdar, A. S., Eds.; 2010; Vol. 1; pp 59–82.
28. Jangam, S. V.; Mujumdar, A. S. Heat Pump Assisted Drying Technology–Overview with Focus on Energy, Environment and Product Quality. In *Modern Drying Technology: Energy Savings*; Tsotsas, E., Mujumdar, A. S., Eds.; John Wiley & Sons: Weinheim, Germany, 2011; Vol. 4; pp 121–162.
29. Jeevarathinam, G.; Pandiselvam, R.; Pandiarajan, T.; Preetha, P.; Balakrishnan, M.; Thirupathi, V.; Kothakota, A. Infrared Assisted Hot Air Dryer for Turmeric Slices: Effect on Drying Kinetics and Quality Parameters. *LWT-Food Sci. Technol.* **2021,** *144*, E-article 111258.
30. Jojo, S.; Mahendran, R. Radio Frequency Heating and its Application in Food Processing: A Review. *Int. J. Curr. Agric. Res.* **2013,** *1* (9), 42–46.

31. Kerr, W. L. Food Drying and Evaporation Processing Operations. Chapter 12. In *Handbook of Farm, Dairy and Food Machinery Engineering*, 2nd ed.; Kerr, W. L., Ed.; Academic Press: Amsterdam, Holland, 2013; pp 317–354.
32. Magoon R. E. Method and Apparatus for Drying Fruit Pulp and the Like. US Patent No. 4,631,837, 1986.
33. Mahmutoğlu, T.; Emir, F.; Saygi, Y. B. Sun/Solar Drying of Differently Treated Grapes and Storage Stability of Dried Grapes. *J. Food Eng.* **1996,** *29* (3–4), 289–300.
34. Mejia-Meza, E. I.; Yanez, J. A.; Davies, N. M.; Rasco, B.; Younce, F.; Remsberg, C. M.; Clary, C. Improving Nutritional Value of Dried Blueberries-Combining Microwave-Vacuum, Hot-Air Drying and Freeze Drying Technologies. *Int. J. Food Eng.* **2008,** *4* (5), 1364–1367.
35. Mermelstein, N. H. Microwave and Radio Frequency Drying. *Food Technol.* **1998,** *1* (1), 84–86.
36. Moreira, R. G. Impingement Drying of Foods using Hot Air and Superheated Steam. *J. Food Eng.* **2001,** *49* (4), 291–295.
37. Moses, J. A.; Norton, T.; Alagusundaram, K.; Tiwari, B. K. Novel Drying Techniques for the Food Industry. *Food Eng. Rev.* **2014,** *6* (3), 43–55.
38. Mujić, I.; BavconKralj, M.; Jokić, S.; Jarni, K.; Jug, T.; Prgomet, Ž. Changes in Aromatic Profile of Fresh and Dried Fig–the Role of Pre-treatments in Drying Process. *Int. J. Food Sci. Technol.*, **2012,** *47* (11), 2282–2288.
39. Mujumdar, A. S. Superheated Steam Drying. In *Handbook of Industrial Drying,* 2nd ed.; Mujumdar, A. S., Ed.; Marcel Dekker Inc.: New York, 1995; pp 1071–1086.
40. Nindo, C. I.; Tang, J. Refractance Window Dehydration Technology: A Novel Contact Drying Method. *Dry. Technol.* **2007,** *25* (1), 37–48.
41. Nindo, C. I.; Tang, J.; Powers, J. R.; Bolland, K. Energy Consumption during Refractance Window Evaporation of Selected Berry Juices. *Int. J. Energy Res.* **2004,** *28* (12), 1089–1100.
42. Nindo, C.; Sun, T.; Wang, S. W.; Tang, J.; Powers, J. R. Evaluation of Drying Technologies for Retention of Physical Quality and Antioxidants in Asparagus (*Asparagus officinalis,* L.). *LWT-Food Sci. Technol.* **2003,** *36* (5), 507–516.
43. Nipkow, J.; Bush, E. In *Promotion of Energy-Efficient Heat Pump Dryers*, Proceedings of the 5th International Conference on Energy Efficiency in Domestic Appliances and Lighting (EEDAL '09), Berlin, Germany; 2009; pp 1215–1225.
44. Nyangena, I.; Owino, W.; Ambuko, J.; Imathiu, S. Effect of Selected Pretreatments Prior to Drying on Physical Quality Attributes of Dried Mango Chips. *J. Food Sci. Technol.* **2019,** *56* (8), 3854–3863.
45. Orsat, V.; Raghavan, G. S. Radio-Frequency Processing. *Bio Resour.* **2005,** *1* (1), 446–450.
46. Ortiz-Jerez, M. J.; Gulati, T.; Datta, A. K., Ochoa-Martínez, C. I. Quantitative Understanding of Refractance Window™ Drying. *Food Bioprod. Process.* **2015,** *95* (1), 237–253.
47. Oyebanji, A. O.; Ajani, A. O.; Adeize, A.; Oyelakin, M. O.; Ikotun, I.; Agboola, A. A.; Alimi, J. Effect of Pre-Drying Treatments on Utilizability of Sweet Potato Tubers for Production of Chips for Confectionery Flour. *Afr. J. Food Sci.* **2013,** *7* (9), 258–263.
48. Ozdemir, M.; Devres, O. Analysis of Color Development during Roasting of Hazelnuts using Response Surface Methodology. *J. Food Eng.* **2000,** *45* (1), 17–24.

49. Pangavhane, D. R.; Sawhney, R. L.; Sarsavadia, P. N. Effect of Various Dipping Pretreatment on Drying Kinetics of Thompson Seedless Grapes. *J. Food Eng.* **1999,** *39* (2), 211–216.
50. Pawar, V. N.; Singh, N. I.; Dev, D. K.; Kulkarni, D. N.; Ingle, U. M. Solar Drying of White Onion Flakes. *Indian Food Packer* **1988,** *42* (1), 15–24.
51. Raghavi, L. M.; Moses, J. A.; Anandharamakrishnan, C. Refractance Window Drying of Foods: A Review. *J. Food Eng.* **2018,** *222* (1), 267–275.
52. Rastogi, N. K.; Raghavarao, K. S. Mass Transfer during Osmotic Dehydration: Determination of Moisture and Solute Diffusion Coefficients from Concentration Profiles. *Food Bioprod. Process.* **2004,** *82* (1), 44–48.
53. Ratti, C.; Mujumdar, A. S. Infrared Drying. In *Handbook of Industrial Drying*, 2nd ed.; Mujumdar, A. S., Ed., Marcel Dekker Inc.: New York, 1995; pp 567–588.
54. Rostami, H.; Dehnad, D.; Jafari, S. M.; Tavakoli, H. R. Evaluation of Physical, Rheological, Microbial and Organoleptic Properties of Meat Powder Produced by Refractance-Window Drying. *Dry. Technol.* **2018,** *36* (9), 1076–1085.
55. Rowley, A. T. Radio Frequency Heating. In *Thermal Technologies in Food Processing*, Richardson, P., Ed.; Woodhead Publishing: Cambridge, UK, 2001; pp 162–177.
56. Sagar, V. R.; Kumar, P. S. Recent Advances in Drying and Dehydration of Fruits and Vegetables: A Review. *J. Food Sci. Technol.* **2010,** *47* (1), 15–26.
57. Sakai, N.; Hanzawa, T. Application and Advances in Far-Infrared Heating in Japan. *Trends Food Sci. Technol.* **1994,** *5* (11), 357–362.
58. Sandu, C. Infrared Radiative Drying in Food Engineering: A Process Analysis. *Biotechnol. Progr.* **1986,** *2* (3), 109–119.
59. Sefa-Dedeh, S. Traditional Food Technology. In *Encyclopedia of Food Science, Technology and Nutrition*; Macrae, R., Robinson, R., Sadler, M., Eds.; Academic Press: New Jersey, USA, 1993; pp 4600–4606.
60. Sharma, G. P.; Verma, R. C.; Pathare, P. B. Thin-Layer Infrared Radiation Drying of Onion Slices. *J. Food Eng.* **2007,** *67* (3), 361–366.
61. Sharma, R.; Joshi, V. K.; Kaushal, M. Effect of Pre-Treatments and Drying Methods on Quality Attributes of Sweet Bell-Pepper (*Capsicum annum*) Powder. *J. Food Sci. Technol.* **2015,** *52* (6), 3433–3439.
62. Shrivastava, M.; Nagomidir, M.; Pandey, P. H.; Salokhe, V. M. In *Design and Development of a Low Cost Solar Chilli Dryer*, Proceeding of the International Agricultural Engineering Conference and Exhibition, Bangkok, Thailand; 1990; pp 505–515.
63. Sosle, V. A Heat Pump Dehumidifier Assisted Dryer for Agri-Foods. PhD dissertation, McGill University: Ontario, Canada, 2002; http://www.redalyc.org/pdf/1698/169823914004.pdf; p 139.
64. Sosle, V. Heat Pump Drying. *Stewart Postharvest Rev.* **2006,** *2* (6), 1–5.
65. Stroem, L. K.; Desai, D. K.; Hoadley, A. F. A. Superheated Steam Drying of Brewer's Spent Grain in a Rotary Drum. *Adv. Powder Technol.* **2009,** *20* (3), 240–244.
66. Strumillo, C.; Zylla, R. Optimization of Heat Pump Dehumidifier. In *Handbook of Industrial Drying*, Mujumdar, A. S., Ed.; Elsevier Science Publishers: Amsterdam, Netherlands, 1985; pp 21–32.
67. Tang, X.; Cronin, D. A.; Brunton, N. P. The Effect of Radio Frequency Heating on Chemical, Physical and Sensory Aspects of Quality in Turkey Breast Rolls. *Food Chem.* **2005,** *93* (1), 1–7.

68. Tiwari, G.; Wang, S.; Tang, J.; Birla, S.L. Analysis of Radio Frequency (RF) Power Distribution in Dry Food Materials. *J. Food Eng.* **2011,** *104* (4), 548–556.
69. Tontand, S.; Therdthai, N. Preliminary Study of Chilli Drying using Microwave Assisted Vacuum Dry. Technol. *Asian J. Food Agro-Indus.* **2009,** *2* (2), 80–86.
70. Vega-Mercado, H.; Góngora-Nieto, M. M.; Barbosa-Cánovas, G. V. Advances in Dehydration of Foods. *J. Food Eng.* **2001,** *49* (4), 271–289.
71. Wang, J.; Luechapattanaporn, K.; Wang, Y.; Tang, J. Radio-Frequency Heating of Heterogeneous Food–Meat Lasagna. *J. Food Eng.* **2012,** *108* (1), 183–193.
72. Zhao, Y.; Flugstad, B. E.; Kolbe, E.; Park, J. W.; Wells, J. H. Using Capacitive (Radio Frequency) Dielectric Heating in Food Processing and Preservation–A Review. *J. Food Process Eng.* **2000,** *23* (1), 25–55.
73. Zotarelli, M. F; Augusto, B.; Carcio, M.; Laurindo, J. B. Effect of Process Variables on the Drying Rate of Mango Pulp by Refractance Window. *Food Res. Int.* **2015,** *69* (1), 410–417.

CHAPTER 5

IRRADIATION OF FOODS AND FOOD PRODUCTS

PAVULURI PAVANI, S. SUBHASHINI, and S. NIVEADHITHA

ABSTRACT

The irradiation process is the exposure of foods to ionizing energy to terminate microbes or pests, which are existing in food and to recover the functional properties of food or to remove contaminants through minimum negotiation on physical and nutritious value. Though, the highly perishable foods and foods with more bacterial content are also main issues concerning to the simple delivery besides harmless intake. The irradiated foods are harmless and price-operative procedure to improve the shelf-life and to improve its physical value. Safety of irradiated foods depends on proper standards offering indication of this irradiation technique is harmless with additional benefits. Marketable use of irradiation is still incomplete. Despite new research and development actions in North America and Europe, the operation of this method is increasing quicker, mostly for hygienic resolutions, in Latin-American and Asian region countries. Development is slower in European Union (EU). In EU, irradiation of food is controlled since 1999 by a General Directive, the Public Report of EU accepted foods that are irradiated comprises only a particular division of stuffs such as "dehydrated fragrant aromatic plant, flavors, and seasonings." Concentration usage of irradiation to foodstuffs was improved resulting from the 1997 U.S. FDA support of irradiation for microbe controller in untreated pink meat and meat products. Irradiation of spice decreases postyield losses, safeguards sanitary quality, and enables trade with food products. This chapter elaborates on history of irradiation, process of food irradiation, emerging applications in

Novel and Alternative Methods in Food Processing: Biotechnological, Physicochemical, and Mathematical Approaches. N. Veena, Megh R. Goyal, & Ritesh B. Watharkar (Eds.)

food sector, legislation, safety of irradiated foods, and consumer acceptance for irradiated foods.

5.1 INTRODUCTION

The development of numerous technologies to enhance food protection and to increase food value includes salting, cooking, pickling, dehydration, and smoking generally used along with sterilization, freezing, refrigeration, canning, and the traditional technologies of preservation. The various techniques of preservation show slight drawbacks related to organoleptic, bacteriological risks, and there may be chance to degrade the nutrients. Many examples are related to preservation techniques, textural weakening in some foods during freezing, changes in sensory quality during heat processing, and loss of flavor and freshness of food in canning technique. Moreover, some boundaries are interlinked with the high-cost apparatus required for chilling, freezing, and different packaging techniques, such as modified atmosphere packaging. Preservation techniques should meet the growing needs of customers for superior quality and bacteriologically harmless nourishments. Adulteration of food with microbes, which are related to public health problems and human suffering all over the world, is mainly due to reproductive forming microbes. To avoid all these issues, researchers developed irradiation technology of foods, which is safe for consumption and quality of foods remains the same.[34]

This chapter focuses on concepts and applications of irradiation technology in food items.

5.2 MILESTONES IN IRRADIATION TECHNOLOGY

Irradiation of food was established as a technically recognized process for safe food through the second half of 20th century.[21] Food irradiation is the final negligible process, which was primarily started and deeply considered, but in European nations, it was not used as much as in other nations. Although food is usually exposed to radiation like microwaves, yet the name irradiation is utilized to label a procedure, where food is irradiated with radiation sources mainly from Radioactive Isotopes (such as Co-60 and Cs-137), thus producing Gamma rays and also X-rays generated using machines with a maximum MeV of 5 and a maximum MeV of 10, in case of Dynamic Energy.[11] The two major types of electromagnetic radiations in use should have excellent diffusion capacity, whereas faster electrons have little permeability.[8] Since mid-20th century, methodical study efforts, numerous nationwide study sequencers, and worldwide collaboration joined hands with practical progresses recognized application of this irradiation technology.

5.2.1 HISTORIC INDICATORS OF FOOD IRRADIATION[12]

- 1921: Inactivation of trichinae in raw pork using radiation with X-rays was published by B. Schwartz.
- 1957: First commercial application on irradiation of spices using an electron beam.
- 1905 (British patent): J. Appleby and A.J. Banks reported the overall quality maintenance of food using radioactive substances in radiation process and to carry quality enhancement in products.
- 1930 (French patent): Use of X-rays for the destruction of microorganisms was done in packed food products by Wüst.
- 1966: Global Conference on Irradiation of foods held in Deutschland.
- 1970–1982: Global Development in the arena of irradiated foods—International Food Irradiation Project—(19/24 nations, International Atomic Energy Agency/Food and Agriculture Organization [IAEA/FAO], World Health Organization [WHO]/Organization for Economic Co-operation and Development).
- 1980: *Codex Expert Commission* for Freshness of Irradiated Foods, Joint Expert Committee on Food Irradiation (JECFI); dosage up to 10 kilo Gray (KGy).
- 1983–1984: JECFI, "Overall Standard for Irradiated Foods" and "Suggested Global Code of Practice for the Process of Radiation Facilities."
- 1979–1990: Helping emerging nations through guidance and demonstration: FAO/IAEA Global Capability for Irradiated Technology in Food—International Facility for Food Irradiation Technology, Wageningen—Netherlands.
- 1983–2004: International Consultative Group on Food Irradiation.

5.3 CONCEPT OF IONIZING RADIATION IN FOODS

It is defined as the process of application of ionizing radiation to food materials, such as gamma and electron beams intentionally to improve its shelf-life and to supply safe food to the consumer. The process applied to certain fruits and vegetables shows delay in browning process and ripening process. Consequently to increase the shelf-life and to avoid the loss of nutrients, along with other innovative food processing methods, irradiation can help in this goal.[29]

5.3.1 PRINCIPLE OF IRRADIATION

Irradiation of food provides the contact of wholesale or prepacked food with gamma (γ) rays produced from emitting radiation isotopes, such as Cesium (Cs-137) and Cobalt (Co-60) or X-rays or high-energy electron rays in a precise manner. Among three kinds of energy, one kind of energy is incorporated in the selected food product, the energy incorporated into food is restricted as it does not interact with the core parts of food to induce radiation.[14] Moreover, irradiation process has a harmful influence on microbes in the food substance when the food is exposed at a precise dosage. The ionizing radiation-induced into microbes inactivates microbes by destructing the most important component in the cell generally the chromosomal genetic material, that is, DNA.[1] The destruction stops reproduction and randomly dismisses maximum cell functioning. The destruction of the genetic material leads to the straight impact among the radioactivity energy and the genomic substance.[33]

5.4 PROCESS OF IRRADIATION

Throughout the irradiation procedure (Figure 5.1), food substances are exposed to radiation source with an accurate dosage, with the objectives: to achieve the desirable power outcome of the cause for each period; to possess a well-distinct geographic association among the food product and source; and to depict the food substance for a precise period. The energy dosage generally utilized in food handling industries ranges from 10 to 50 kGy and is dependent on the type of food.[41] The design and physical prearrangement of food irradiation process plant depends on the regular usage. There are two types of processes for irradiation food: batch and continuous[41]:

- **In batch process,** the food is irradiated for a period at required dosage. After irradiation process, the food is removed from the vessel and another batch of food that is to be irradiated is loaded.

- **In continuous process,** the food irradiated features the portion of food that is delivered through the vessel at a precise frequency designed to safeguard all the food takes accurately the proper dosage.

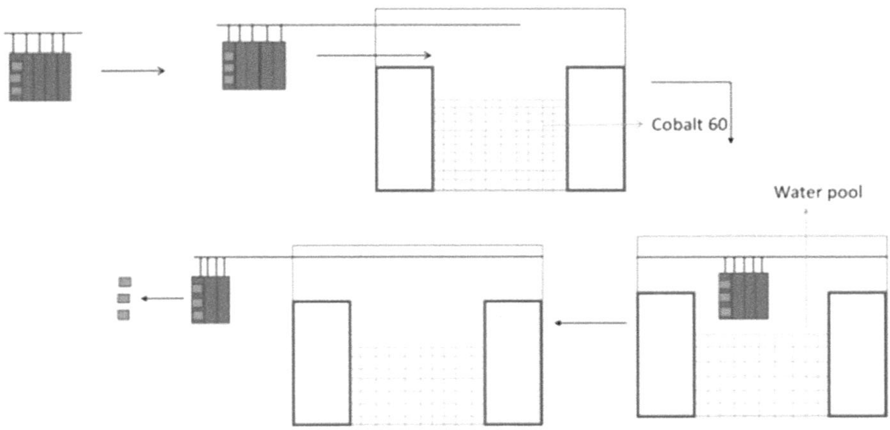

FIGURE 5.1 Flow diagram of the irradiation process.

The process is simple and easy to function in the batch process when compared with continuous process, moreover batch process is more feasible. The number of doses can be active and adapted to research. On the other hand, continuous process is better to produce larger capacities of the produce, particularly when a single food is treated at a specified dosage. Continuous processes are generally chosen in the food processing units, since they give substantial amount of produce.[41]

5.4.1 COMPONENTS OF AN IRRADIATION SYSTEM[5]

- **Radiation chamber:** Persons can securely arrive into the irradiation room when the Co-60 is in the liquids like water.
- **Source of radiation:** Co-60 is a source of radiation and it is covered in water in subversive tank after the use and when not in use.
- **Controller:** Process is measured by the speediness of the conveyer belt. Measure of energy is required and is differs by the mass of the batch.
- **Loading:** For irradiation treatment, the food is passed into vessel through conveyor belt.

- **Unloading of treated produce:** Handling of treated food is done immediately. Separation of processed and unprocessed food is done after treatment.
- **Radiation protect:** Walls made from concrete stops gamma rays from leaking into the surroundings.

5.4.2 IRRADIATION SOURCES

Irradiation of food process includes subjecting the foodstuff to ionized radiations.[42] The radiation resources that are being used for human consumption are

- Cobalt 60 or Caesium137 are the sources of gamma-rays.
- Manufacturing of X-rays from machinery sources operate at a source level of 5 MeV.
- Manufacturing of electron beam from machinery sources operate at 10 MeV.

5.4.3 DOSAGE

Throughout the cycle of food irradiation, the food material is subjected to ionizing radiation up to a precise value. The immersed dosage is defined as "the amount of energy imparted by ionizing radiation to the substance in a size component distributed by the substance in a capacity portion." In the SI unit, the quantity of captivated radio activity is expressed in Gray (Gy). One Gy is defined as one Joule of immersed power per kg. Based on the type of food and the intended result, the dosage amounts are split into three groups:

- **Low dosage** (1 kGy) is utilized in products, such as fresh vegetables and fruits to delay maturing and germination; also to regulate beetles and fleas in breakfast cereal and pulses, dehydrated fruits, dehydrated fish, meat, etc.[34]
- **Medium dosage** (1–10 kGy) is utilized to decrease decomposition and harmful microbes on mushrooms, strawberries, fresh fish, fresh, and frozen poultry, meat products, fresh, and ice-covered seafood, to improve technological properties, such as dip cooking time in dry tubers and to maximize the shelf-life of food products.[34]
- **High dosage** (10–50 kGy) is utilized for fumigation of various products, such as seafood, meat, poultry, and supplementary ready foodstuffs in mixture with slight reheating for the production of enzymes;

bacteriological fumigation of food products and components like flavors, colors, and for enzyme inactivation.[34]

5.4.4 VARIATIONS IN FOOD MATERIAL AFTER TREATMENT

Food irradiation brings firm changes in the DNA fragment of microbes.[34] Various changes that may occur in food composition are

- Direct radiolysis is ionization of particles, which leads to the development of the electrically excited ions due to an electron loss.
- Application of this treatment for materials is also done by manufacture of nonaligned fragments with free radicals.
- Biological bonds occur in food particles by cleavage formation.
- Indirect radiolysis for products results in the reorganization of biological links of sensitive radicals with agricultural products and is manufactured by new-fangled units.
- The DNA of the bacteria induces destruction of the biochemical linkages and damage to the cells capacity to replicate. Comparatively minor modification of the DNA of the microbial cell may extinguish the cell, the main effect of food irradiation enabling flea inactivation, slowing maturation, pest destruction, and inhibition of germination in rhizomes and roots.

Other minor variations in irradiated food composition leads to the growth of rotten off-flavors; and odor occurs to the resulting biochemical responses and radiolytic compounds, correspondingly:

- Development of fatty acid oxides during irradiation in fat including foods.
- In fruits and vegetables, the splitting of higher molecular sugars is converted into lower molecular elements and the formation of the proteinaceous compounds occurs.
- Reduction of natural antioxidants such as ascorbic acid as well as tocopherol.

5.4.5 EFFECTS ON MICROORGANISMS

Microbe's chiefly Gram-negative organisms (such as Salmonella) are diminished by irradiation treatment. Bacteriological spores, however, are destroyed by applying high dosage, because they cause extremely toxic foodborne illness. Radiation cannot prevent botulism. Amount of irradiation dosage can destroy only some part of the infectious population when exposed to radiation, irrespective of the number of microbes that exist.[40]

5.5 SAFETY OF RADIOACTIVITY

On an essential level, all nourishments are radioactive to some degree of treatment because of regular substance radiation. The food exposed to gamma rays will not be dangerous because the discharge of neutrons from cobalt-60 is not imparted. In this way, changes in nuclear are not created in the cores of food particles. There is no logical proof that exposed food will have concentrations of radiation more than those in the nonradiated foods.[19,37,41]

5.5.1 SAFETY REGARDING TOXICITY

Self-governing examination on specialized research on irradiation of foods and educated by *Codex commission* on the Healthfulness of light of food products, FAO, International Consultative Group on Irradiation of Food (ICGIF), WHO, United States Food and Drug Administration, and IAEA have concluded that the irradiation process of food in expressed circumstances is hazardous and that "radiation of any food item is a worldwide normal measurements of 10 kGy speaks to no toxicological hazard, hence toxicological test to nourishments so rewarded is not, at this point required."

Most extreme studies and complete examination on safety issues of food irradiation[30,34] by Raltech research (1976) have endured 7 years in United States—Army Medical Department. It involved 20 study plans for examining the finish of expending chicken meat cleaned using electron beams and gamma rays by an average dose of 58 kGy in hamsters, rodents, hounds, mice, and hares in terms of nutritive worth, destructiveness, cancer-causing properties, propagative execution, and genomic poisonousness. In view of the findings of the in vitro genotoxicity contemplated by Thayer et al.[39] inferred that radioactivity execution of poultry meat had no adverse impacts in the creatures.

5.6 LABELING OF IRRADIATED FOODS

Foods that are irradiated must have extensive label, except the declaration according to US Food and Drug Administration that "Foods exposed to irradiation are accepted to expose again to radiation. The trade tag accepts the subsequent logo along other declarations “treated with radiation,” or “treated by irradiation.”[26] The symbol of irradiation internationally named as Radura (Figure 5.2).

FIGURE 5.2 Irradiation symbol.

5.7 APPLICATIONS OF IRRADIATION TECHNOLOGY IN FOODS

Irradiation process in food application basically categorized into three groups (low, medium, or high), based on the dosage applied to the target food.

5.7.1 LOW-DOSE: RADICIDATION

5.7.1.1 GERMINATION INHIBITION

Due to the demand of consumer’s all-time distribution of numerous budding/germinated vegetables such as tubers like onions, garlic, yams, and potatoes, storing periods are ahead up to numerous months.[16] Inhibition of sprouting is avoided by preservation using cooling technique and using hydrazide, which is used for preharvest and isopropyl chlorocarbamate used for postharvest, these two are considered as chemical preservatives. The cooling technique is very costly and primarily used in humid regions, while the hazardous chemical preservatives are comparatively effective as well as inexpensive. But these chemicals remain in the food substances. So, nowadays, the technique recommended was irradiation. Germination inhibition and decreases decaying and loss of net weight were detected in tuberous vegetables such as onions, garlic, yams, and potatoes in the span of 50–150 Gy when irradiation technique is used.[23]

5.7.1.2 PEST DISINFESTATIONS

Ethylene dibromide or ethylene oxide are chemical preservatives used for fumigation in pest control in grain and grain products. Fruits and vegetables are fumigated using chemical preservative still 1984. In some countries, the usage of chemical preservatives is prohibited due to its bad effect on human well-being and environmental surroundings. When chemical preservatives are used for fumigation, degradation of product quality like changes in flavor and taste[23,35] can occur. For fumigation process, recommended method is irradiation for disinfestation. Disinfestations are generally done to avoid loss produced by pests in godowns for products like dry fish, nuts, cereals, pulses, coffee beans, flour, etc.[2,11] Based on the techniques applied it is understood that the essential dosage of radiation should be in the scale of 150–700 Gy. For the treatment of fruit flies, the dosage level should be minimum 250 Gy if they are treated separately, whereas dosage levels of radiation nearly 500 Gy can eliminate maximum pests in all phases.[11,20]

5.7.2 MEDIUM DOSE: RADURIZATION

5.7.2.1 PATHOGENS IN FOOD

All these products are known to be the main cause of foodborne diseases: pork, beef, seafood, poultry, eggs, and dairy. The most noxious strains are *Escherichia coli, Listeria*, and a cattle tapeworm. *Salmonella* and *Campylobacter* are the primary bacteria for poultry products. With dosage in the range of 1–3 KGy, effective controller of all such bacteria can be achieved.[34,44]

5.7.2.2 SHELF-LIFE EXTENSION

The similar dosage concentrations suitable for management of foodborne disease-causing bacteria may also substantially expand the shelf-life of food products, which are simply talked about by decreasing microbial populations of decay microbes, fungi, and molds. A dosage of 2.5 kGy, for example, may expand the shelf-life time of poultry as well as the beef by a few weeks. With 5 kGy, the storage of minimal-fatty fish can be increased from usually 3 to 4 days with no radiation technique to many weeks.[34] In addition to various cheese shelf-life may be greatly increased by eliminating yeast at 0.5 kGy or below concentrations. The expansion of shelf-life for packaged leafy vegetables, strawberries, mushrooms, carrots, and papaya seems to be encouraging at dosage levels of a few kGy or less.[15,16] Application of irradiation in mushrooms at 2–3 kGy decreases the opening of the cap and increases the elongation of the stem, which can be improved at least twice when storage can be found at 10°C. Healing of strawberries with a dosage of 2–3 kGy, followed by storing at 10°C can lead to an improvement in storage time to nearly 14 days. The process of maturation in mangoes, bananas, and papaya's may be delayed by 0.25–1 kGy irradiation. It is essential to irradiate them prior to the whole process of ripening begins.[15, 23]

5.7.2.3 IRRADIATION OF SPICES

The new plants from which flavors are derived are generally polluted due to microorganisms from soil, wind-propelled traces, and winged insect fecal matter. During the drying cycle, these microorganisms will grow population densities in excess of 106 life forms per gram of substance.[23] At a point when used as seasonings in the production of prepared food for which the assembling procedure excludes a sanitizing step, these life forms may result in quick food deterioration and may lead to foodborne disease. Because clammy thermal processing is not usually appropriate for such parched foods, zest makers earlier regularly utilized as disinfectants for dispersion. The creators are gradually moving towards ionizing radiation. To be told, the market lighting of flavors has been sponsored and refined in a variety of nations for a long time. Dosage of 5–10 kGy for the most part give particularly good outcomes (end of microorganisms, shape spores, and creepy crawlies) without negative effect on concoction or tactile.[17, 23]

5.7.3 HIGH-DOSE: RADAPPERTIZATION

Only certain foods, for example in raw products like fresh vegetables and fruits, when subjected to high dosage of radiation portions result in damages, other products like poultry products, meat, and some shellfishes, may continue to be of high quality when specific precautionary measures are taken.

Accordingly, it is conceivable to adequately sanitize these nourishments with dosages in the scope of 25–45 kGy.[35] To avoid off-flavors occurring due to fat oxidation, oxygen needs to be removed via vacuum bundling; and lighting must be carried out at a very less temperatures from 20 to 40°C.

The food products that are already heated to deactivate compounds can be commercially cleaned, like canned foods. Such products may be easily kept at room temperature. Although this additional technique and portions boost overall costs, yet these items are in any case important to serious hospitalized people with a smothered resistant framework and for space travelers and their agencies.[32,35] Instructions for the treatment of food irradiation and their dosages are indicated in Table 5.1.

TABLE 5.1 Instructions for the Treatment of Food Irradiation and their Dosages.[6,11]

Preserving impacts and categories of application	Dosage (kGy)
Decreasing the number of viruses by 10^6	10–40
Decreasing the number of fungi, vegetative bacteria, and molds by 10^6	1–10
Decreasing the number of dry frozen vegetative microbes, spores, and fungi by 10^6	2–20
Extension of food shelf-life by decreasing bacteriological communities	0.5–5.0
Food sterilization	20–45
Food sterilization for parasites and insects	0.03–0.2
Hindering the ageing of vegetables and fruits to reduce ripening	0.5–5.0
Inactivating and disinfecting pests	0.2–0.8
Inhibition of germination in onions and potatoes	0.03–0.1
Inhibition of the replica of food-borne organisms	0.1–3.0
Killing parasites	0.05–5
Removal of the possible disease-causing bacteria, found in fresh and chilled food	1.0–7.0
Removal or decrease of microbial load in parched-food materials	3–10

5.8 ADVANTAGES/DISADVANTAGES OF IRRADIATION TECHNOLOGY

5.8.1 ADVANTAGES OF IRRADIATION METHODS[3,10,11,25,38,40,43]

- Enhanced hygiene: Reduction of hazards in food by decreasing food-borne illness by an effective method, which is irradiation by regulating foodborne illness microbes such as *Salmonella, Listeria monocytogens, Toxoplasma, E. coli, Vibrio,* and *Campylobacter*. Ready-to-eat meat and meat products, cereals, seafoods, vegetables, fruits, and leguminous products are preserved by this method in both raw and processed forms also. Sanitization of food enhances the sanitary value of the foods by preventing it from harmful bacteria, mould, yeast, and pests.

- Irradiation improves the keeping quality of food materials by increasing the storage period of food items by killing pests and delaying the process of softening, which increases the nutrition source.
- Irradiation is also known as cold processing hence, it can be utilized in farming, regulating the freshness value.
- Irradiation process is atmosphere-friendly and harmless to employees and the peoples nearby the process.
- Preservation of nutritive quality and organoleptic value of food nourishments. Slight irrelevant loss of vitamins and minor variations in color, odor, and textural when related other protection methods during storing.
- The three major sources of Irradiation sources used in the process are electron beams, gamma rays, and X-rays, when used in standard doses. Dose will not cause any contamination of radiation in the food product and its packing material.
- Wide-range isolation process: The isolation method utilized for reducing pests in diseased food materials is by irradiation method, therefore by enhancing the supply of essential fresh groceries from rising states in industrialized countries in tropical and subtropical regions. Reduction of postharvest loss is better by irradiation process when compared with the use of chemical substances. The chemical substances (such as dibromide, ethyl oxide, and ethylene) used for the regulation of flies are prohibited for usage because of their harmful effects on human wellbeing and the environment.

5.8.2 DISADVANTAGES OF FOOD IRRADIATION[3,11,10,25,40,41]

- Application of irradiation to the foods leads to changes in chemical properties, which leads to sensory changes. Dairy and milk foods are inappropriate for irradiation treating because it causes a rotten flavor even at radiation with 0.1 kGy.
- Application of radiation causes decrease in vitamins level such as tocopherol, vitamin B_2, ascorbic acid, vitamin B_6, and cyanocobalamin.
- Change in food sensory properties mostly causing rotten flavor and change in color in meat and meat products.
- Damage of cell walls in fruits and vegetables leads to softening when irradiation is applied.
- Irradiation is ineffective against foodborne prions and viruses.

- Irradiation of food can lead to the formation of hazardous oxidative substances such as ketones, aldehydes, alcohols, etc.
- The initial cost of investment and processing is very high.
- The major disadvantage of irradiation processing is that, it is not suitable for application to all food products. Quality of fruits is degraded when this method of processing is used on them.

5.9 CONSUMER ACCEPTABILITY OF IRRADIATED FOODS

Based on a survey, irradiated food products have acceptance rates by customers nearly from 45–90% dependent on the food product.[13] View polls reports excellence data offered on the pack and the awareness program specified. The data on irradiation process inclines to encourage acceptability. Currently, Nayga et al.[24] stated that the customer's procurement of irradiated foods is highly demanding. Those report results highlight the status of educating the community on the disagreement on process and advantages of the process, expressly the community have presented to highly open to the damaging argument. Labeling of irradiated foods is major issue to reach consumers.[4,9] The labeling should be informative that consumers should be willing to buy the foods which are irradiated along with labeling information; nutritive information should be included. The data that defines the need of the process should encourages customers' acceptance. For example, chicken that is irradiated should be properly mentioned in the sentence "preserved with irradiation treatment to regulate Salmonella pathogenic bacteria."[28] Extra information for customer understanding needs to be presented in the residence wherever the produce is promoted.

5.10 OUTLINE OF IRRADIATION OF FOODS AROUND THE GLOBE

In 2005, the total volume of grains irradiated globally reached 405,000 tons. The main nations for using this technique for products like dried vegetables and sanitization of spices (P1) are China with 52,000 tons, United Nations with 80,000 tons, South Africa with 16,000 tons, and Brazil with 20,000 tons and for products like fruits and disinfestation of grains (P2), Ukraine leads by 70,000 tons, Vietnam by 14,000 tons, USA by 8000 tons, and Belgium by 5500 tons.

The major countries for sea foods and meat disinfection (P3) are Japan with 8100 tons and China with 10,000 tons and main country for processing foods for inhibition of sprouts is China with a capacity of 10,000 tons and for P5 food products like honey and mushrooms are led by China holds a prominent place (Figure 5.3). The three countries with maximum quantities of food irradiation around the entire globe are China, Ukraine, and United States; and Ukraine add up to three residences of the entire quantities of irradiated food around the world. In countries like US, Brazil, China, and South Africa, they mainly preserved vegetables and flavors; and in Japan, Vietnam, and Ukraine, the radiated products are mainly focused on potatoes, grains, and seafood products.[18]

5.11 PROSPECTS OF IRRADIATION TECHNOLOGY

The irradiation technology development helps in decreasing of originated diseases, decomposition as well as longer storage-life in addition to the commitment to be paid to the processed for food safety.[22] New advances in food irradiation laws and regulations, mainly in the European Union, would promote a larger recognition of the process through all appropriate stakeholders. Consumer education, identification is an issue as well as proper communication reducing unethical process by which irradiation of food is an atomic technology.[36] Misrepresentation on irradiated food is considered as a method of violence, avoiding the consumption of a harmless and beneficial method and obstructing customers to take a decision amongst alternate process. Marketing's trials represented that growing quantity of customers are enthusiastic to obtaining irradiated food when they are appropriately knowledgeable approximately about the procedure and its properties on food composition. Additional advances in equipment design and reworking of usages of apparatus energy sources such as electron rays and X-ray machines[1,7,27] can also support changing the image of the procedure into one kind of electrical procedure.

5.12 SUMMARY

Food irradiation is regarded as harmless, as well as the most efficient method to control bacterial infection without leaving chemical residues, which is used to enhance the protection of processed food. The sensory value of irradiated exposed foodstuffs, mostly seasonings, should be concentrated more in future studies. In this view, it is required to regulate analytic methods for clear detection of the radiation therapy along with its impact on the nutritional value. Irradiated food products are essential to make up an international agreement on

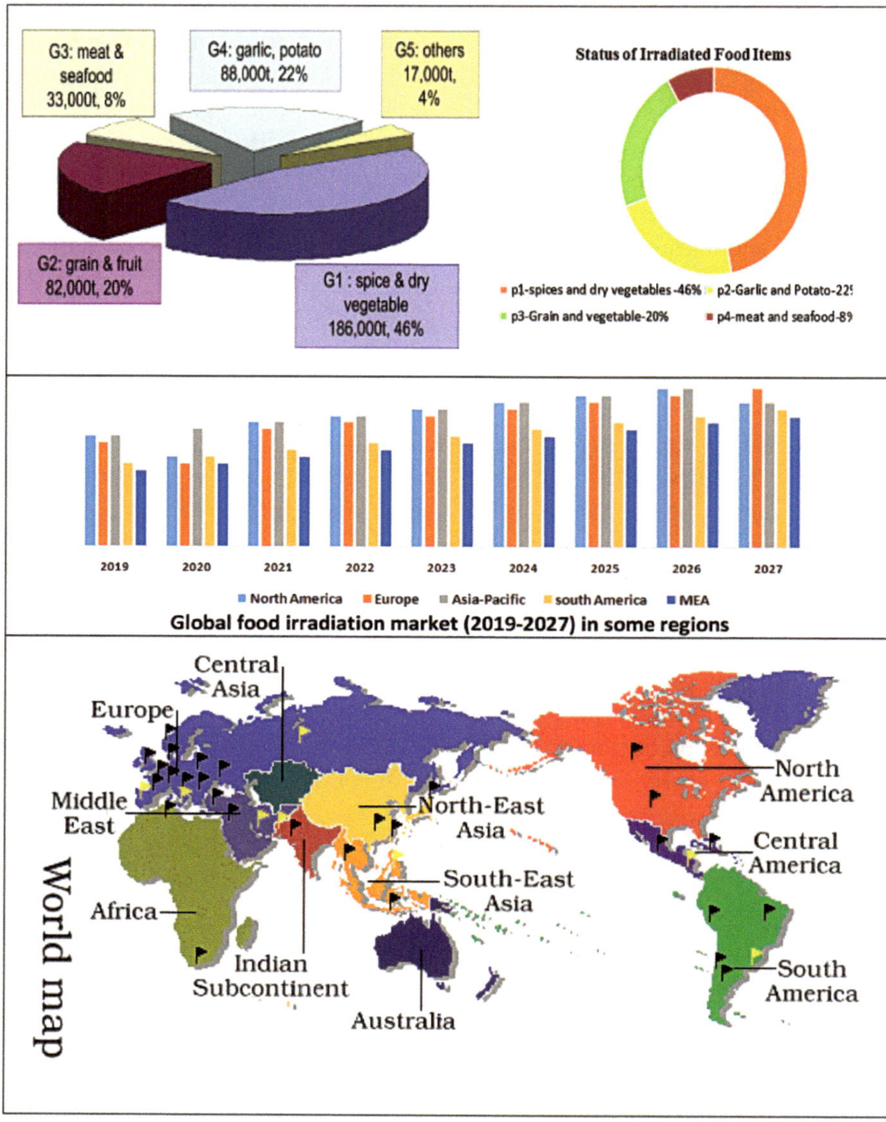

FIGURE 5.3 Status of irradiated food items (countries with commercial irradiation of food).[18]

a sensible and stable product marking. Ultimately for the improved protection and target for customer knowledge, regarding to their freedom of choice. In recent times, International Council on Food Irradiation in 2004 joined hands with researchers, professors, government representatives, and dealers of food irradiation apparatus and facilities to collect information and bring awareness about the security and advantages of irradiated food. Researchers have a duty to assist the customer to comprehend the radiation exposure as well as its ability to enhance human life and defend our wellbeing, in the same way as the pasteurization of dairy products, preserving, and freezing.[31]

KEYWORDS

- **ionizing energy**
- **irradiation**
- **pasteurization**
- **regulations**
- **safety**

REFERENCES

1. Acheson, D.; Steele, J. H. Food Irradiation: A Public Health Challenge for the 21st Century. *Clin. Infect. Dis.* **2001,** *33*(3), 376–377.
2. Arthur, T. M.; Wheeler, T. L.; Shackelford, S. D.; Bosilevac, J. M.; Nou, X.; Koohmaraie, M. Effects of Low-Dose, Low-Penetration Electron Beam Irradiation of Chilled Beef Carcass Surface Cuts on *Escherichia coli* O157:H7 and Meat Quality. *J. Food Protect.* **2005,** *68*(4), 666–672.
3. Borsa, J.; Lacroix, M.; Ouattara, B.; Chiasson, F. Radio-Sensitization: Enhancing the Radiation Inactivation of Foodborne Bacteria. *Radiat. Phys. Chem.* **2004,** *71*(1), 137–141.
4. Bruhn, C. M.; Eustice, R. F. Consumer Acceptance and Marketing of Irradiated Foods, Chapter 10. In *Food Irradiation Research and Technology*; Xuetong, F., Christopher H. S., Eds.; Blackwell Publishing: Chicago, 2013; pp 173–195.
5. Bruhn, C. M. How does Food Irradiation Work? [Online] 2017. https://ccr.ucdavis.edu/food-irradiation/how-does-food-irradiation-work (accessed July 17, 2021).
6. Chadwick, K. H.; Ehlermann, D. A. E.; McLaughlin, W. L., Eds.; *Manual of Food Irradiation Dosimetry. Technical Reports Series No. 178*; International Atomic Energy Agency: Vienna: 1977; p 180.
7. Cleland, M. R. Advances in Electron Beam and X-Ray Technologies for Food Irradiation, Chapter 2. In *Food Irradiation Research and Technology*; Xuetong, F.; Christopher, H. S., Eds.; Blackwell Publishing: Chicago, 2013; pp 9–27.
8. Diehl, J. F. *Safety of Irradiated Foods*, 2nd ed.; Marcel Dekker, Inc.: New York, 1995; p 464.
9. Farkas, J. Five Years' Experience of the International Facility for Food Irradiation Technology. *Radiat. Phys. Chem.* **1977,** *25*(1–3), 227–231.

10. Farkas, J. Irradiation as a Method for Decontaminating Food: A Review. *Int. J. Food Microbiol.* **1998,** *44*(3), 189–204.
11. Farkas, J. Irradiation for Better Foods. *Trends Food Sci. Technol.* **2006,** *17*(4), 148–152.
12. Farkas, J.; Mohácsi-Farkas, C. History and Future of Food Irradiation. *Trends Food Sci. Technol.* **2011,** *22*(2–3), 121–126.
13. Fox, J. A. Influences on Purchase of Irradiated Foods. *Food Technol.* **2002,** *56*(11), 34–37.
14. Grandison, A. S. Irradiation, Chapter 5. In *Food Processing Handbook*; Brennan, J. G., Grandison, A. S., Eds.; Wiley-VCH: Germany, 2012; pp 153–177.
15. Hammad, A. A.; Abo, E.; Salah, A. International Atomic Energy Agency, Food and Environmental Protection Section. In *Use of Irradiation to Ensure Hygienic Quality of Fresh, Pre-Cut Fruits and Vegetables and Other Minimally Processed Foods of Plant Origin*; Bandekar, J. R., Dhokane, V. S., Shashidhar, R., Hajare, S., Sharma, A., Saroj, S., Eds.; IAEA (International Atomic Energy Agency): Islamabad, Pakistan, 2006; pp 106–129.
16. IAEA (International Atomic Energy Agency)–FAO (Food and Agriculture Organization). *Use of Irradiation to Ensure the Hygienic Quality of Fresh, Pre-Cut Fruits and Vegetables and Other Minimally Processed Food of Plant Origin, 2005.* Islamabad, Pakistan: International Atomic Energy Agency; 2006; p 328.
17. Koopmans, M.; Duizer, E. Foodborne Viruses: An Emerging Problem. *Int. J. Food Microbiol.* **2004,** *90*(1), 23–41.
18. Kume, T.; Furuta, M.; Todoriki, S.; Uenoyama, N.; Kobayashi, Y. Status of Food Irradiation in the World. *Radiat. Phys. Chem.* **2009,** *78*(3), 222–226.
19. Loaharanu, P. *Irradiated Foods*, 6th ed.; American Council on Science and Health, Inc.: New York, 2007; p 50.
20. Miller, R. B. *Electronic Irradiation of Foods: An Introduction to the Technology*; Springer Science & Business Media: Boston, 2006; p 293.
21. Molins, R. A. *Food Irradiation: Principles and Applications*. John Wiley & Sons, Inc.: New York, 2001; p 457.
22. Mossel, D. A. A.; Drake, D. M. Processing Food for Safety and Reassuring the Consumer. *Food Technol.* **1990,** *44*(12), 63–67.
23. Mostafavi, H. A.; Fathollahi, H.; Motamedi, F. Food Irradiation: Applications, Public Acceptance and Global Trade. *Afr. J. Biotechnol.* **2010,** *9*(20), 2826–2833.
24. Nayga Jr, R. M. Will Consumers Accept Irradiated Food Products? *Int. J. Consum. Stud.* **2003,** *27*(3), 220–220.
25. O'bryan, C. A.; Crandall, P. G.; Ricke, S. C.; Olson, D. G. Impact of Irradiation on the Safety and Quality of Poultry and Meat Products: A Review. *Crit. Rev. Food Sci. Nutr.* **2008,** *48*(5), 442–457.
26. Pauli, G. H.; Takeguchi, C. A. Irradiation of Foods—An FDA Perspective. *Food Rev. Int.* **1986,** *2*(1), 79–107.
27. Pillai, S. D.; Braby, L.; Maxim, J. Technical Challenges and Research Directions in Electronic Food Pasteurization, Chapter 23. In *Food Irradiation Research and Technology*; Xuetong, F., Christopher, H. S. Eds.; IFT Press: Iowa, **2012**; pp 425–433.
28. Pszczola, D. Irradiated Poultry Makes US Debut in Midwest and Florida Markets. *J. Food Technol.* **1993,** *47*(11), 89–96.
29. Robertson, R. E.; Hoy, J. *Food Irradiation: Available Research Indicates that Benefits Outweigh Risks*; Report to Congressional Requesters, United States General Accounting Office: Washington, DC, 2000, pp 1–36.

30. Ronning, D. C. *A Chronic Toxicity, Oncogenicity, and Multigeneration Reproductive Study Using CD-1 Mice to Evaluate Frozen, Thermally Sterilized, Colbalted-60 Irradiated, and 10 MeV Electron Irradiated Chicken Meat: Final Report*; St. Louis, MO: Ralston Purina Company to USDA - ARS, Eastern Regional Research Center, in partial fulfilment of Contract 53-3K06-1-2; 1983; vol. 1, pp 112–113.
31. Sádecká, J. Irradiation of Spices—A Review. *Czech J. Food Sci.* **2007,** *25*(5), 231–242.
32. Smith, J. S.; Pillai, S. Irradiation and Food Safety. *J. Food Technol.* **2004,** *58*(11), 48–55.
33. Sommers, C. H.; Fan, X. *Food Irradiation Research and Technology*, 1st ed.; Wiley-Blackwell: Iowa, 2006; p 336.
34. Stefanova, R.; Vasilev, N. V.; Spassov, S. L. Irradiation of Food, Current Legislation Framework, and Detection of Irradiated Foods. *Food Anal. Methods* **2010,** *3*(3), 225–252.
35. Stewart, E. M. Food Irradiation: More Pros than Cons? Part 2. *Biologist* **2004,** *51*(2) 141–144.
36. Teisl, M. F.; Fein, S. B.; Levy, A. S. Information Effects on Consumer Attitudes Toward Three Food Technologies: Organic Production, Biotechnology, and Irradiation. *Food Qual. Prefer.* **2009,** *20*(8), 586–596.
37. Terry, A. J.; McColl, N. P. Radiological Consequences of Food Irradiation. *National Radiological Protection Board* **1992,** *23*(13), 1–16.
38. Thayer, D. W.; Christopher, J. P.; Campbell, L. A.; Ronning, D. C.; Dahlgren, R. R.; Thomson, G. M.; Wierbicki, E. Toxicology Studies of Irradiation-Sterilized Chicken. *J. Food Protect.* **1987,** *50*(4), 278–288.
39. Thayer, D. W.; Rajkowski, K. T. Developments in Irradiation of Fresh Fruits and Vegetables. *J. Food Technol.* **1999,** *53*(11), 62–65.
40. Urbain, W. M. Food Irradiation: The Past Fifty Years as Prologue to Tomorrow. *Food Technol.* **1989,** *43*(7), 76–92.
41. Urbain, W. M. Government Regulation of Irradiated Foods, Chapter 14. In *Food Irradiation*; Academic Press Inc.: London, 1986; pp 276–287.
42. Van Kooij, J.; Leveling, H.; Schubert, J. Food Preservation by Irradiation. *Int. At. Energy Agency Bull.* **1981,** *23*(3), 33–36.
43. WHO (World Health Organization). *Safety and Nutritional Adequacy of Irradiated Food*; World Health Organization: Geneva, Switzerland, 1994; p 182.
44. Zienkewicz, L. S. H.; Penner, K. P. Consumers' Perceptions of Irradiated Ground Beef after Education and Product Exposure. *Food Protect. Trends* **2004,** *24*(10), 740–745.

PART II

ALTERNATE METHODS IN FOOD PROCESSING

CHAPTER 6

MATHEMATICAL MODELING IN FOOD PROCESSING

SAGAR V. GOKHALE, JYOTI S. GOKHALE, and SMITA S. LELE

ABSTRACT

Food systems generally consist of a mixture of complex molecules having biochemical and functional roles that are often linked with sensory attributes, which make it difficult to fully define these systems mathematically. Good model is an approximate representation of a system in terms of mathematical equations, which then helps in the process optimization, scale up, instrumentation, equipment sizing, reducing waste, energy consumption, etc. Mathematical modeling is a well-established branch in chemical and other process engineering, but its application is not fully explored in food processing. Modeling helps in reducing the number of experiments and hence time and cost. A good mathematical model should be based on unified approach to encompass many parameters in a wide range so that it could be used for many similar food systems.

6.1 INTRODUCTION

Mathematical modeling is well-established in food process engineering but is not that common in the food industry. Mostly researchers from academics and big companies have used modeling techniques, which is an interdisciplinary branch of study as it includes the understanding of biology, microbiology, biochemistry, statistics, etc., in addition to chemical engineering (unit operations, kinetics, transport phenomena, numerical methods, etc.). Although

Novel and Alternative Methods in Food Processing: Biotechnological, Physicochemical, and Mathematical Approaches. N. Veena, Megh R. Goyal, & Ritesh B. Watharkar (Eds.)

fundamental principles of unit operations are applied to foods, yet the nature of raw materials, and how they interact is different and sometimes may be more complicated than the chemical industry.

The purpose of this book chapter is to introduce readers to mathematical modeling with a few case studies in food processing.

Food processing is essentially an interdisciplinary area that involves knowledge of biochemistry, microbiology, food chemistry, nutrition along with the application of food engineering principles to get desired product with minimum energy consumption, low or no waste production and improved nutrition. Food processing makes the food in edible form, improves shelf-life and safety, brings convenience, and have an impact on nutritional quality of food. A unique feature of any food system needs special attention, when classical chemical engineering unit operations are applied. Unlike pharmaceutical products which look for sterile conditions of the drug, for a food product, a bigger challenge is to minimize heat treatment time so that nutritional, sensory, and textural damages are minimum, yet the desired microbial safety is achieved. Most foods have 80–95% moisture and retaining color, vitamins, taste of raw material could demand slower rate of low-temperature drying. Thus, producing healthy, tasty, and affordable products is the crust of the challenge.

Several food additives and ingredients are used in a single food formulation and positive and negative synergy on sensory attributes and molecular interactions may significantly influence flow characteristics, viscosity, and other physicochemical properties of the food product. Another challenge for food systems is its heterogeneous nature. For example, a classical dessert like ice cream is an emulsion of air, fat, water with milk proteins and sugars. In fact milk itself is also an emulsion. The food science and technology is also changed commodity-wise. Fruits, vegetables, grains, eggs, fish may behave completely different in a simple unit operation like drying—may it be a different equipment design, or it could need a pretreatment to avoid undesirable biochemical changes like Maillard reaction. Thus, mathematical modeling of food processing is a very challenging and interesting area.

In simple terms, a mathematical modeling is an approximate representation of the system in several mathematical terms. The fundamental principles of the modeling techniques can be found in many books by Bird et al.,[6] Luyben,[24] Tijskens et al.,[38] Özilgen,[30] etc. Typically, inputs and outputs or independent and dependent parameters or cause and results are related to a mathematical equation. Negligible inputs and outputs are generally neglected. More and more complex model by considering all these factors can be built but then it becomes more system specific instead of generalizing its application and which should not be the case. If one can write 100 models for 100 systems, then we can say it is okay, if 10 models can be written for 100 systems, it is better and if one can propose a single model for 100 systems, then that can be considered as best one. The aim of the modeling should be to reduce the experimentation and hence time and money rather than building complex mathematical equations. The best rule of thumb is given by Glasscock and Hale[13] and Özilgen,[30] which is an 80-20 rule: you get 80% of the benefit from the first 20% of the model complexity.

Let us consider a process with $x_1, x_2, \ldots, x_n$ as independent parameters or inputs and $y_1, y_2, \ldots, y_m$ as dependent parameters or outputs. Then, a mathematical model is built, which took $x_1, x_2, \ldots, x_n$ as inputs and gave $y_1', y_2', \ldots, y_m'$ as outputs (Figure 6.1). Then comparison of experimental data with the model output data through graph for visual comparison and also error calculations as "$y_m'-y_m$" are done with minimization of errors using numerical methods. The typical process of mathematical modeling is shown in Figure 6.2.

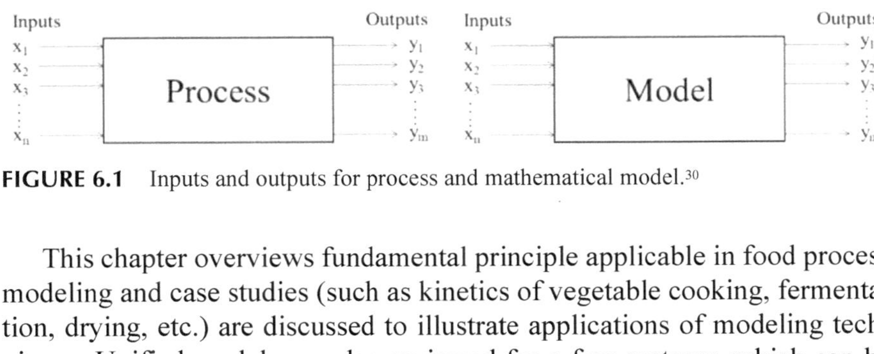

FIGURE 6.1 Inputs and outputs for process and mathematical model.[30]

This chapter overviews fundamental principle applicable in food process modeling and case studies (such as kinetics of vegetable cooking, fermentation, drying, etc.) are discussed to illustrate applications of modeling techniques. Unified models are also reviewed for a few systems, which can be applied to multiple products.

6.2 APPROACHES FOR MATHEMATICAL MODELING

The approaches for mathematical modeling can be described as follows:

- Fundamental approaches: Kinetic modeling and fluid mechanics, heat transfer, and mass transfer modeling.
- Empirical approaches: Black box modeling like artificial neural network (ANN), genetic algorithm (GA), fuzzy logic, etc.; and data mining using statistical methods.

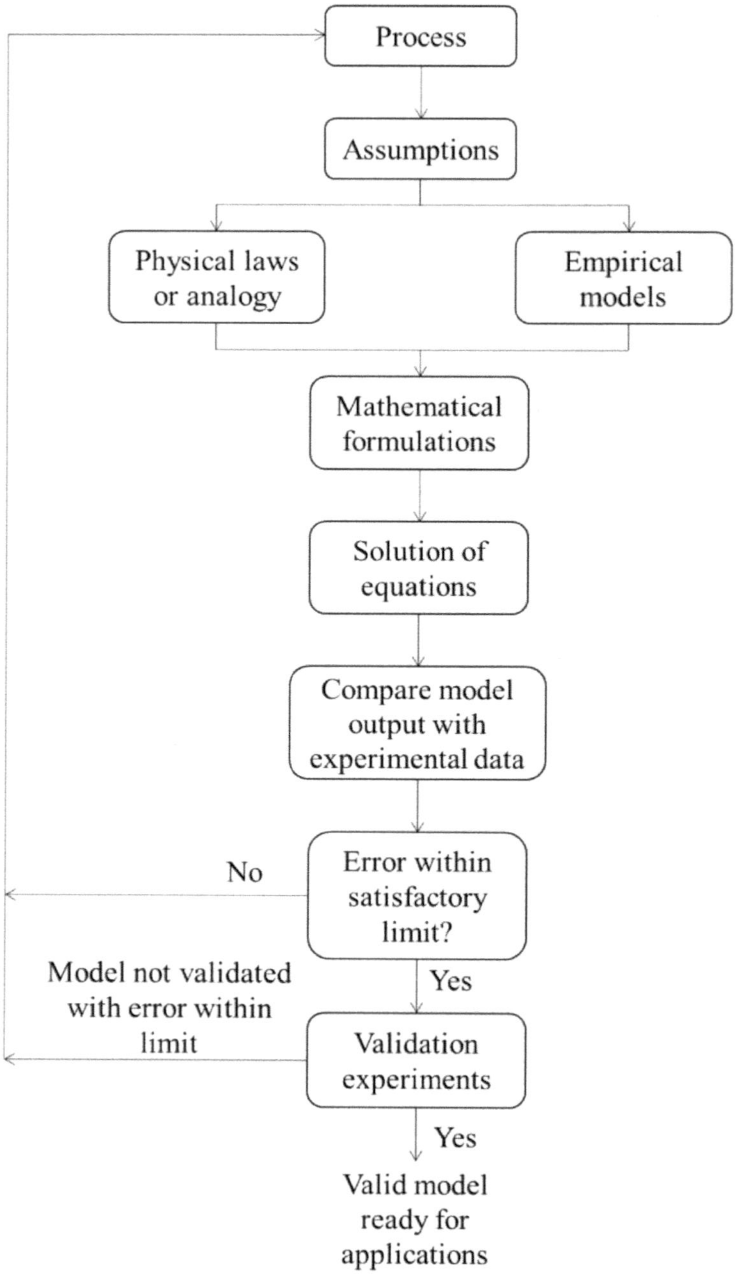

FIGURE 6.2 Flow chart for mathematical modeling.[30]

Fundamental modeling is done using physical laws and hence these models are consistent. The parameters, which are unrelated to the system, can be omitted from the model. At the same time, additional complexities can be systematically incorporated. And in this process, understanding of deeper fundamental science is established. It then can be easily adapted to different processing systems and can be made universal. Since it is established from fundamental laws, universality is the major advantage of the fundamental approach over the empirical approach. As we know, biological systems are complex systems and sometimes explanation with physical laws becomes too complex, therefore empirical models are adapted, which are easy to establish. Also, when datasets are too large (such as sensory evaluations, shelf-life studies, consumer feedbacks, consumer trends, etc.), then the empirical modeling, statistical data mining techniques are more useful.

In this chapter, three kinds of case studies are discussed:

- **Kinetic modeling**: Fundamental modeling based on chemical reaction kinetics. The similar principle can be used for microbial destruction, cooking of vegetables, drying of food materials, etc. Also, microbial fermentation kinetics is explained, as fermentation is a complex process and cannot be described with just one kinetic equation.
- **Unified modeling**: It is semiempirical approach to develop a universal model, which can be used for many systems.
- **Modeling for product development**: Product development is a combination of art and science, as it involves sensory evaluation as a dependent parameter and parameters of unit operations, ingredient properties, etc., as independent parameters. There are large number of academic research papers are published in peer-reviewed Journals.

Table 6.1 reviews a few of the representative reports on the application of fundamental and empirical approaches in different food systems.

6.3 KINETIC MODELING

Kinetic modeling plays an important role in the food processing and food quality as it enables to describe the food quality changes and their rates quantitatively. It also unravels basic reaction mechanisms. Common examples of applications of kinetic modeling in food processing are microbial destruction kinetics, drying kinetics, cooking kinetics, kinetics of various reactions happening during processing (e.g., acrylamide formation during frying), enzymatic reaction kinetics, fermentation kinetics, etc.

TABLE 6.1 Fundamental and Empirical Modeling Approaches for Food Processing Systems.

Food	Unit operation	Model type	Independent parameters	Dependent parameters	References
			Fundamental modeling		
–	Radio-frequency (RF) tempering/thawing	–	Temperature and electric field changes, movement of food product, movement of RF electrode	Power uniformity index	10
Beet	Hot air dryer	Kinetic modeling, heat, and mass transfer modeling	Time, drying temperature, mass flow rate of air	Moisture content, drying rate, effective diffusivity, overall heat transfer coefficient	14
Biscuit	Baking oven	Kinetic modeling	Time	Concentration of acrylamide and 5-hydroxymethylfurfural (HMF)	29
Blackcurrant juices	Heating in water bath	Kinetic modeling	Time	Anthocyanin	18
Canned celery stew	Retort process	Conductive heat transfer using CFD	Size and shape of container, holding time, retort settings	Temperature	5
Cylindrical sections of potato	Freezing	Heat transfer modeling	Time, freezing temperature	Freezing progression, product temperature, enthalpy change	20
Frozen foods	Frying	Heat and mass transfer modeling	Time, frying temperature	Product temperature, cooking of food, frying time, moisture loss	37
Ice cream	Freezing		Initial temperature, freezing rate, and different compositions with and without stabilizers	Ice crystal size	11
Large particle food items	Fluidized bed freezing	Heat transfer modeling	Particle diameter, time, air flow rate	Heat transfer coefficient, freezing time, moisture loss	34
Mashed potatoes	Microwave heating	Heat transfer modeling	Initial temperature, time, final temperature	Dielectric properties of mashed potatoes to predict temperature profile of product	31

TABLE 6.1 *(Continued)*

Food	Unit operation	Model type	Independent parameters	Dependent parameters	References
Morcilla, a typical cooked sausage	Vacuum packaging and shelf-life	Kinetic modeling, shelf-life prediction modeling	Time (shelf-life), initial microbial count (artificially inoculated), temperature	Lactic acid bacteria (LAB) count	26
Mushroom	Microwave vacuum drying, freeze drying	Kinetic modeling, heat transfer modeling	Time, temperature, pressure	Moisture content, diffusivity coefficient, quality of mushroom	
Protein-stabilized emulsions	–	Rheological modeling	–	Bulk rheological properties, interfacial rheology of proteins adsorbed at oil-in-water (O/W) interface, and the links between interfacial and bulk rheology of proteins/polysaccharide mixtures	12
Tomato paste	Heating in water bath	Kinetic modeling	Time	Color of paste (L, a, b, ΔE)	4
Winter mushrooms (*Flammulina velutipes*)	Heating in water bath	Kinetic modeling	Time	Texture (hardness, springiness, fracturability)	21
			Empirical modeling		
Canned peas	Retort process	Artificial neural network (ANN)	Time, temperature	F_0	17
Green chickpea	Hydration	ANN	Hydration time, hydration temperature	Moisture ratio, moisture content, and hydration ratio	23
Pea based extruded snack	Twin screw extrusion	Genetic algorithm (GA)	Screw speed, temperature	Radial expansion, density, WAI, WSI, pressure, motor torque, SME, and color	22

TABLE 6.1 *(Continued)*

Food	Unit operation	Model type	Independent parameters	Dependent parameters	References
Plum spread	Product development/sensory optimization	ANN	Concentrations of plum pomace lyophilisate, sugar, amidated low-methoxyl pectin	Chemical, textural, and color	2
Sugarcane juice	Micro fluidization	GA	Pressure, number of passes	Total soluble solids, total phenolic content, free radical scavenging activity, total flavonoid content, chlorophyll, color, polyphenol oxidase activity, microbiological quality	37
	Canning	Fuzzy logic	Retort temperature, time, pressure	F_0, time–temperature profile	7
Tomato soup and model salt solution		Multinomial logit model	Salt concentration	Sensory and molded modeling clay into a shape that best reflected their sensory experience	3
Wheat flour dough	Dough proofing	Fuzzy logic	Yeast concentration, temperature	Volume of dough ball	41

Many researchers have studied food quality parameters (such as color, nutrient content, microbial content, physical changes, etc.) during the processing and shelf-life using kinetic models for these parameters.[39] Mostly chemical reaction kinetics are developed based on simple elementary reactions, whereas reactions in foods involve many parameters. Hence better model development is needed. For solid or semisolid foods, diffusion of the reactants to encounter with each other for reaction to happen can be a rate-limiting factor, whereas in many cases, rate of reaction itself can be rate-limiting step versus diffusion rate. The simple kinetic model for n^{th} order reaction is as follows:

$$r = -\frac{dC}{dt} = kC^n \tag{6.1}$$

where C is the concentration of reactant; k is reactions rate constant; t is time; and n is order of the reaction.

When diffusion is the rate-limiting step, rate constant k can be defined as follows:

$$k_{diff} = \frac{8\times 10^3 RT}{3\eta} \tag{6.2}$$

where R is the universal gas constant; T is absolute temperature; and η is viscosity.

Further dependence of the rate of reaction/process on temperature can be defined using Arrhenius equation:

$$k = k_0 e^{\frac{-E_a}{RT}} \tag{6.3}$$

Zero-order or first-order reactions have more commonly been reported in food matrix.[39] The graphs for different orders of reaction are shown in Figure 6.3. Few of the case studies are presented in the following subsections.

6.3.1 CASE STUDY 1: KINETICS OF COOKING OF VEGETABLES

An extensive research has been done on microbial destruction kinetics; and equations for D value and Z value are well-established. Gokhale has applied similar approach to the cooking of vegetables to establish cooking kinetics and its application in retort process modeling.[16] First-order kinetics was applied for the cooking of vegetables at different temperatures. Table 6.2

represents the first-order rate constants and D values for rice and vegetables used in *Pulav*. The basic model for the rate of cooking is first-order kinetics, as follows:

$$\frac{dc}{dt} = -k_r c \tag{6.4}$$

where k_r is first-order rate constant for cooking.

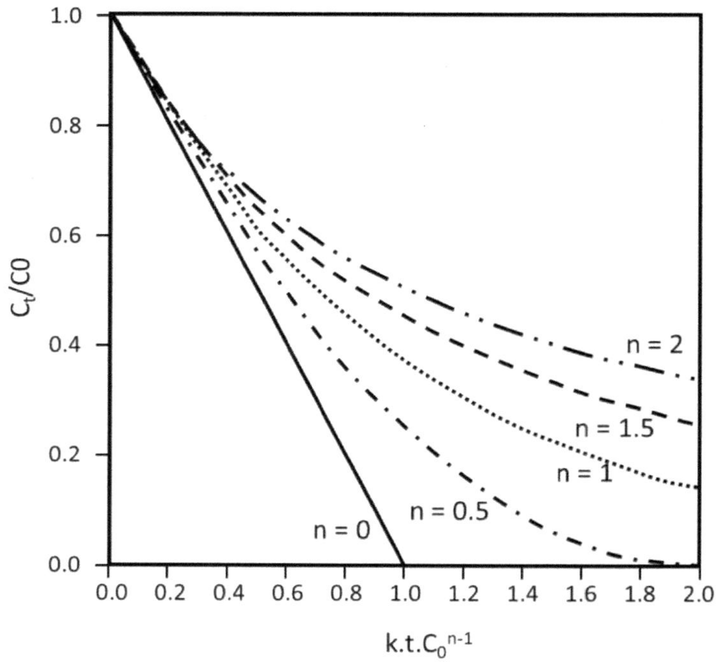

FIGURE 6.3 Decrease of a reactant with different order of reactions.[39]

The degree of cooking was measured on the texture analyzer (Stable Microsystems TXT2i Texture analyzer) using 5-mm cylindrical probe. The maximum force required was taken as an index of hardness (higher value indicates greater hardness and lower degree of cooking) that is denoted by "c" in eq 6.4. Decimal reduction time (D_c) and thermal resistance constant (z_c) were defined like microbial destruction. Also, like lethal rate and F-value for microbial destruction, cooking rate and cook value (C) were calculated with the reference temperature of $T_R = 121.1°C$ and the "z_c" value was calculated as follows:

$$D_c = \frac{t}{\log(C_0/C)} \text{ or } D_c = \frac{2.303}{k_r}$$

$$z_c = \frac{2.303R}{E_a}(T_{A1} \times T_{A2}) \text{ or } z_c = \frac{T_2 - T_1}{\log D_{c_{T1}} - \log D_{c_{T2}}}$$

$$C = \int_{10} \frac{T - T_R}{z_c} dt \quad (6.5)$$

In these equations, the required C value was calculated as follows:

$$C_{R,121} = \log\left(\frac{C_0}{C}\right) D_{c,121} \quad (6.6)$$

A set of experiments was conducted to obtain the cooking rate of carrot under isothermal conditions at different temperatures (90–120°C). Figure 6.4 represents time versus temperature relationship for cooking of rice and vegetables. Samples were taken after every 2 min; and time required for cooking the vegetable/dal (grain) for respective temperature was noted. The z_c value was calculated from the data for each vegetable and rice (ingredients for ready-to-eat *Pulav*). Table 6.3 shows z_c values for different ingredients of *Pulav*.

TABLE 6.2 First-Order Rate Constants and D-Values for Rice and Vegetables Used in *Pulav*.

Grain/vegetable	Temperature (°C)	k_r (min^{-1})	D_c (min)
Rice	120	0.125	18.38
	110	0.090	25.73
	100	0.063	36.79
	90	0.020	115.15
Green peas	120	0.141	16.33
	110	0.091	25.31
	100	0.074	31.12
	90	0.032	71.52
Carrot	120	0.131	17.62
	110	0.089	25.96
	100	0.067	34.32
	90	0.029	78.87
Corn	120	0.121	19.08
	110	0.097	23.67
	100	0.061	37.82
	90	0.028	82.25

TABLE 6.3 Thermal Resistance Constant (z_c) Values for *Pulav*.

Grain/vegetable	z_c value
Carrot	48.31
Corn	47.62
Green peas	49.50
Rice	39.37

These z_c values (Table 6.3) can then be used in calculations of retort processing in a very similar way to calculate the required cook value, C_0 (similar to required F_0) values, and optimize the retort process for the said product.

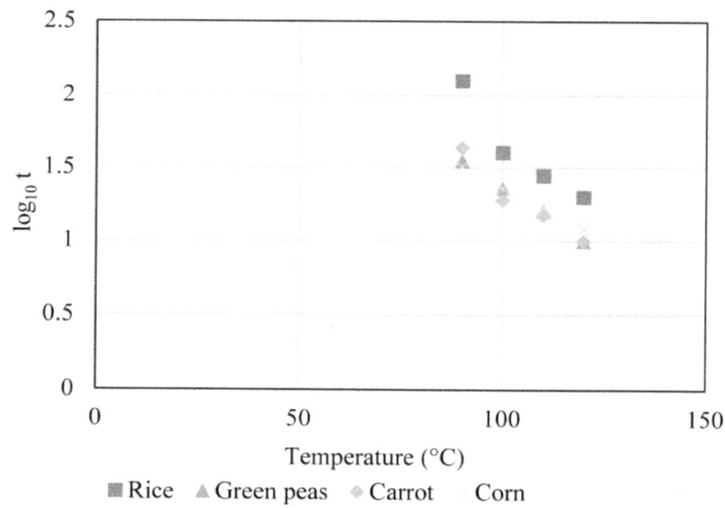

FIGURE 6.4 Time–temperature relationship for cooking of rice and vegetables.

6.3.2 *CASE STUDY 2: LEAVENING REACTION KINETICS*

Leavening agents are used mainly in the bakery food systems to leaven or to raise the products wherein a gas is produced because of chemical reaction between leavening acid and leavening base. This case study for leavening reaction kinetics is at conceptual level only as the experimental data are unpublished and proprietary to a company. The set of experiment was carried out to explore novel leavening acids. Leavening reaction is the acid–base reaction, typically used in bakery products where the base is usually baking soda or sodium bicarbonate. To achieve different textures in cakes, breads, etc., different leavening acids are used (such as mono calcium phosphate, sodium aluminum phosphate, di-calcium phosphate, etc.), which have different rates of reaction along with different activation energies. The typical reaction during leavening is as follows:

$$HX \text{ (acid salt)} + NaHCO_3 \xrightarrow{\text{Heat}} NaX \text{ (neutral salt)} + H_2O + CO_2 \quad (6.7)$$

The reaction was carried out in a cake batter matrix (using the standard recipe for cake) and a leavening system in a closed stirred reactor. Carbon dioxide (CO_2) concentration was measured using CO_2 analyzer probe. The reactor was placed in an oil bath to maintain reactions at different temperatures. Leavening acid was mixed with the batter and measured quantity was poured in the reactor. Sodium bicarbonate ($NaHCO_3$) was added at time t = 0. Therefore to measure rate of reaction stated above, rate of production of CO_2 was assessed and it was found that most of the reactions followed first-order kinetics, as follows:

$$\frac{dC_c}{dt} = kC_c \quad (6.8)$$

where C_c is concentration of CO_2; t is time; k is the reaction constant.

Since the rate of production of CO_2 is measured, the equation has a positive sign. The effect of temperature can be represented by Arrhenius equation:

$$k = k_0 e^{\frac{-E_a}{RT}} \quad (6.9)$$

where, k_0 is pre-exponential factor; E_a is activation energy; R is the universal gas constant; and T is the absolute temperature.

- In the *first step*, commonly used leavening acids are monocalcium phosphate, dicalcium phosphate, sodium acid pyrophosphate, calcium acid pyrophosphate, sodium aluminum phosphate, sodium aluminum sulfate, etc. The landscape of rates of reactions of all these acids at different temperatures was developed. Functions of leavening are gas bubbles formation, uniform cell structure, thin cell walls, improve crumb color, and improve eating quality. To achieve it, combinations of fast acting, slow acting, and temperature activated leavening acids are used. Now the aim of the study was to find alternative novel leavening acids instead of acid salts listed above.
- The acids tested in the *second step* were citric acid, lactic acid, malic acid, glucono delta lactone, cream of tartar, vinegar, tamarind extract, etc. Then, it was found that many of these acids are fast acting.

- Therefore in the *third stage*, the rate of reaction was controlled using different processes, that is, encapsulation, etc.

The first-order rate of reaction is then shifted to the complex kinetic model considering the diffusion step added to the first-order reaction rate. Therefore, rate of reaction was altered and matched to the chemical leavening acids, and replacers were developed for same. The alternatives were then tested in actual product for validation (unpublished work by Gokhale S. V.).

6.3.3 CASE STUDY 3: FERMENTATION KINETICS

Food fermentation is a traditional process and study of fermentation kinetics is important in many aspects of processing, such as process design, equipment sizing, process scale-up, process control, etc. Fermentation is a complex process and hence its kinetics cannot be as simple as using just reactant kinetics, which is the case for simple chemical reactions. One has to measure the consumption of nutrients, increase in biomass, and increase in product concentration with time and establish relations. This is because part of the nutrient can be used for the cell growth and some part can go toward production of various biochemical out of which one biochemical can be of our interest, which is our product. Here, one such case study is presented on the production of natto kinase in shake flask and a 5 L fermenter.[25] The kinetic modeling was done to represent increase in biomass (X, dry cell weight g/L), increase in product (P, natto kinase activity U/L), and decrease in substrate (S, maltodextrin g/L) with respect to time (t).

The most commonly used models in fermentation kinetics are Monod kinetics, logistic equation, and the Haldane model. To describe microbial growth in various biological systems, logistic model is commonly used, which takes into consideration simple calculations of fermentation parameters having geometrical and biological significance using sigmoidal profiles while ignoring substrate concentration. Thus, biomass growth kinetics can be written as follows:

$$X = \frac{X_0 X_{max} e^{\mu_{max} t}}{X_{max} - X_0 + X_0 e^{\mu_{max} t}} \tag{6.10}$$

$$\grave{\imath}_{max} = \frac{\text{initial volumetric rate of biomass formation}}{\text{initial biomass concentration}(X_0)}$$

where t is time (h); X is biomass concentration (g/L); X_{max} is maximum concentration of biomass (g/L); and X_0 is the initial biomass concentration (g/L).

The product formation kinetics can be written as proposed by Mercier et al.[27]:

$$P = \frac{P_0 P_{max} e^{P_r t}}{P_{max} - P_0 + P_0 e^{P_r t}} \tag{6.11}$$

where P_r is defined similar to μ_{max} as follows:

$$P_r = \frac{\text{Initial volumetric rate of product formation}(r_p)}{\text{initial product concentration}(P_0)} \tag{6.12}$$

where t is time (h); P is enzyme activity (U/L); P_{max} is maximum enzyme activity (U/L); and P_0 is the initial product concentration (U/L).

Substrates are consumed by the cells to form its components, production of metabolites, and for the maintenance of cells. Substrate consumption (maltodextrin in the present study) kinetics can be written as:

$$S = S_0 - \frac{1}{Y_{P/S}}(P - P_0) - \frac{1}{Y_{X/S}}(X - X_0) \tag{6.13}$$

where $Y_{P/S}$ (U/g) and $Y_{X/S}$ (g/g) are the substrate consumed to form product (enzyme) and biomass, respectively; S_0, P_0, X_0, S, P, and X are initial and final concentrations of substrate (g/L), enzyme (U/L), and biomass (g/L).

Figure 6.5a–c shows comparison of experimental and model predicted data for shake flask and fermenter for biomass growth (R^2 = 0.9586 and 0.9788), enzyme production (R^2 = 0.9940 and 0.9896), and substrate utilization (R^2 = 0.9724 and 0.9627), respectively. A correlation regression coefficient of more than 0.95 shows that models are giving agreeable prediction with less than 5% error.

6.4 UNIFIED MODELS

As mentioned earlier in this chapter, the aim of the mathematical modeling is to minimize the number of experiments and hence the time and cost. In many cases, there is little difference in the product composition or seasonal variation in the same ingredient. Then to minimize the experimentation, one can simulate the product and develop the process and a mathematical model for it. Then with the help of mathematical model, process development for each product variant with minimum number of experiment or in later case can be done, and one can predict what will be the effect of change in ingredient (seasonal change or change in variety) on the product. This approach of mathematical modeling is known as *unified modeling*.

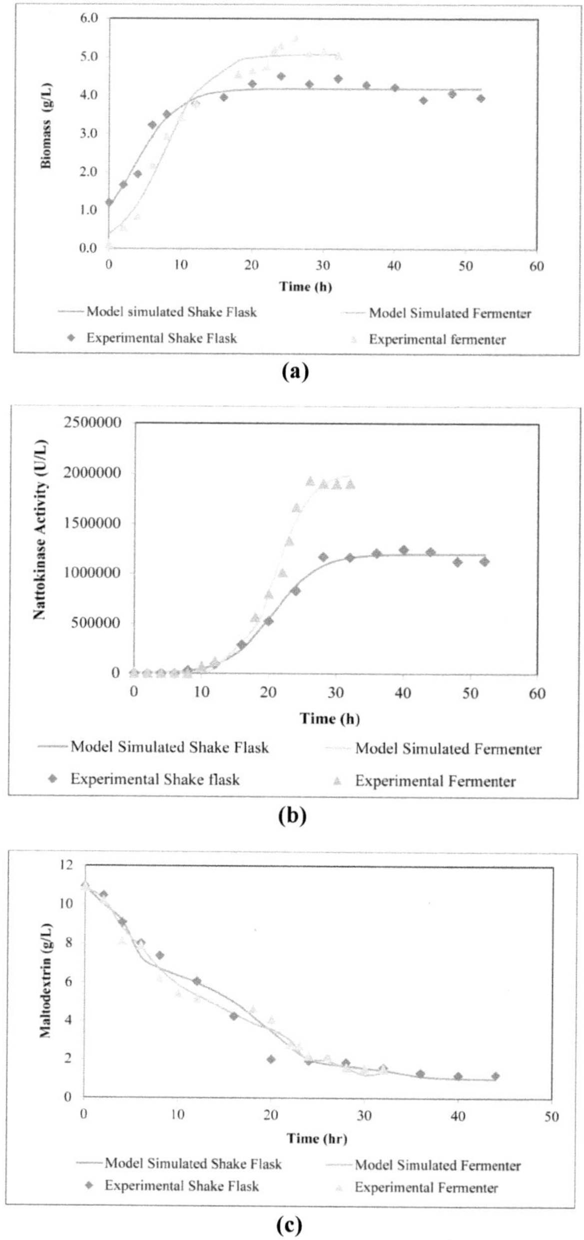

FIGURE 6.5 Comparison of the calculated values and experimental data of (a) biomass formation (g/L); (b) natto kinase production (U/L); and (c) substrate utilization (g/L) at shake flask and fermenter scale with respect to time.

Suppose one want to develop a drying process for different fruits and vegetables (say 10 each). To get optimum drying conditions with minimum energy consumption, one need to carry out drying of each one at different temperatures to check what is its effect on drying rate, mass transfer, heat transfer, etc., and develop a drying process with variable temperature resulting in optimum energy consumption. If one need to study five different temperatures, then 5 × 20 = 100 experiments are to be carried out. Now let us think in different ways. See what are fruits and vegetables made up of and drying rate will depend on what factors? Therefore, fruits and vegetables mainly contain water, starch, fibers, sugars, vitamins, minerals, etc. Out of these, water, starch, fiber, and sugar are the parameters, which will affect drying rate. Then can one simulate a mixture with various compositions of these and carry out drying experiments to find the effect of each set on drying rate. Then, one will have a model correlating process parameters and composition of vegetable/fruit. Thus, take example of any vegetable or fruit, one can estimate how it will behave in the dryer and thus one may get results only based on half the number of experiments. Of course, minor tweaking in the process parameters may require from product to product, but number of experiments will be reduced.

The other example can be variation in the same ingredient. For example, if one is using potato powder in some product and source potatoes are from the local market to make a powder. Then there can be variation in the properties of the potato powder with variety, with season and harvesting time, etc. With each batch of potatoes, one will need to optimize the product as per variation in properties in potato powder made in that particular batch. In this way, one will end up with conducting many experiments for each batch. To avoid this, if one wants to find out what are the important properties of potato powder to be considered (gelatinization, color, etc.) and which parameters (such as starch content, skin content, protein content, etc.) are responsible for the said result; then one can run a set of experiments wherein simulation of the potato powder with different percentage of starch, protein, fibers, etc., can be done and related them with properties (such as gelatinization or rheology of a paste, etc.). Thus, one will end up developing a mathematical model, which will tell how potato powder is going to behave if there is a change in its composition. Then using the optimization of such parameters, one can develop the product. Then every time for the new batch of potatoes, one will have two options: (1) separate starch, protein, fibers of the potato, dry, and mix them again in the required proportion to get the optimum potato powder; (2) by knowing starch, protein, fiber content, one can predict how it is going to affect the product and one can vary its use accordingly. Therefore, now one need not worry about the change in variety of a potato or seasonal variation, etc.

Two different case studies are discussed in this section, that is, (1) simulating the product for process development (retort processing); and (2) simulating the raw material for product development (wheat flour).

6.4.1 CASE STUDY 1: UNIFIED MODEL FOR RETORT PROCESSING

Retort processing is a hurdle technology in which in pouch sterilization of foods takes place because of heating with pressurized steam to extend shelf-life ranging between 6 and 18 months depending on nature of the product. Recently, retort foods are preferred by consumers due to dietary habit changes and its advantages (such as convenience, portability, time saving, etc.). Gokhale and Lele[15] have developed a unified model for retort process for traditional Indian foods. Food product processed using retort technique is typically a heterogeneous system with some slurry (curry) and solid pieces (vegetables or meat) in it. The products considered in this report were vegetarian products typically prepared with raw materials like vegetables/lentils/grains/cereals, and spices, etc. In the present study, instead of using actual food product, simulated system was used for mathematical model development and model was then validated using four different kinds of vegetarian products. To develop a simulated food system, food product is imagined to be made with four class of components:

- Dry solids contained in vegetables, grains, lentils, spices, etc. It also includes sugar.
- Water, either externally added while making the product or contained by vegetables.
- Fats.
- Salts and minerals.

Out of these four categories, salt, sugar, and fats are added in minute quantities and hence can be neglected in the present study. Thus, the two major components remaining are considered as a part of the simulated food and those are dry vegetable solids and water. It was assumed that the thermal properties of the food product will depend largely on ratio of total solids to water in the food product. Due to cultural diversity across India, there are over two thousand different recipes of staple and snack foods. If we go product by product for process development, then many set experiments will be involved. Hence unified model development approach was used in this study so that most of the products could be viewed as the combination of total solids (carbohydrates, fibers, sugars, etc.) and water; based on which one can predict process parameters for the same equipment.

The objective of this study was to develop a mathematical model to predict time–temperature profile of the product during retort processing; provided one knows the process parameters like retort temperature, initial temperature of the product, and property of a food product like ratio of solid content to water in the product. A semiempirical model was proposed by Gokhale and Lele[19] to describe temperature profile of the product with respect to time. The driving force for the heat transfer and hence rise in product temperature is defined as follows:

$$\Delta T = T_R - T_p \tag{6.14}$$

where T_R is the desired retort temperature (°C); and T_P is the product temperature (°C) at any time, t (min).

In differential form, the model was described as follows:

$$\frac{dT_P}{dt} = k_h T_P \left(\frac{T_R - T_P}{T_R} \right) \tag{6.15}$$

Integrating the eq 6.15 and rearranging, we get:

$$T_P = \frac{T_{P0} e^{k_h t}}{\left[1 - \left(T_{P0} / T_R\right)\left(1 - e^{kt}\right)\right]} \tag{6.16}$$

where T_{P0} is initial product temperature (°C); and k_h is the empirical model constant.

The heating and cooling process is considered as two separate parts that were modeled differently and two separate model constants k_h and k_c were estimated.

Further as a simulated system, mixture of dry vegetable solids and water was used. As a first step, it was checked that all vegetable solids have similar thermal properties. Three simulated food systems were used with different ratios of vegetable solids to water (vegetable solid content, S_s) at different initial temperatures (T_{P0}, °C) processed in a retort at different temperatures (T_R, °C). The time–temperature profile for different combinations of these three parameters (S_s, T_{P0}, and T_R) was recorded and dependence of model constants k_h (for the heating part) and k_c for the cooling part was studied on these three parameters. Figure 6.6 shows the effect of solid content and initial product temperature on k_h. The equations relating k_h and k_c and process parameters are as follows:

$$k_h^3 = 0.038 - 0.048\ S_s + 0.03\ T_{P0}\ S_s + 0.016\ T_R^2 - 0.016\ T_{P0}^2 + 0.061\ S_s^2$$

$$k_c^2 = 18.91 - 0.36\ T_R + 0.14\ T_{cw} + 0.01\ S - 5.57\times10^{-4}\ T_R T_{cw} - 2.1\times10^{-4}\ T_R\ S_s + 1.86\times10^{-4}\ T_{cw}\ S_s + 1.63\times10^{-3}\ T_R^2 - 9.95\times10^{-4}\ T_{cw}^2 + 1.63\times10^{-4}\ S_s^2 \quad (6.17)$$

where k_h and k_c are model constants for heating and cooling equations, respectively; S is solid content in the product pouch; T_R is target retort temperature; T_{P0} is initial product temperature; T_{cw} is cooling water temperature.

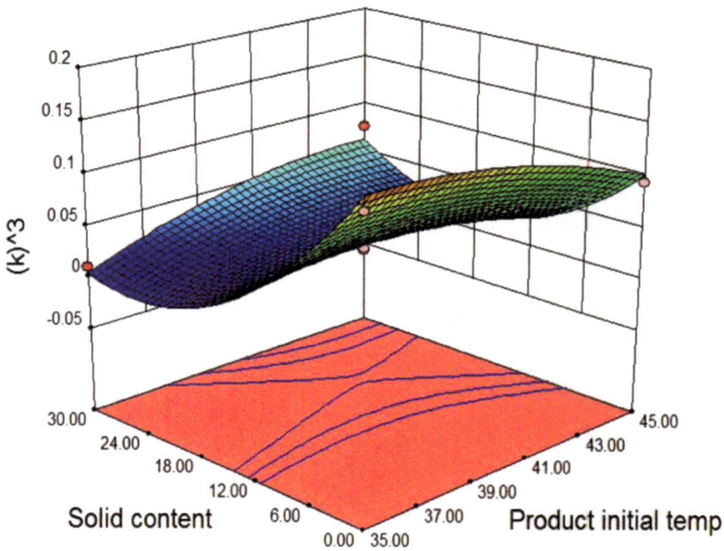

FIGURE 6.6 k_h as a function of solid content of pouch and initial product temperature; adapted from Gokhale and Lele.[15]

Source: Reprinted with permission from Ref. [15]. ©Association of Food Scientists & Technologists (India) 2012.

Now one may have thought that solid content in the pouch may be just one parameter, which is affecting the rise in temperature; and size and shape of the pieces in the pouch are also important parameters, which affect the heat transfer. The model stated above is just the first step and there can be a few more such steps to make it complete. The objective here was just to showcase the approach of the unified model to develop a process.

Thus, if three parameters (such as S, T_{P0}, and T_R) for any vegetarian food product are known, then we can calculate the model constants, k_h and k_c and which can lead us to the time temperature profile for the desired food product. Using the predicted temperature, one can then calculate F_0 value and thus decide the time cycle for the retort process. The validation of this model was done by Gokhale and Lele[15] using four traditional Indian products having different product nature:

- *Pulav*, which is a rice variety with vegetables incorporated and has no curry (slurry) in it.
- *Sambar*, which is a curry product with high water content and lentil along with some vegetable pieces.
- *Gajar Halawa*, which is desert with higher fat and sugar content.
- *Upama*, which is breakfast product made with Suji and some vegetables.

The products so chosen that they have different solid-to-water ratios and different recipes so as to cross-check if it is applicable to variety of products. A sample validation with *Upama* is shown here. *Upama* contained 29.9% solids, retort temperature was kept 120°C and initial product temperature was 36°C. Using these conditions, model constants k_h and k_c were predicted as 0.25 and 0.027, respectively, and experimental values obtained were 0.27 and 0.030. Thus, model was validated with relative error of 7.4% and 10%, respectively. The predicted cumulative F_0 value was 2.26 and experimental value found was 2.08 having relative error of 8.67%. The predicted and experimental time–temperature profile is shown in Figure 6.7.

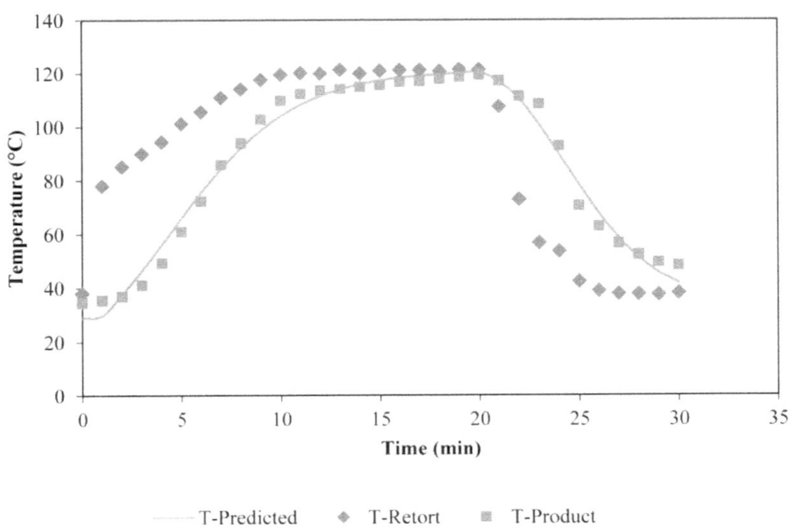

FIGURE 6.7 Experimental and predicted time–temperature profiles for *Upama*.

Source: Modified from Ref. [15].

6.4.2 CASE STUDY 2: UNIFIED MODEL FOR WHEAT FLOUR DOUGH

Wheat flour dough is an industrially important material as bread (leavened or flatbread). The manufacturing process and product quality largely depend on dough rheology, which is a complex fluid consisting of hydrated starch granules (~5–40 μm) suspended in a cohesive viscoelastic network of gluten matrix. The main ingredients in *chapatti* dough are wheat flour, oil, water, and salt. Further, on a dry weight basis, the major constituents of wheat flour to make the Indian breads are 75–90% starch, 6–15% gluten, and 1–3% lipids. Now, there can be variations at two levels, which have been considered by Desai[8]: (1) different water and oil content while making dough; (2) variation in wheat flour because of different varieties/seasons. The first one is quite straight forward, whereas for later one, simulated flour approach was used. It was proposed by Desai[8] that the starch and gluten are the two important factors in wheat flour that affects the dough properties and if we can simulate the dough with different combinations of starch and gluten, and relate it with rheological properties of dough through a mathematical model, then one gets any variety of wheat, and get it tested for starch and gluten percentage, then the model can tell what are the rheological properties of it and predicts how dough will behave during the processing.

Shear modulii (G' and G") and complex viscosity (η) were studied on a rheometer by Desai[8] with different water content, oil content, and various additives at first level and then different combinations of starch and gluten at second level. This makes a landscape of all the combinations that one can use in the flatbread making; and then for any formulation of the dough, one can predict how dough will behave during processing. Further, the rheological properties of the dough can be related to sensory properties of the product, such as softness, chewability, etc.

Figure 6.8 shows variation of rheological properties with different water, oil, and salt content. Further, one can also study the effect of different additives like enzymes, etc., on rheological properties to have extensive studies with additives. Figure 6.8 shows that the variation of rheological properties with variation in water was not linear whereas with oil and salt it was linear. The further model dough was made with different proportions of wheat starch, gluten, and water. Although we know that there is no wheat flour with 0% or 100% gluten, the proportions were varied between these two values for model development. Six two-component mixtures, with varying quantities of starch and extracted gluten (SG), were prepared as shown in Table 6.4.

Each model dough was prepared in mixograph (TMCO) for optimum dough development time (~8 min) (Figure 6.9).

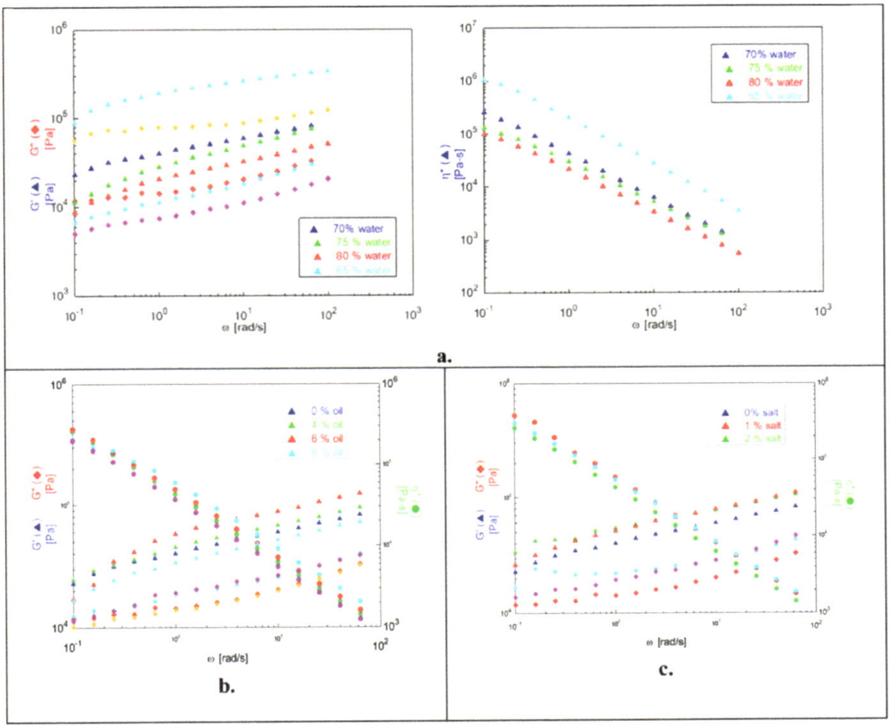

FIGURE 6.8 Effect of (a) water; (b) oil; and (c) salt content in dough on rheological properties of dough.

TABLE 6.4 Nomenclature, Composition, and Mixograph Characteristics for Model Dough.

Model dough	Gluten % (dry basis)	Gluten volume fraction	Starch volume fraction	Water volume fraction
SG0	0	0	52.9	47.1
SG10	10	5.9	47.6	46
SG20	20	11.7	42.3	45.9
SG40	40	23.5	31.8	44.7
SG60	60	35.3	21.2	43.5
SG100	100	58.8	0	41.2

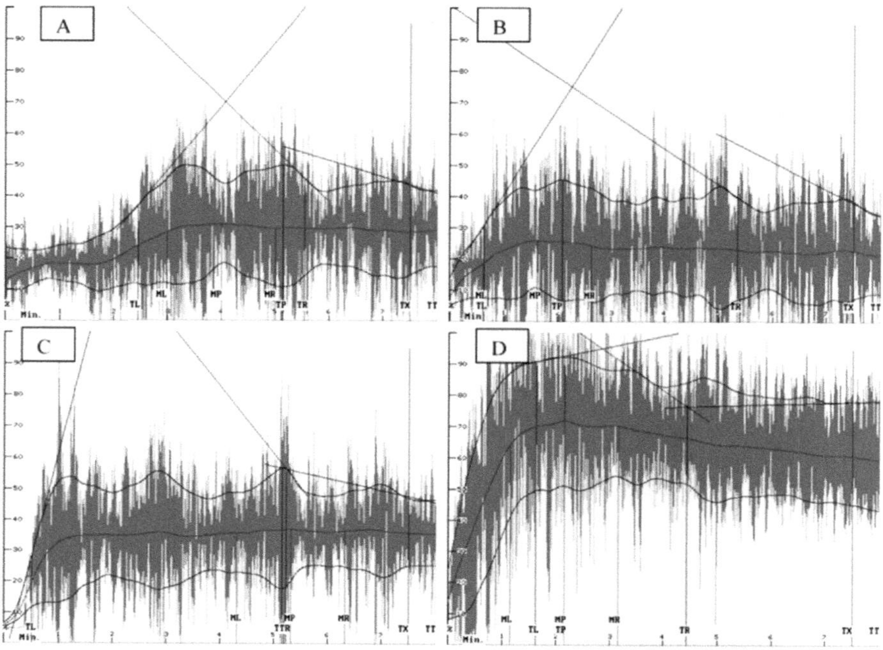

FIGURE 6.9 Mixographs of model dough: (A) SG0; (B) SG10; (C) SG20; and (D) SG40.

Further, the rheological studies of the model dough are shown in Figure 6.10. The study of model dough given in this book chapter is just a small part of the detailed study carried out by Desai and Lele[8,9] and is just to demonstrate how a simulated ingredient can be made to study its characteristics, which later can be related to machinable properties of dough or sensory parameters of finished products as discussed earlier. Therefore, mathematical modeling here helps us to minimize the number of experiments, every time we get a new variety of wheat. Therefore, now if depending on product attributes, we have decided on optimum rheological parameters of dough (G', G", etc.) and now we got some new variety of wheat, then how modeling will help us here? Let us analyze that a new variety of wheat essentially means different starch and gluten content and here now we have complete landscape of how dough will behave at different starch and gluten content.

Therefore, firstly we will have to see what will be rheological parameters for new starch and gluten content. If they are within range of our optimum rheological parameters that we have decided as per product attributes, then we need not optimize our dough formulation. But if they are out of range, then further we have graphs for different water, oil content in the dough, etc. From that, we can find optimum water, oil content, etc., to result in the required rheological properties that we need on machine and to get the required product attributes. Thus, there will be no or very little experimentation (for validation) required when we get a new variety of wheat in the factory. Thus, reduced experimentation will reduce time and hence money.

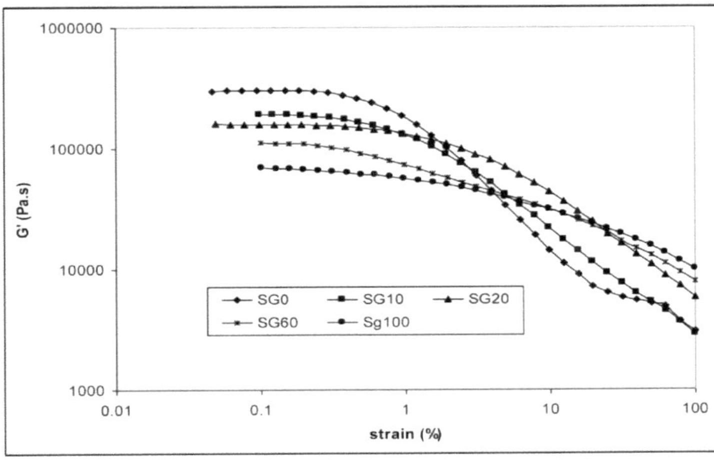

(a)

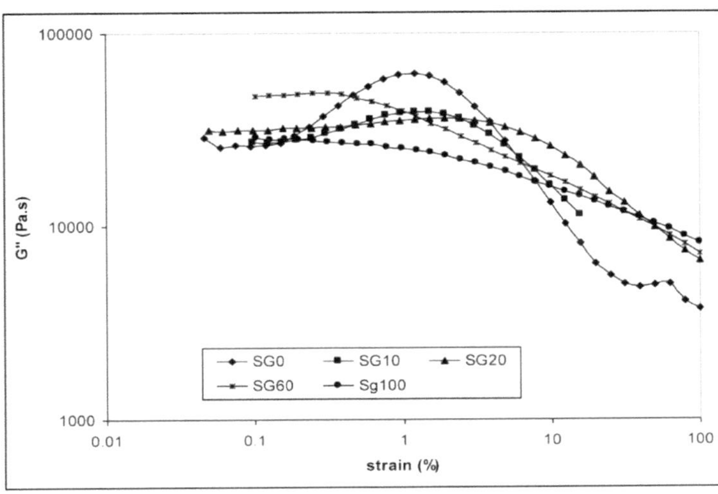

(b)

FIGURE 6.10 Strain sweep for model dough formulations: (a) storage modulus (G'); (b) loss modulus (G") of model dough.

6.5 PRODUCT QUALITY OPTIMIZATION/NEW PRODUCT DEVELOPMENT

Modeling and optimization of a unit operation (like drying, freezing, retorting, extrusion) is the first step among the many aspects to be considered in the commercialization of a product. Food product should have good customer appeal as well as desired shelf-life. Mathematical modeling helps in many ways in the new product development, product quality optimization, and product renovation. Many researchers have used the response surface methodology in product development, which has been considered in this category.

6.5.1 CASE STUDY 1: SCREENING OF INGREDIENTS FOR SNACK BARS

The demand for convenient and healthy ready-to-go snack food has influenced the sale of snack bars all over the world. A case study in this section consists of screening of binder ingredients for snack bars. This study is discussed at conceptual level only as the experimental data is unpublished and proprietary to a company (Unpublished work by Gokhale S. V.). Assume that a snack bar contains binder mixture (containing corn syrup/sugar, etc.), which binds with the inclusions (like oats/peanuts, etc.) and holds the entire bar together. Now one wants to screen number of ingredients, which can bind the inclusions and then make an optimum binder slurry. Then modeling is an important tool, in this case, to screen large number of ingredients and explore their potential use as binder, screen them, and shortlist them to the actual trial in a product. Let us see how this problem was solved as follows:

a. Known binding ingredients like corn syrup, sugar syrup, date syrup, etc., were analyzed for various properties listed in Figure 6.11. Examples from different disciplines were studied to understand how binding properties of materials are studied for polymers, packaging materials, etc.; and then a few properties of the material were shortlisted for analysis.
b. After analysis of ingredients, snack bars were made using these known binders analyzed in step 1 and standard inclusions (e.g., rolled oats). These bars were analyzed on the texture analyzer to measure the

force required for extension/breaking, compression test, etc. Sensory scoring for chewiness/crunchiness was also done. This gave the target bar properties.

c. A black box modeling technique, such as ANN, GA, etc., was then applied to find correlations among the properties of known binder ingredients and properties of bars made from them (Figure 6.12).
d. Test number of potential ingredients and their combinations, which can be used in binding slurry to find similar properties, as discussed in step 1 and shown in Figure 6.11.
e. Based on the properties of potential ingredients and using a model developed in step 3, properties of bar were estimated if a bar is made using binder slurry containing that ingredient/combination of ingredient. And these predicted bar properties were compared with the target bar properties. If they are in the target range, then an ingredient is shortlisted otherwise it is discarded from the list.
f. The shortlisted ingredients/combinations of ingredients were then used to make a binding slurry and then snack bar to validate their binding ability in the snack bar system and were then taken for further product development.

Polymer adhesive POV	Binder slurry properties	Molecular properties	Powder properties
• Glass transition temperature • Adhesiveness on texture analyzer	• Viscosity • Rheology (G', G") • Surface tension • Contact angle • Wettability • Consistency • Stringiness • Stickiness/ cohesiveness	• Zeta potential • Thermal conductivity • Crystalline structure	• Water activity • Powder flow ability • Powder density

FIGURE 6.11 Properties of binder ingredients/slurry.

Thus, in this case study, once a model is developed, then a number of experiments were reduced to check whether any ingredient can be used in binder slurry or not. The testing of different ingredients for their properties is much less time consuming and cheaper than testing them in actual snack bar system. Thus, modeling saved time and money here.

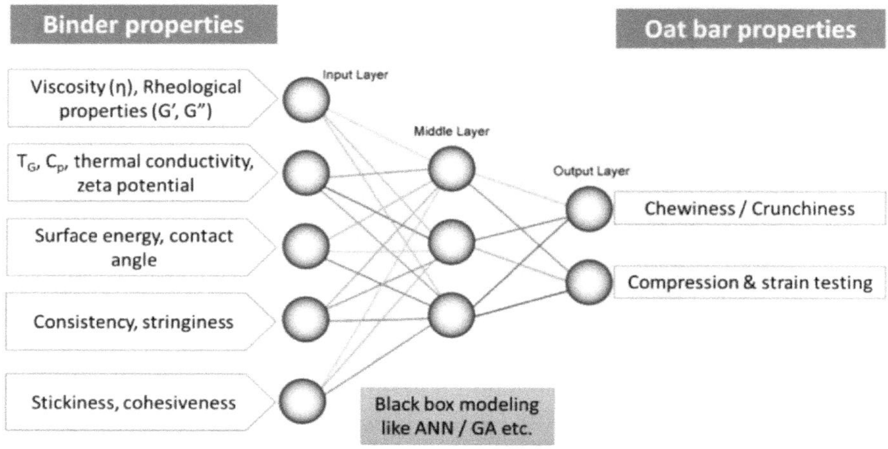

FIGURE 6.12 Relating binder and bar properties (ANN modeling).

6.5.2 *CASE STUDY 2: ICE CREAM CRYSTALLIZATION*

Crystallization and viscosity of ice cream are widely studied phenomena by many researchers with extensive use of modeling techniques.[1,33,40] Creamy mouth-feel and smooth texture of the ice cream are dependent on viscosity of the ice cream and crystal size in the ice cream. The case study in this section describes the rheological modeling of ice cream mix and frozen ice cream.

Sherman[35] used a model emulsion system of liquid paraffin in water. The emulsion was stabilized using skimmed milk powder at 2.5–10.0% level. When ice cream mix freezes, no remarkable change in size and size distribution of globules was observed. But when it is remelted, globules come together to coalesce. When agitation is incorporated during freezing, the process of globule formation during freezing and remelting may influence the subjective assessment of texture of ice cream on palate. The ice cream mix and frozen ice-cream or remelted ice cream show viscoelastic behavior. The shearing deformation is neither completely conserved as that in solids, nor completely dissipated as in fluids. The creep compliance is defined as:

$$J(t) = J_0 + J_1\left(1 - e^{-t/\tau_1}\right) + J_2\left(1 - e^{-t/\tau_2}\right) + \frac{t}{\eta^N} \tag{6.18}$$

where J(t) is total compliance at time t (cm²/dyne); J_0 is instantaneous elastic compliance (cm²/dyne); J_1 and J_2 are compliances associated with retarded elastic compliance (cm²/dyne); τ_1 and τ_2 are retardation times associated with retarded elasticity (s); η is viscosity components associated with retarded elasticity (poise); and N is the number of fat globules per unit volume of ice cream mix at time t. Theoretical curves for model emulsion system derived using above equation and along with calculated values of all model constants showed nonsignificant error, when compared with experimental creep compliance–time curve.

To describe the creep behavior of frozen ice cream, six-element model was presented by Shama and Sherman,[33] which relates creep behavior and instantaneous elasticity (E_0) of ice and fat content, elastic moduli (E_1 and E_2) of weak stabilizer gel, and protein enveloped air cells, respectively, viscoelastic components associated with retarded elasticity, η_1 and η_2 of fat crystals and weak stabilizer gel and protein enveloped air cells, respectively, and Newtonian viscosity (η_N) of ice and fat crystals. The model is represented in the form of resistances and capacitance, as shown in Figure 6.13.

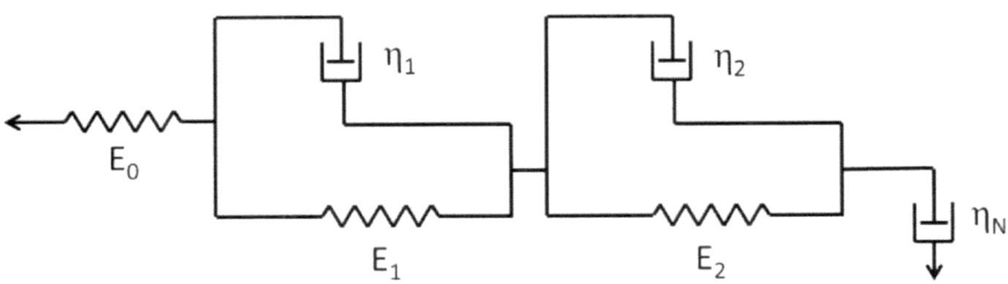

FIGURE 6.13 Six element model for ice cream showing rheological associations with structural components; adopted from Shama and Sherman.[33]

The relationship among fat content, overrun, and temperature with E_1, E_2, η_1, η_2, τ_1, τ_2 can be observed in the original article by Shama and Sherman.[33] As compared to pure ice, values of E_1, E_2, η_1, η_2, and η_N are lower by many orders of magnitude for ice cream. If ice cream is considered as network of ice crystals with fat and air in void spaces, then ice crystal parameters are by far altered because of air content in the ice cream. Size of air cells, ice crystals, and fat cells are inversely proportional to the overrun.[33] Size of air cells, ice crystals, and fat cells decrease and hence their number per unit volume increases with increase in overrun and more bonds are established, so that the structure becomes stronger. This is reflected in the rise of E_1, E_2, η_1, η_2. The influence of temperature can be described using an equation similar to the Arrhenius equation:

$$\eta_N = Ge^{\frac{E}{RT}} \tag{6.19}$$

6.6 SUMMARY

As seen from the discussions above throughout this chapter, application of food process modeling can be found in three broad areas: processing of foods, processes happening in the food, and processes applied to model food behavior. As seen from the case studies explained earlier, mathematical model describes the process approximately, and one should focus on building a model which can help in reducing the experimentation which will in turn save time and money; rather than build complex mathematical equations which will describe all small aspects of the process. The use of model systems to develop unified models is a common practice in chemical engineering and not much common in food engineering. Two such case studies described in this chapter can be adopted to more processes like high-pressure processing, nonthermal processing, etc. The use of statistical optimization (e.g., response surface methodology) is now common in optimization of process and product development. More mathematical modeling techniques can be used for screening of functional ingredients, relate properties of ingredients with their effect on sensory attributes, etc.

KEYWORDS

- **case studies**
- **kinetic modeling**
- **mathematical modeling**
- **product quality optimization**
- **unified models**

REFERENCES

1. Adapa, S.; Dingeldein, H.; Schmidt, K. A.; Herald, T. J. Rheological Properties of Ice Cream Mixes and Frozen Ice Creams Containing Fat and Fat Replacers. *J. Dairy Sci.* **2000,** *83*(10), 2224–2229.
2. Azar, A. B.; Ramezan, Y.; Khashehchi, M. Numerical Simulation of Conductive Heat Transfer in Canned Celery Stew and Retort Program Adjustment by Computational Fluid Dynamics (CFD). *Int. J. Food Eng.* **2020,** *16*(9), 20190303.

3. Bajić, A.; Pezo, L.; Stuper, A.; Filipčev, B.; Cvetković, B.; Horecki, A. T.; Mastilović, J. Application of Lyophilized Plum Pomace as a Functional Ingredient in a Plum Spread: Optimizing Texture, Color and Phenol Antioxidants by ANN Modelling. *LWT- Food Sci. Technol.* **2020,** *130*, E-article 109588.
4. Barnett, S. M.; Diako, C.; Ross, C. F. From Abstract to Recognizable: Modelling Tendencies of a Basic Salt Solution and a Tomato Soup Based on Affective Reactions. *J. Sens. Stud.* **2019,** *34*(5), E-article 12510.
5. Barreiro, J. A.; Milano, M.; Sandoval, A. J. Kinetics of Colour Change of Double Concentrated Tomato Paste during Thermal Treatment. *J. Food Eng.* **1997,** *33*(3–4), 359–371.
6. Bird, R. B.; Steward, W. E.; Lightfoot, E. N. *Transport Phenomena*, 2nd ed.; John Wiley & Sons, Inc.: New York, 2006; p 905.
7. Chung, C. C.; Chen, H.; Ting, C. H. Fuzzy Logic for Accurate Control of Heating Temperature and Duration in Canned Food Sterilization. *Eng. Agric. Environ. Food* **2016,** *9*(2), 187–194.
8. Desai, K. M. Biopolymer Processing: Modelling And Machinery Design. Ph.D. Degree, Thesis; University of Mumbai, Mumbai, India 2009, pp 83–147.
9. Desai, K. M.; Lele, S. S.; Lele, A. K. In *Rheology of Model Dough Formulation.* In: The XV International Congress on Rheology, The Society of Rheology 80th Annual Meeting, Monterey, California; Co, A.; Leal, L. G.; Colby, R. H. and Giacomm, A. J., Eds.; Melville, New York: American Institute of Physics; 2008; pp 1226–1228.
10. Erdogdu, F.; Altin, O.; Marra, F.; Bedane, T. F. A Computational Study to Design Process Conditions in Industrial Radio-Frequency Tempering/Thawing Process. *J. Food Eng.* **2017,** *213*, 99–112.
11. Faydi, E.; Andrieu, J.; Laurent, P. Experimental Study and Modelling of the Ice Crystal Morphology of Model Standard Ice Cream, Part I: Direct Characterization Method and Experimental Data. *J. Food Eng.* **2001,** *48*(4), 283–291.
12. Félix, M.; Carrera, C.; Romero, A.; Bengoechea, C.; Guerrero, A. Rheological Approaches as a Tool for the Development and Stability Behavior of Protein-Stabilized Emulsions. *Food Hydrocoll.* **2020,** *104*, E-article 105719.
13. Glasscock, D. A.; Hale, J. C. Process Simulation: The Art and Science of Modelling. *Chem. Eng.* **1994,** *101*(11), 81–89.
14. Gokhale, S. V. Process Development and Modelling of Preservation Techniques for Indian Traditional Foods. Ph.D. Thesis, University of Mumbai, Mumbai, India, 2010, pp 91–129.
15. Gokhale, S. V.; Lele, S. S. Retort Process Modelling for Indian Traditional Foods. *J. Food Sci. Technol.* **2014,** *51*(11), 3134–3143.
16. Gokhale, S. V.; Lele, S. S. Dehydration of Red Beet Root (*Beta vulgaris*) by Hot Air Drying: Process Optimization and Mathematical Modelling. *Food Sci. Biotechnol.* **2011,** *20*(4), 955–964.
17. Guldas, M.; Kurtulmus, F.; Gurbuz, O. Prediction of Lethality by Nonlinear Artificial Neural Network (ANN) Modelling. *J. Food Process Eng.* **2017,** *40*(3), E-article 12457.
18. Harbourne, N.; Jacquier, J. C.; Morgan, D. J.; Lyng, J. G. Determination of the Degradation Kinetics of Anthocyanins in a Model Juice System using Isothermal and Non-Isothermal Methods. *Food Chem.* **2008,** *111*(1), 204–208.
19. Horak, F. P. Über die Reaktionskinetik der Sporenabtö tung und Chemischer Veranderungen bei der thermischen Haltbarmachung von Milch zur Optimieruung

von Erhitzungsverfahren (About the reaction kinetics of spore killing and chemical changes in the thermal preservation of milk to optimize heating processes). PhD. Thesis, Technical University of Munich, Germany, 1980, p 219.

20. Kerr, W. L.; Kauten, R. J.; Ozilgen, M.; McCarthy, M. J.; Reid, D. S. The NMR Imaging, Calorimetric, and Mathematical Modelling Studies of Food Freezing. *J. Food Process Eng.* **1996,** *19*(4), 363–384.
21. Ko, W. C.; Liu, W. C.; Tsang, Y. T.; Hsieh, C. W. Kinetics of Winter Mushrooms (*Flammulina velutipes*) Microstructure and Quality Changes during Thermal Processing. *J. Food Eng.* **2007,** *81*(3), 587–598.
22. Kowalski, R. J.; Li, C.; Ganjyal, G. M. Optimizing Twin-Screw Food Extrusion Processing Through Regression Modelling and Genetic Algorithms. *J. Food Eng.* **2018,** *234*, 50–56.
23. Kumar, Y.; Singh, L.; Sharanagat, V. S.; Tarafdar. A. Artificial Neural Network (ANN) and Mathematical Modelling of Hydration of Green Chickpea. *Inf. Proc. Agric.* **2021,** *8*(1), 75–86.
24. Luyben, W. L. *Process Modelling, Simulation and Control for Chemical Engineers*, 2nd ed.; McGraw-Hill Inc.: Boca Raton, 1989; p 725.
25. Mahajan, P. M., Gokhale, S. V., Lele, S. S. Production of Nattokinase Using *B. natto* NRRL 3666: Media Optimization, Scale Up and Kinetic Modelling. *Food Sci. Biotechnol.* **2010,** *19*(6), 1593–1603.
26. Martins, W. F.; Longhi, D. A.; Falcao de Aragão, G. M.; Melero, B.; Rovira, J.; Diez, A. M. Mathematical Modelling Approach to the Quantification of Lactic Acid Bacteria in Vacuum-Packaged Samples of Cooked Meat: Combining the TaqMan-based Quantitative PCR Method with the Plate-Count Method. *Int. J. Food Microbiol.* **2020,** *318*, E-article 108466.
27. Mercier, P.; Yerushalmi, L.; Rouleau, D.; Dochain, D. Kinetics of Lactic Acid Fermentation on Glucose and Corn by *Lactobacillus amylophilus*. *J. Chem. Technol. Biotechnol.* **1992,** *55*(2), 111–121.
28. Milliatti, M. C.; da Silva Lannes, S. C. Impact of Stabilizers on the Rheological Properties of Ice Creams. *Food Sci. Technol.* **2018,** *38*(4), 733–739.
29. Nguyen, H. T.; van der Fels-Klerx, H. J. I.; van Boekel, M. A. Acrylamide and 5-Hydroxymethylfurfural Formation During Biscuit Baking, Part II: Effect of the Ratio of Reducing Sugars and Asparagine. *Food Chem.* **2017,** *230*, 14–23.
30. Özilgen, M. *Handbook of Food Process Modelling and Statistical Quality Control*, 2nd ed.; CRC Press: Boca Raton, 2011; p 704.
31. Regier, M.; Housova, J.; Hoke, K. Dielectric Properties of Mashed Potatoes. *Int. J. Food Prop.* **2001,** *4*(3), 431–439.
32. Rodríguez, R.; Lombraña, J. I.; Kamel, M. de Elvira, C. Kinetic and Quality Study of Mushroom Drying under Microwave and Vacuum. *Dry. Technol.* **2005,** *23*(9–11), 2197–2212.
33. Shama, F.; Sherman, P. The Texture of Ice Cream, Part 2: Rheological Properties of Frozen Ice Cream. *J. Food Sci.* **1966,** *31*(5), 699–706.
34. Sheen, S.; Whitney, L. F. Modelling Heat Transfer in Fluidized Beds of Large Particles and its Applications in the Freezing of Large Food Items. *J. Food Eng.* **1990,** *12*(4), 249–265.
35. Sherman, P. The Texture of Ice Cream, Part 3: Rheological Properties of Mix and Melted Ice Cream. *J. Food Sci.* **1966,** *31*(5), 707–716.

36. Singh, R. P.; Vijayan, J. Advances in Predictive Modelling in Food Frying and Cooking Processes. *Latin Am. Appl. Res.* **1998,** *28*(4), 217–221.
37. Tarafdar, A.; Nair, S. G.; Kaur, B. P. Identification of Microfluidization Processing Conditions for Quality Retention of Sugarcane Juice Using Genetic Algorithm. *Food Bioproc. Tech.* **2019,** *12*, 1874–1886.
38. Tijskens, L. M. M.; Hertog, M. L. A. T. M.; Nicolai B. M. *Food Process Modelling*; Woodhead Publishing Ltd.: Cambridge, England, 2001; p 496.
39. Van Boekel, M. A. J. S. Kinetic Modelling of Food Quality: A Critical Review. *Compr. Rev. Food Sci. Food Saf.* **2008,** *7*(1), 144–158.
40. Vilas, C.; Balsa-Canto, E.; Arellano, M.; Benkhelifa, H.; Alvarezb, G.; Flick, D.; Leducq, D. Alonso, A. *Model Identification of the Ice Cream Crystallization Process*, Food Process Engineering in Changing World: Proceedings of 11th International Congress on Engineering and Food (ICEF11), Athens, Greece; Taoukis, P. S., Stoforos, N. G., Karathanos, V. T., Saravacos, G. D., Eds.; Cosmosware: Athens, Greece, 2011, 1, p 6.
41. Yousefi-Darani, A.; Paquet-Durand, O.; Hitzmann, B. Application of Fuzzy Logic Control for the Dough Proofing Process. *Food Bioprod. Process.* **2019,** *115*, 36–46.

CHAPTER 7

STRUCTURAL DESIGN OF FOODS BASED ON LAWS OF PHYSICS

PALAK MAHAJAN, SHUBHAM SARASWAT, and MANAB B. BERA

ABSTRACT

Understanding of food materials from the classical realm of physics including soft condensed matter physics has been an area of interest, especially in structural design engineering of food products. The contents of this chapter would help the reader in understanding the thermodynamics of food polymer, structural design principles, structural hierarchy, steps involved in food structuring, and structure measurement techniques. The method of construction and understanding of the state diagram of a food system is explained by using various theories and models. SD acts as a tool for manipulating process parameters to obtain the desired food product, and in the prediction of food structural changes during storage.

7.1 INTRODUCTION

The term "food physics" might seem a strange combination of two immiscible entities. Let us understand food and physics separately. A food material is any substance derived from plant or animal origin that supplies nutrients to our body to sustain life. In contrast, physics is a science dealing with the properties of a matter and the energies associated with them. We practice and observe food physics in our day-to-day life, but without even knowing it. Also, a few physicists and food technologists have been able to mention or highlight this weird relationship.[70] Food physics does not involve the concept

Novel and Alternative Methods in Food Processing: Biotechnological, Physicochemical, and Mathematical Approaches. N. Veena, Megh R. Goyal, & Ritesh B. Watharkar (Eds.)

of chemical reactions to explain the food system while it just considers that the molecules remain unchanged but behave differently under different conditions. It can be observed during the foam aeration, material flow, even during water/moisture content determination or in melting or evaporation processes.

Various unit operations of food processing (such as size reduction, concentration, cooking, baking, extraction, leaching, etc.) could also be explained by applying many fundamentals of physics, such as heat-mass transfers and energy exchange. Similarly, during food consumption, the destruction of structured products, that is, breakage or unjamming of food into pieces by the application of small forces involves fundamentals of applied mechanics.[10] Physics governs the food system, especially its components, such as fats, proteins, and carbohydrates, which together with water, give a structure to food products. It helps in understanding food system at molecular level and also helps in predicting structure–texture–property–flavor relationship.

Mechanism of food breakdown in our mouths and release of the flavor molecules from the matrix, bioaccessibility and bioavailability of nutrients at a specific site of the gastrointestinal tract have given a new direction to the food product designing and fabrication and have become an interesting area of study.[58] All such accumulated knowledge indicates the importance of fundamentals of physics and its application in understanding food structure (FS).

However, some physicists considered food as a soft condensed matter (SCM), where the atoms and molecules interact to form denser solids and liquids since their attractive energies are stronger than the thermal energy. Such matter includes complex fluids having a continuous phase wherein mesoscale assemblies (colloidal particles, gas/air bubbles, amphiphilic, and polymers) are dispersed. Properties of (edible) soft matters (SMs) at mesoscale and beyond it mostly do not depend on the chemical characteristics of their components. However, their properties vary with length scale and are prone to externally applied fields (mechanical stresses) which provides them soft appearance.

Understanding food structural designing (FSD) requires thorough knowledge on phase transformation, equilibrium, internal properties of material, and changes occurring in the food system or FS with respect to the changes in conditions such as temperature and moisture or solid content. Understanding of thermodynamics, phase transition (PT), and moisture sorption behavior is the prerequisites of the FSD process. Data related to changes are mapped as a state diagram (SD) and help in understanding food behavior during processing and storage; establish strategies to optimize design parameters of FS having desired functionalities, nutrient uptake, and release properties.

This chapter highlights the concepts of food physics based on thermodynamics, FSD, and equations to develop the SD for studying the structural changes in food products/materials.

7.2 UNDERSTANDING THE FOOD AS A SOFT CONDENSED MATTER (SCM)

Food as an SCM is still a rarely investigated topic.[26] Basically, a food material can be perceived as a three-dimensional matrix having both continuous and dispersed phases that sustain life. The simplest form of food material could be tissues of biological origin. A tissue contains a large number of cells. A cell has a cytoplasm that has a continuous phase and salt, water, polymer (carbohydrates, proteins, fat) as dispersed phase, besides, other nuclear components. They (SMs) are easily deformed by the application of thermal fluctuations, external stress, and magnetic or electric fields.[66]

SCM physics helps the food scientists to understand the main structural elements responsible for building up of food materials. SM occupies a middle space between the two extremes: ideal solid and fluid states. In addition to these features, SMs are accompanied by higher viscosity than the simple liquids and exhibit viscoelastic properties. Most food products are a multiphase system of colloids possessing liquid or solid aggregation state with crystalline or glassy phases and made up of the self-assembled structures.[3] The building blocks of SM are organic molecules having macro- or supramolecular entities have complicated architectures, shape anisometric, and are bounded by weak interactions.[22]

Unit operations involved in the food processing (structuring) go through a series of energy exchanges. The study of forces required at every single step needs a considerate understanding of the basic thermodynamic concepts related to food polymers, discussed in this chapter.

7.3 THERMODYNAMICS OF FOOD POLYMERS

Food is considered a biopolymer; therefore, the theories including thermodynamics that are used in studying synthetic polymers may be used in understanding food polymers. Thermodynamics deals with energy exchanges between a system and surrounding or among the components of a system. During food processing, there are series of energy exchanges occurring between systems causing thermal effects. These exchanges influence the components involved and cause heating or cooling, activation, or repression of any reaction.[73] Many phase changes occurring during processing or storage of food and degradation rate of a nutrient can be described at a macroscopic scale with the help of principle thermodynamics including chemical kinetics. Once the molecular makeup of polymers is known, one can predict controlling and interactions taking place among the molecules.[1] Therefore, thermodynamics gives useful information as to which direction a system will move and what would be the result of processing variables for instance, temperature, pH, and concentration.

A key thermodynamic parameter important for studying the phases at the equilibrium is the Gibbs free energy, G, and is expressed as follows:

$$\Delta G_{mix} = \Delta H_{mix} - T\Delta S_{mix} \quad (7.1)$$

where for any mixing, ΔG_{mix} = Gibbs free energy; ΔH_{mix} = enthalpy; T = temperature; and ΔS_{mix} = entropy.

Thus, mixing involves both changes in enthalpy and entropy. In an ideal case (a mixture of ideal molecules), interaction among the molecules is the same and thus the change in volume after mixing is zero (the continuity equation). For an ideal case, $\Delta H_{mix} = 0$, therefore the sign of ΔG_{mix} depends only on the entropic term. For the regular solution, ΔH_{mix} is finite and is defined as follows:

$$\frac{\Delta G_{mix}}{N} = \Delta H_{mix} + RT(x_1 \ln x_1 + x_2 \ln x_2) \quad (7.2)$$

where x represents the molar fraction of solvent (7.1) and solute (7.2); and N is the total number of moles.

However, food thermodynamics could be understood by following the Flory-Huggins theory (FHT). Basically, FHT is a lattice model, based on dissimilarity among molecules depending on their sizes. Each lattice site is occupied either by a polymer segment or by solvent molecule.[1] FHT is based on some simplifying assumptions, such as volume change does not occur upon mixing of polymer and solvent. It can be used to study the phase behavior of polymer mixture and the thermodynamics involved, such as gelatinization of starches.[65] It deals with the free energy of a polymer solution as given below:

$$\Delta G_{mix} = RT\left(\chi_{12}\varnothing_1\varnothing_2 + \varnothing_1 \ln\varnothing_1 + \frac{\varnothing_2}{x} \ln\varnothing_2 \right) \quad (7.3)$$

where Ø is the volume fraction of solvent (7.1) and solute (7.2). The terms used in eqs. 7.2 and 7.3 are the same, but molar fractions have been replaced by volume fractions.

Enthalpy term is associated with interaction parameter known as Flory-Huggins Interaction Parameter (FHIP) (χ_{12}) as follows:

$$\chi_{12} = \frac{\Delta H_{mix}}{RTN\varnothing_2} \tag{7.4}$$

where H_{mix} is enthalpy of the mixture; and RT is a type of "thermal energy" related to the intermolecular bonding. Thus, χ is the ratio of energy involved in the interaction of neighboring molecules and type of thermal energy. The positive value of χ_{12} indicates repulsion while the negative value represents the attraction or miscibility.[1]

Depending on the nature of the interaction, phase behavior of the mixture can be predicted by using FHT. Repulsive interactions are indicated by upper critical solution temperature (UCST) behavior and attractive interactions are shown by lower critical solution temperature (LCST). Complete miscibility may exist between the polymer and solvent above UCST and below LCST. These thermodynamic interactions among the mixture components influence their morphological properties and also predict incompatibility if any, among them.[36] Few research articles are available on its application in food product development.[61,62]

Van der Sman[66] used the SM model to study the heat and mass transport phenomenon occurring during the meat cooking, which was based upon the Flory-Rehner theory[63,69] for the mass transfer in the swelling and shrinkage of the gels, and water holding capacity (WHC) of food products, such as jam, jellies, meats, and vegetables.[62,67–69] Based on thermodynamic property of food product like WHC of gels, one can predict the amount of water that could be held by the food material without compromising the mechanical load taken as food quality/sensory indicator (such as hardness and juiciness).

7.4 FOOD STRUCTURE (FS)

FS is a network that supports its components and its own weight within itself when force applied on it. The food structuring principle is very similar to architecture of building structures and uses its components (fat, protein, and carbohydrate) as construction materials including some prefabricated materials (egg, lecithin, hydrocolloids, etc.) for the design and fabrication of food products. Food microstructure focuses on three key features: existence of recognizable distinct domains, organization of spaces between discrete domains, and the presence of some kind of interaction between them. For example, hydrogels, when viewed, are formed of the arrangement of the polymeric network, and the spaces between them are filled with the water.

The relationship among the FS and its physical properties forms the center point FSD process. FS influences the sensory properties and delivery of nutrients or nutraceuticals when ingested. Researchers are now engaged in developing food products similar to that of controlled drug release concepts where is food is supposed to deform at specific sites in the gastrointestinal tract for bioaccessibility and bioavailability of nutrients.[58] Novel FS can be created by applying the knowledge of both traditional concepts of processing and fundamental knowledge of physics including SCM. All SMs are thermodynamically unstable, so their stabilization is required. This can be achieved by modifying their surface characteristics or by bringing them to jammed/arrested state.

7.4.1 STRUCTURAL STABILITY AND INCOMPATIBILITY AMONG FOOD POLYMERS

Proteins, carbohydrate, fats, and water are the structural components or structural elements of the food[57] and a minute dissimilarity in these structural components brings incompatibility in the structure. Other factors such as salt concentration and pH strongly influence their interaction and incompatibility that can be seen in the formation of food gels, emulsion, or any colloidal system.[14] Structural stability of a food system requires the compatibility among its ingredients. This compatibility is related to strength of attractive and repulsive forces among the segments of species.

Theoretically when FHIP (χ) shows a positive value, then it represents immiscibility or repulsion while a negative value shows miscibility or attraction. Similarly, when ΔG_{mix} is positive, then the biopolymers are thermodynamically incompatible and in reverse, the polymer forms a single-phase mixed solution. It means that each polymer prefers to be surrounded by its own molecule instead of any other molecules and when such condition is formed it brings thermodynamic incompatibility. This gives rise to the concept of free volume (FV) and excluded volume (EV). These concepts are based on understanding of the interactions among the molecules as "strong short-range repulsive force" and "weak long-range attractive forces."

These repulsive forces are responsible for the formation of FV where desired new particles can be inserted.[66] Around each particle, there is a volume called EV where no new particle can be inserted (Figure 7.1).

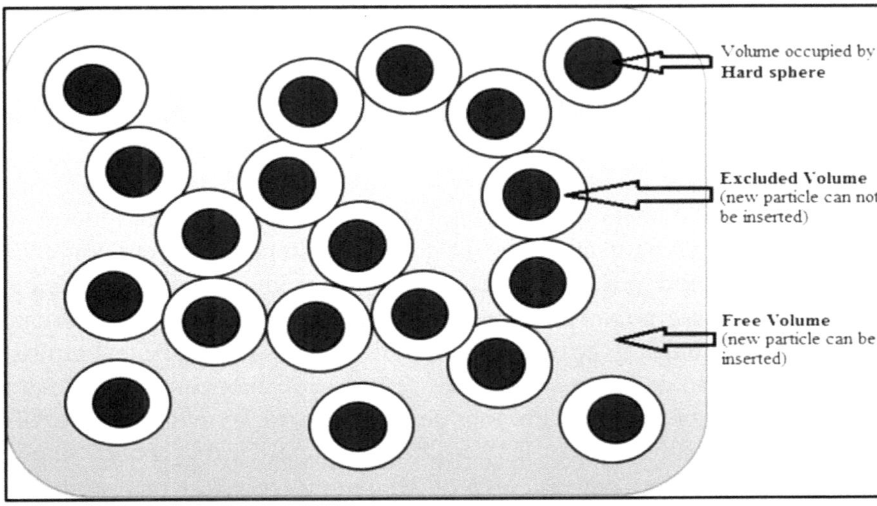

FIGURE 7.1 Excluded and free volume around hard spheres.

Source: Adapted and modified from Ref. [66].

7.4.1.1 FREE VOLUME (FV)

Every polymer depending upon the size of its molecules contains an FV. It forms as a result of the uneven packing, instability of density, and topographical restrictions among the molecules. It determines the structural properties and also dynamics of polymeric materials. FV can be determined experimentally using confocal microscopy (CFM) and theoretically by Widom insertion method.[15] The magnitude of FV depends both on temperature and moisture. At the glass transition temperature (GTT) of a material, FV is basically zero and this can be shown with the extreme change in its thermal and physical properties. FV is smallest at maximum densities and at low moisture content. This effect is known as antiplasticization. When the material is near to its GTT at higher water concentration, then its FV becomes larger.

7.4.1.2 EXCLUDED VOLUME (EV)

EV is also known as steric exclusion effects. This occurs when the food components that are capable of occupying large volume in a system exclude other components by occupying the same volume. EV increases when the effective volume of molecules or particles in the system increases or if their molar concentration increases.[33] Higher molecular weight (MW), unfolding of molecules, and less interaction among them increase their incompatibility and formation of EV. EV shows enhancement of biopolymeric association, reduction in solubility, and cosolubility of the macromolecular solutes, and also destabilization or unjamming of the polymeric suspensions.

An increment in EV means a decrease in FV of a solution.[50] Such understanding helps in determining the space occupancy in a biopolymer solution and can be useful in FSD. Series of literature on enrichment or fortification food products/functional food products and development of biodegradable polymeric films involving various types of plasticizer show that the majority of the process optimizations are based on empirical assumptions. In such processes, desired biomolecules/micronutrients or plasticizers are inserted in the food matrix without knowing the available FV of the matrix and may lead to the development of products having high or poor loading per unit area of desired molecules, which ultimately affects economics of the product. Similarly, numerous literatures on the optimization of process variables using empirical models only predict the dependent variables or responses (quality attributes/ nutrition) of the products at an optimum level of independent variables but fail to quantify changes taking place at molecular levels. On the other hand, models based on fundamentals of physics and the use of sophisticated software like molecular dynamic simulation (MDS) and few other software help in scanning the process at the molecular level while predicting various responses at length and time scale.

7.4.2 ASSEMBLING OF FOOD STRUCTURE

7.4.2.1 STICKING OF FOOD MOLECULES AND MOLECULAR GLUES

The basic unit of food molecules is atom and depending upon the electron arrangement, food molecules can be categorized as polar, nonpolar, and ionic. The occurrence of full or partial charge on molecule's surface determines their adhesiveness or stickiness. No attraction among food molecules means the impossible structural building of foods.

The forces that connect or assemble the food ingredients together can be referred to as molecular glues or food cement. Based on the unique electrical properties of the food molecules, the molecular glues are formed in the food system from the various forces, such as hydrogen, van der Waals, electrostatic forces, and hydrophobic bonding.[33] These molecular glues hold different kind of molecules, such as water, proteins, carbohydrates, fats, etc. For example, in cheese and yoghurt production, control over prevailing electrostatic forces between protein molecules is required. Casein, a milk protein contains negative charge, results in repulsion among the molecules and make obstacle in making links among them. However, when the lactic acid bacteria are added to milk, lactose (milk sugar) converts into lactic acid, the protein molecules lose its charge on them, facilitate attraction or linkage among molecules via hydrophobic and van der Waals forces. This linkage results in formation of 3D delicate network of the coagulated proteins which provides a silky and delicate texture to yoghurt.

7.4.3 LEVELS OF FOOD STRUCTURE (FS) FORMATION

FS occurs over a wide length scale (Figure 7.2). On the basis of their characteristic length scale, the biopolymer interactions can be divided into four structural levels: super macromolecular, macromolecular, submacromolecular, and macroscopic levels. These levels can also be referred to as conformational potential (CP) of food macromolecules. CP means the capacity of biopolymer to form an intermolecular junction zone that gives the required properties of food systems, such as rheological, structural, and other physicochemical properties. The size of the junction zone, strength, and nature of the intermacromolecular forces are an important aspect of the structural hierarchy of the food products and has a great influence on the food properties.[55]

7.4.4 PRINCIPLES OF STRUCTURAL DESIGN

Attributes of food products (physicochemical, functional, and sensory) are described by their characteristics, structural organization, and the interaction among various ingredients/constituents they contain. There are a large number of lipophilic components, such as bioactive lipids, antioxidants, vitamins, antimicrobials, and flavors, which are required to be delivered in edible form.[33] Usually they are delivered in aqueous systems as they increase the palatability, stability, and bioactivity in food. The incorporation of active lipophilic components into the aqueous delivery system poses varieties of technical challenges. Biopolymers of the food system show variety of conformations and characteristics, such as molar mass, charge, flexibility, and hydrophobicity. On altering the environment (temperature, ionic strength, pH, or solvent composition), biopolymers undergo a transition from one aggregation state to the other or from one conformation to another. An understanding of molecular forces (hydrogen bonding, electrostatic interaction, hydrophobic interactions, and steric exclusion), conformation and interactions of the biopolymer, mechanism of assembly of the components having some specific structures, and its delivery in the food system plays a crucial role.[32,33] Some design principles that can form novel FS using the food components are discussed in this section.

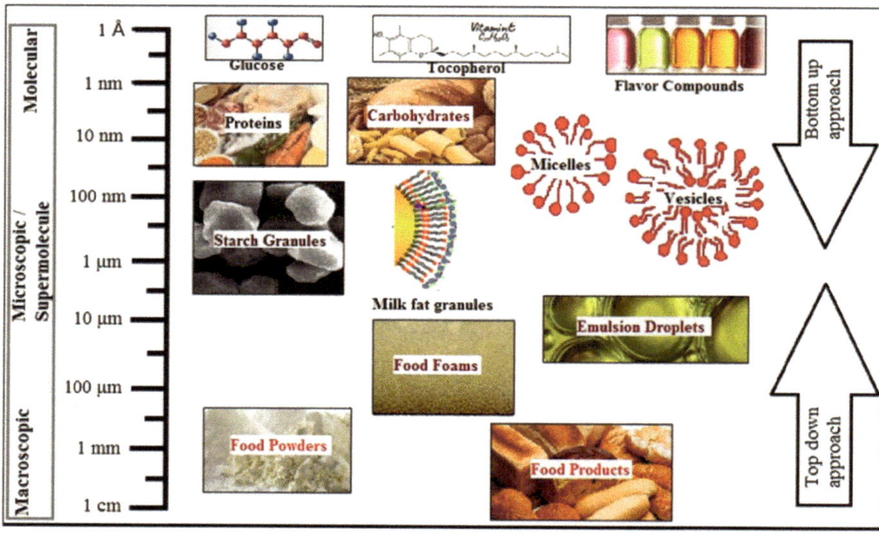

FIGURE 7.2 Characteristic length scale of food items with their respective examples.
Source: Adapted and modified from Ref. [59].

7.4.4.1 PHASE SEPARATION

Whenever two unlike or dissimilar materials are mixed, any one of two phenomena may occur, either they can be completely miscible forming a single phase or they may separate forming different phases (Figure 7.3a). This phenomenon depends upon interaction among different molecules present and their strength. Separation may occur as a result of strong thermodynamic unfavorable interactions (steric exclusion/electrostatic repulsion) between the different biopolymers.[32,33]

7.4.4.2 SELF-ASSEMBLY

Under some circumstances, the mesoscopic structure tries to self-assemble and to become stable. Structural elements are bonded by sufficiently weak attractive interactions that alter their positions and initiate orientations to facilitate more favorable self-assembled configurations. Self-assembly occurs when there is a balance between attractive and repulsive forces. If the interactions are too strong, then the system will most likely end up in an arrested state. The structuring of microentities depends upon the colloidal interaction between the components, their chemical and physical properties and the way they assemble. These microentities are thermodynamically unstable, exhibiting kinetically trapped nonequilibrium state as their free energy decreases with the time. However, on the other hand, self-assembled nanostructure entities are thermodynamically stable until their molecule degrade/react with the environment.[34]

7.4.4.3 SPONTANEOUS SELF-ASSEMBLY

Some food components assemble themselves to form a well-defined structure under appropriate environmental conditions to minimize the free energy of the system. Self-assembly mostly involves hydrogen bonding, hydrophobic attraction, and/or electrostatic interactions.[32,33] Some examples of self-assembled structures are micelles, microemulsions, and vesicles, as shown in Figure 7.3b.

7.4.4.4 DIRECTED SELF-ASSEMBLY

This type of self-assembly does not occur spontaneously upon simply mixing all the components. However, it requires controlled preparation conditions, such as temperature, ionic strength, and time profile, to carefully direct the assembly of components to form a metastable structure, as shown in Figure 7.3c.

7.4.5 STEPS INVOLVED IN FOOD STRUCTURAL DESIGNING (FSD) PROCESS

FSD involves the use of external fields resulting in deformation at mesoscale and jamming or unjamming of structural elements, leading to new structures via transitions and stabilizations. FSD requires a careful balance between various forces at length and time scales and is often classified into three steps:

- destruction of the native structure of the food materials,
- development of new FSs, and
- arresting the FS in the jammed state.[12]

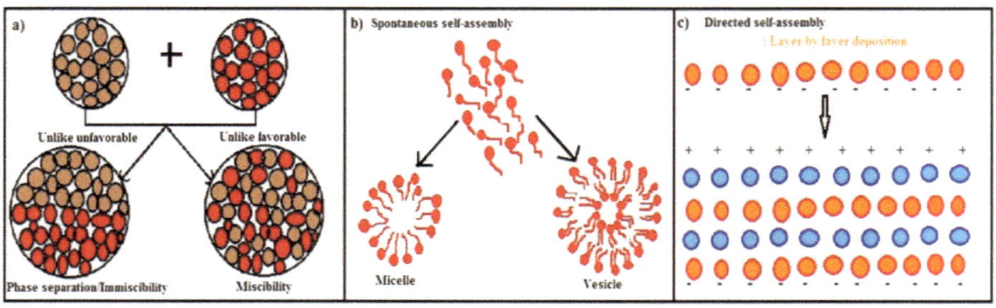

FIGURE 7.3 Structural design principles for assembling the food ingredients into a specific structure: (a) phase separation; (b) spontaneous self-assembly; (c) directed self-assembly.

Source: Adapted and modified from Ref. [33].

For example, the formation of dairy gelled products includes (1) destruction of the native colloidal stability via the application of the external field as temperature, pH; (2) formation of the new structure via the aggregation of proteins; and then finally (3) preservation of newly formed FS via their dispersion.[53]

Figure 7.4b shows the formation of a hydrogel as a nutraceutical delivery system and its destruction upon consumption or with respect to alterations in the environmental conditions, such as changes in temperature and pH. There are four main transformations that are involved in FSD: (1) development or destruction of a phase; (2) heat and mass transfer; (3) PT; and (4) biochemical changes occurring by pH, ions, and enzymes.[16] These transformations could be understood from the view of SM physics, employing the FSD steps, as discussed in Figure 7.4a.

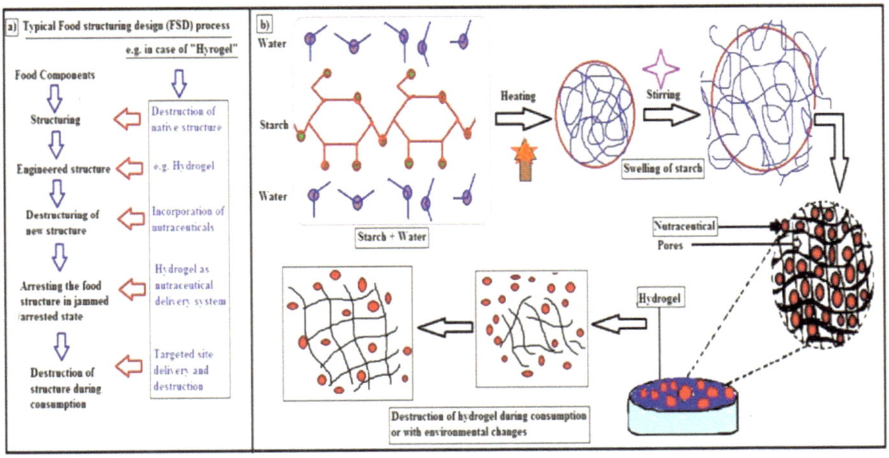

FIGURE 7.4 Food structuring process: (a) basic steps involved in food structuring; and (b) formation of the starch hydrogel as a nutraceutical delivery system.

7.4.5.1 DESTABILIZATION OF NATIVE AND DEVELOPMENT OF NEW PHASES

During FSD, the native FS is deliberately destabilized to form a new structure by the application of various external parameters, such as pH, temperature, enzymes, or ionic strength (making of cross-links). In the preparation of dairy products such as yoghurt, cheese, or ice cream, the firmness and stability of milk colloids are damaged to form a new structure. Colloids have short-range interactions and can be stabilized by creating a condition where these interactions are more than the thermal fluctuations.[64]

Food colloids aggregate to form new phases after destabilization, via various methods such as PT[60] and aggregation. Aggregation can be flocculation or coagulation depending upon strength of interaction potential (IP). If IP is more than thermal fluctuation then it leads to coagulation and if less or equals then it is aggregation. Aggregation can be seen in soy protein products, where protein aggregates via fermentation. Similarly, starch gelatinization involves melting of crystalline phase followed by leaching of amylose and forming gels (stabilization).[64]

Size distribution profile of newly formed dispersed phases plays an important role in sensory acceptance of the food products. The required sizing can be performed predominantly by deforming, diffusing, and dispersing actions of the fluid stream.[74] The dispersed phase must be in the fluid state in order to deform,[30] as it requires unjamming, destabilization, or melting of the crystalline dispersed phase, as seen in the case of starch and fat. This deforming action leads to the formation of new phases.

7.4.5.2 STABILIZATION OR JAMMING OF NEWLY FORMED STRUCTURES

Deformation of newly formed FS needs to be preserved until consumption. It can be achieved by converting food into gel state or in jammed/arrested states (also known as glassy), and macroscopic structures involves immobilized percolating structure.[66] In this state, food materials are considered shelf stable, despite being vulnerable to slow deterioration or exhibit slow dynamics process upon storage.[11] In slow dynamics, thermal fluctuations are small and a cage is formed by neighboring dispersed elements, from where they are not able to escape as seen in bread crumbs and staling.

Liu and Nagel[27] proposed the jamming/arresting state transition concepts in the SD. The transition depends on the volume fraction φ, strength of interaction among the particles, and on ratio of mechanical stress and yield stress (i.e., $\sigma/\sum$).[27] Unjamming of structure can be obtained by increasing the temperature above the GTT (T_g), applying mechanical stress via shearing and hydrostatic pressures beyond yield stress and by changing the volume fraction by diluting with the solvent or concentrating with the solute.

7.4.6 MAPPING TECHNIQUES FOR FS

Food is complex and it involved various dynamics in its structural development during production,[2,3] storage and consumption,[28] and digestion process.[7] Structure measurement techniques include both direct and indirect approaches. Direct approaches include tomography, scanning electron microscopy (SEM), optical microscopy (OM), and CFM while indirect measurements include spectroscopy or mechanical response topographical techniques, for example, X-ray tomography and magnetic resonance imaging allow full 3D reconstruction of sample structure.[59] Such techniques possess low resolution and acquisition time. OM can even be used under highly controlled conditions and can trace the changes occurring when the sample is cooled or is prone to mechanical deformation on the application of tensile stresses. Transmission electron microscopy can be used to study the food ultrastructure. The electron microscopic technique offers microscopic images with a high resolution (mainly SEM).

Environmental scanning electron microscopy enables the analysis of samples at a required relative humidity. Atomic force microscopy has gained enormous significance in SCM physics, physical chemistry and biophysics, over the past few decades. It allows the mapping of 3D surface structure with high resolution and also determines the interaction forces and elastic properties of the surface. It has been widely used for imaging the biopolymers and biological systems for their structure as well as functional properties.[19]

A new technique, positron annihilation lifetime spectroscopy, is used to determine the molecular structure including packing of crystalline and amorphous matrices.[21] However, it does not provide topographical information directly but it gives quantitative detail such as molecular holes/voids present among the molecules in compressed matrix.[35] These structural techniques when combined with the SDs help in better understanding of transitional changes and structural changes occurring over the temperature ranges, with respect to the changing in the moisture content of the system. This helps one to find the optimum conditions required by the food system during its processing and storage.

7.5 STATE DIAGRAM (SD)

SD is a graphically representation of several physical states of a food product with respect to temperature and moisture content (Figure 7.5).[5,9] SD is combined with the phase diagram.[43]

SD relies on the principle that in the presence of moisture and temperature, a food polymer undergoes a complicated multiple PTs. A PT line in SD separates the regions that differ distinctively in their structures and properties.[60,61] One of the strong points of SD is that it can explain the aggregation states along with the arrested/jammed state (glassy or gel state), in area of food science and SM physics. Prediction of food stability during various phases and state changes occurring during processing, storage, and consumption can be made using SD. Dependence of glass transition of sugars on the water content was first established for glucose and sucrose.[29] After that many SDs for other sugars such as arabinose, fructose, galactose, mannose, etc., were reported.[49]

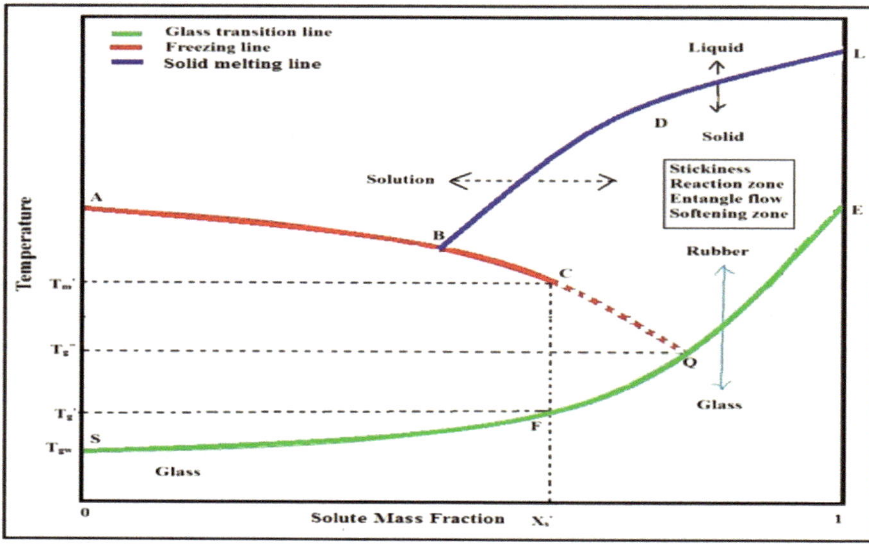

FIGURE 7.5 Typical state diagram (SD) of food materials.

Source: Adapted and modified from Ref. [40].

7.5.1 CONSTRUCTION OF SD CURVES

SD curve for a food material can be developed by using experimental data of maximum freeze concentration (MFC), freezing temperature depression, and glass transitions. These form the three main curves of SD as (1) glass transition curve mainly obtained from the Gordon Taylor (GT) equation (eq 7.5), (2) solid melting curve estimated from the Flory-Huggins equation (eq 7.7), and (3) the freezing curve is obtained from the Chen model (eq 7.10) which is a modified Clausius Clapeyron model. SD also gives an idea about the conditions for MFC equivalent to onset ice melting temperature (T_m') and GTT for the MFC (T_g'). These models are briefly discussed below.

7.5.1.1 GLASS TRANSITION TEMPERATURE (GTT) CURVE

T_g of mixture depends upon its water content, composition, MW of the components, and polymerization degree.[56] Water has a plasticizing effect.[31] It increases the FV and thereby causes molecular relaxation. Moreover, it has small MW and a T_g (−135°C) and thus it depresses the overall T_g of food system.[49,56] The food system having high interaction among its constituents has a higher GTT. Components such as simple sugars, with small MW have GTT (10–20°C) while those with large MW have higher GTT (50°C).[42,49] There are various equations for the prediction of GTT as Couchman and Karasz equation, GT equation, and Fox equation. But the GT equation has been widely used as a tool for predicting the GTT of mixtures with varying water content.[17]

GT equation is related to the linear change in the volume with respect to temperature. This equation (eq 7.6) can be obtained by modifying the Couchman and Karasz equation (eq 7.5).

$$\ln T_{gm} = \frac{x_1 \Delta C_{p1} lnT_{g1} + x_2 \Delta C_{p2} lnT_{g2}}{x_1 \Delta C_{p1} + x_2 \Delta C_{p2}} \tag{7.5}$$

where x is the mole fraction concentration; ΔC_{pi} is the difference in heat capacity of liquid (C^{L}_{Pi}), and glass (C^{g}_{Pi}) forms of the compound *i*.

Upon dividing both of its numerator and denominator by ΔC_{p1}, and substituting $k = \Delta C_{p2} / \Delta C_{p1}$ in the above expression, then we will get the GT eq 7.6.

$$T_{gm} = \frac{w_1 T_{g1} + k w_2 T_{g2}}{w_1 + k w_2} \tag{7.6}$$

where w is the weight fraction concentration. The parameter k depends upon the variation in thermal expansion coefficient (α) of the components when they change in to liquid (rubbery) state from their glassy (amorphous) state.

The curvature of GTT and solid content represents the strength of interaction among water and solid content present.[23] For any binary mixture, large value of *k* means that the solid content presents are highly plasticized with water and vice-versa. Increment in *k* parameter presents a decrease in the GTT. If $k = 1$, then GTT and solid content relation give a straight line. If $k<1$, then the curve will be convex where as for the $k > 1$, it will be concave. Starch has smaller value of *k* in comparison to sugar or proteins.[49] The model parameters (*k* and T_{gs}) of GT theory using $T_{gw} = -135°C$, can be estimated using nonlinear regression analysis.[4] Moreover, GT model can also be used to calculate GTT of a mixture at any water concentration, if the T_{gs} and *k* are available. For example, glucose–water solutions have its T_{gs} and *k* as 31°C and 4.07, respectively.[25] Now, using GT equation, GTT for the mixture containing solid content 0.9 and water content 0.1 can be determined as 20.7°C.

7.5.1.2 SOLID-CRYSTAL MELTING TEMPERATURE CURVE

According to FHT, the crystals melting peak temperature can be modeled as follows:

$$\frac{1}{T_{mp}}-\frac{1}{T_{sp}}=-\left(\frac{R}{\Delta H_u}\right)\left(\frac{V_u}{V_w}\right)\left(\varepsilon_w-\chi\varepsilon_w^2\right) \tag{7.7}$$

where R is the gas constant; T_{mp} and T_{sp} are the melting temperature peaks for the polymer and pure solvent, respectively; ΔH_u is the latent heat of fusion of the polymer units in the diluent; V_u and V_w are molar volume of polymer units and diluents, respectively; and χ is FHIP.

The volume fraction of water can be determined by using their mass fractions and density values as:

$$\varepsilon_w=\frac{\left(X_w/\rho_w\right)}{\left(X_w/\rho_w\right)+\left(X_s/\rho_s\right)} \tag{7.8}$$

where X_w and X_s represent the mass fraction of water and solid content, respectively; ρ_w and ρ_s are, respectively, the densities of water and dry solids. The density of the solid can be estimated from the estimation of proximate composition.

7.5.1.3 FREEZING CURVE

For determining the melting and freezing point of food systems, DSC and the cooling curve method are widely used. In the DSC system, the temperature at which first ice crystal starts melting is referred as initial freezing temperature (T_f). Moreover, ice-melting temperature (T_m) is a temperature where last ice crystals melt upon cooling them to adequate low temperature. The cooling curve method was also used widely as for the foods.[39,46] It is highly preferred over the DSC, as it is inexpensive and sample preparation is easy. But the result obtained is less accurate when applied to the materials containing smaller amount of water.[49]

Theoretically, freezing point of a food product is mainly estimated from the Chen model which is the modification of Clausius Clapeyron equation (eq 7.9), as follows:

$$\delta = -\frac{\beta}{\lambda_w} \ln\left[\frac{1 - X_s^o}{1 - X_s^o + EX_s^o}\right] \tag{7.9}$$

where $\delta = (T_w - T_f)$ is freezing point depression, that is, difference between T_f and T_w (freezing point of food and water) (°C); λ_w: MW of water; β is the molar freezing temperature constant for water, that is, 1860 kgK/kg mol; and E is the ratio of molecular mass of water to solids (λ_w/λ_s).

Chen equation can be applied to estimate the freezing temperature for the nonideal systems as:

$$\delta = -\frac{\beta}{\lambda_w} \ln\left[\frac{1 - X_s^o - BX_s^o}{1 - X_s^o - BX_s^o + EX_s^o}\right] \tag{7.10}$$

It includes the amount of water not available for ice formation (B), which is ratio of unfrozen water to the concentration of total solids.

7.5.1.4 MAXIMUM FREEZE CONCENTRATION CONDITION

It corresponds to formation of maximum ice at low temperature below 0°C. It is characterized by the two temperature: the GTT for MFC (T_g') and for the onset of melting (T_m'), both are below 0°C.[49] The difference between both temperatures is very less and may coincide with the high MW compounds.[42] Both temperatures are directly related to MW of the components. The amount/quantity of water present at T_g' is referred to as unfrozen water. Generally, the temperature for the onset of melting (T_m') is evaluated by plotting the linear baseline to the endotherm. For material containing freezable water, annealing is performed at temperature ($T_m' - 1$), which results in maximum ice formation. After the annealing step, the materials are again cooled down as per the previous set conditions. The value of T_g' is determined mostly by extending the freezing curve to the glass transition line in the SD. However, the accuracy depends upon several conditions, such as annealing and heating–cooling rates.[37]

7.5.2 APPLICATIONS OF STATE DIAGRAMS

Various researches have estimated the GT constants (T_{gs} and k) from the relationship of glass temperature and water content or solid content via glass transition line, Chen model constants (E and B) via freezing line, and MFC conditions (T_g', T_m') of foods, required for SD development. Some of food products and their estimated parameter are shown in Table 7.1. The physical characteristics of food materials are associated with their glass transition. SDs are found useful in monitoring and predicting the changes occurring in the food system upon thermal processing. Structural transformations such as stickiness, ice formation, and collapses are seen with the change in temperature or relative humidity. SDs have predictive approach toward the food materials or products as they can explain effect of temperature and moisture content through which a material undergoes changes. For example, SD for extrusion cooking explains the specific moisture contents and temperature, where proteins present undergoes the cross-linking reactions, providing crispiness to the extruded products.

Food products may incur various deteriorative reactions during their processing and storage, like stickiness of food powders while drying,[41] cake formation (absorption of moisture) upon storage,[13] sugar crystallization,[24] and loss in crispiness.[44] The main reasons behind such behavior are low GTTs and their high hygroscopicity.[46] For example, many fruit solids have a T_g' value ranging from −10 to −60°C, and moreover, their collapse temperatures also lie in the same range. Therefore, freeze-drying temperature need to maintained below T_g' or adding high MW ingredients having high T_g and via monitoring their processing and storage temperature below GTT to avoid collapse.[49] SD helps the food engineers and technologists in designing a product having the desired structure with sensory properties, besides preventing food wastage during manufacturing, transportation, and storage.

7.6 SUMMARY

Food structuring design has become pivotal in developing newer food products based on consumer's preference, nutritional need, and bioavailability of the functional compounds upon ingestion. Concepts of SM physics could be used for understanding of structural design engineering of the food products. FHT could be used to study the interaction and incompatibility behavior among the constituents of the food product. State/PTs can be constructed using FHT, along with GT equation, and Chen's model. SD could be used as predicting tool for FS and the quantitative and qualitative changes occurring in food properties with the time scale. MDSs could further open up less traveled area of food physics where one could visualize the effect of chemical bonds

TABLE 7.1 State Diagram (SD) Data of Some Food Products.

Products	Glass transition line				Solid melting curve		Freezing curve		References
	T_{gs} (°C)	K	T_g' (°C)	X_s'	B	E	X_s'	T_m' (°C)	
Freeze-dried products									
Mango	12.2	4.487	–54.6	–	0.16	0.0695	–	–33.0	76
Agaricus bisporus	–5.42	3.06	–77.9	0.704	–	–	0.704	–31.9	51
Raspberry	42.6	4.73	–61.6	0.780	–	–	0.780	–28	53
Plums	102.7	3.76	–57.5	–	–	–	–	–	54
Garlic	40.1	3.7	–38.6	0.18	–	–	–	–26	39
Apple	–13.30	2.7	–	–	–	–	–	–	72
Apple juice	–16.3	2.4	–	–	–	–	–	–	
Fruit juices									
Cactus pear	18	3.55	–55.1	0.75	0.1657	0.0778	0.75	–41.8	18
Orange	23.5	3.32	–52.4	0.773	0.1139	0.0735	0.773	–35.3	
Mango	40.8	4.93	–50.5	0.772	0.1003	0.0081	0.772	–35.9	
Strawberry	22	3.02	–53.2	0.762	0.0862	0.0834	0.762	–32.5	
Pineapple	17.5	3.29	–54.2	0.749	0.1287	0.6972	0.749	–39.9	
Low molecular weight sugars									
100% G	31.8	3.79	55.7	0.774	0.0353	0.1181	0.774	–41.8	45
100% F	10.4	2.9	55.9	0.776	0.0048	0.1263	0.776	–42.8	
100% S	64.8	4.72	43.1	0.799	0.0518	0.0652	0.799	–32.4	
50% G+ 50% F	19.1	3.52	54.6	0.793	0.0659	0.1045	0.793	–42.6	
50% G+ 50% S	40.5	3.92	52	0.777	0.0841	0.0858	0.777	–38.8	
50% F+ 50% S	24.7	3.52	–50.2	0.80	0.0246	0.0977	0.800	–38.4	

TABLE 7.1 *(Continued)*

Products		Glass transition line				Solid melting curve		Freezing curve		References
		T_{gs} (°C)	K	T_g' (°C)	X_s'	B	E	X_s'	T_m' (°C)	
					Cereals					
IRS		56.45	2.6202	−42.50	0.71	0.1244	0.0262	0.71	−18.5	71
IRS–SDF		33.85	1.6308	−31.50	0.72	0.1949	0.0278	0.72	−13.50	
Rice		158.5	6.38	−11.6	–	0.329	0.0117	–	–	48
					Fruits and vegetables					
Pumpkin		168	6.30	−59	0.69	0.30	0.000049	0.69	−35	6
Mango pulp										17
Varying weight fraction of maltodextrin	0	48.9	3.13	–	–	–	–	–	–	
	0.2	58.1	4.55	–	–	–	–	–	–	
	0.4	69.3	5.56	–	–	–	–	–	–	
	0.6	76.1	6.08	–	–	–	–	–	–	
	0.8	108.5	6.44	–	–	–	–	–	–	
	1.0	162.6	9.00	–	–	–	–	–	–	
Apple slice		41.3	3.59	−57.8	0.736	–	–	0.736	−50.3	8
Dates		57.4	3.2	–	–	0.429	–	–	–	20
					Meat products					
Tuna meat		95.1	2.89	−54.2	0.719	0.383	0.033	0.610	−13.3	38
Abalone		–	–	−44.3	–	0.307	–	–	−18	47
					Confectionary products					
Date syrup		14.9	1.01	–	–	0.020	0.177	0.73	−42.8	5

TABLE 7.1 *(Continued)*

Products		Glass transition line				Solid melting curve		Freezing curve		References
		T_{gs} (°C)	K	T_g' (°C)	X_s'	B	E	X_s'	T_m' (°C)	
Candied orange peels	COP 100	47.56	7.17	−65.9	0.815	−0.279	0.217	0.815	−36.9	75
	COP 70	38.02	4.71	−54.6	0.804	−0.204	0.188	0.804	−36.1	
	COP 30	42.55	4.32	−54.2	0.782	−0.170	0.183	0.782	−35.4	
	COP 0	45.98	4.73	−53.3	0.788	−0.169	0.175	0.788	−34.8	

Products: G, Glucose; F, Fructose; S, Sucrose, IRS, Indica rice starch; IRS-SDF, Indica rice starch-soluble dietary fiber; COP 100, Candied orange peel in 100% S syrup; COP 70, Candied orange peel in 70% S syrup and 30% G-F syrup; COP 30, Candied orange peel in 30% S and 70% G-F syrup; COP 0, Candied orange peel in 100% G-F syrup

Parameters: T_g', GTT of MFC solution; T_m', onset ice melting temperature; X_s', solid concentration at MFC condition; T_{gs}, glass temperature of dry solids; k, GT parameter; B, unfreezable water (g water/ g solid); E, ratio of MW of water to solids

energies to predict various engineering properties of food blends/materials used for the fabrication of novel food products as per the requirement of consumer.

KEYWORDS

- **Flory-Huggins**
- **food structuring**
- **glass transition**
- **jamming state**
- **soft matter**
- **state diagram**

REFERENCES

1. Aguilera, J. M.; Stanley, D. W., Eds. In *Microstructural Principles of Food Processing and Engineering*, 2nd ed.; Springer: Verlag, 1999; p 432.
2. Aguilera, J. M.; Stanley, D. W.; Baker, K. W. New Dimensions in Microstructure of Food Products. *Trends Food Sci. Technol.* **2000,** *11* (1), 3–9.
3. Aguilera, J. M. Seligman Lecture 2005 on Food Product Engineering: Building the Right Structures. *J. Sci. Food Agric.* **2006,** *86* (8), 1147–1155.
4. Ahmed, J.; Rahman, M. S. Glass Transition in Foods. *Eng. Prop. Foods* **2014,** *2014,* 93–120.
5. Al-Farsi, K. A.; Al-Habsi, N. A.; Rahman, M. S. State Diagram of Crystallized Date-Syrup: Freezing Curve, Glass Transition, Crystals-Melting and Maximal-Freeze-Concentration Condition. *Thermochim. Acta* **2018,** *666,* 166–173.
6. Al-Ghamdi, S.; Hong, Y. K.; Qu, Z.; Sablani, S. S. State Diagram, Water Sorption Isotherms and Color Stability of Pumpkin (*Cucurbita pepo* L.). *J. Food Eng.* **2020,** *273,* E-article 109820.
7. Armand, M.; Borel, P.; Pasquier, B.; Dubois, C.; Senft, M.; Andre, M.; Peyrot, J.; Salducci, J.; Lairon, D. Physicochemical Characteristics of Emulsions during Fat Digestion in Human Stomach and Duodenum. *Am. J. Physiol.* **1996,** *271* (1), G172–G183.
8. Bai, Y; Rahman, M. S.; Perera, C. O.; Smith, B.; Melton, L. D. State Diagram of Apple Slices: Glass Transition and Freezing Curves. *Food Res. Int.* **2001,** *34* (2–3), 89–95.
9. Boonrod, C.; Srikaeo, K. State Diagrams of Two Thai Rice Varieties Differing in Amylose Contents. In *Applied Mechanics and Materials*; Yimnirun, R., Ed.; Trans Tech Publications, Ltd.: Switzerland, 2016; Vol. 848; pp 89–94.
10. Chen, J. Food Oral Processing—A Review. *Food Hydrocoll.* **2009,** *23* (1), 1–25.

11. Cipelletti, L.; Ramos, L. Slow Dynamics in Glassy Soft Matter. *J. Phys.* **2005,** *17* (6), R253–R285.
12. de Kruif, K. G.; Marion, A. M.; van Marle, M. E.; van Mil, P. J.; Sebastianus, P. F. M. Gelation of Proteins from Milk. *Faraday Discuss.* **1995,** *101*, 185–200.
13. Descamps, N.; Palzer, S.; Roos, Y. H.; Fitzpatrick, J. J. Glass Transition and Flowability/Caking Behaviour of Maltodextrin DE 21. *J. Food Eng.* **2013,** *119* (4), 809–813.
14. Dickinson, E. Hydrocolloids at Interfaces and the Influence on the Properties of Dispersed Systems. *Food Hydrocoll.* **2003,** *17* (1), 25–39.
15. Dullens, R. P.; Aarts, D. G.; Kegel, W. K. Direct Measurement of the Free Energy by Optical Microscopy. *Proc. Natl. Acad. Sci.* **2006,** *103* (3), 529–531.
16. Fito, P.; LeMaguer, M.; Betoret, N.; Fito, P. J. Advanced Food Process Engineering to Model Real Foods and Processes: The "SAFES" Methodology. *J. Food Eng.* **2007,** *83* (2), 173–185.
17. Fongin, S.; Kawai, K.; Harnkarnsujarit, N.; Hagura, Y. Effects of Water and Maltodextrin on the Glass Transition Temperature of Freeze-Dried Mango Pulp and An Empirical Model to Predict Plasticizing Effect of Water on Dried Fruits. *J. Food Eng.* **2017,** *210*, 91–97.
18. Grajales-Lagunes, A.; Rivera-Bautista, C.; Loredo-García, I. O.; González-García, R.; González-Chávez, M. M.; Schmidt, S. J.; Ruiz-Cabrera, M. A. Using Model Food Systems to Develop Mathematical Models for Construction of State Diagrams of Fruit Products. *J. Food Eng.* **2018,** *230*, 72–81.
19. Gunning, A. P.; Mackie, A. R.; Wilde, P. J.; Morris, V. J. Atomic Force Microscopy of Emulsion Droplets: Probing Droplet–Droplet Interactions. *Langmuir* **2004,** *20* (1), 116–122.
20. Kasapis, S.; Rahman, M. S.; Guizani, N.; Al-Aamri, M. State Diagram of Temperature vs Date Solids Obtained from the Mature Fruit. *J. Agric. Food Chem.* **2000,** *48* (9), 3779–3784.
21. Kilburn, D.; Claude, J.; Mezzenga, R.; Dlubek, G.; Alam, A.; Ubbink, J. Water in Glassy Carbohydrates: Opening It Up at the Nanolevel. *J. Phys. Chem. B* **2004,** *108* (33), 12436–12441.
22. Kleman, M.; Laverntovich, O. D., Eds. In *Soft Matter Physics: An Introduction*, Springer Science & Business Media: Switzerland, 2003; p 637.
23. Kurozawa, L. E.; Park, K. J.; Hubinger, M. D. Effect of Maltodextrin and Gum Arabic on Water Sorption and Glass Transition Temperature of Spray Dried Chicken Meat Hydrolysate Protein. *J. Food Eng.* **2009,** *91* (2), 287–296.
24. Langrish, T. A. G.; Wang, S. Crystallization Rates for Amorphous Sucrose and Lactose Powders from Spray Drying: A Comparison. *Dry. Technol.* **2009,** *27* (4), 606–614.
25. Levine, H.; Truong, V.; Bhandari, B. R.; Howes, T.; Adhikari, B. Analytical Model for the Prediction of Glass Transition Temperature of Food Systems. Chapter 3. In *Amorphous Food and Pharmaceutical Systems*; Levine, H., Ed.; The Royal Society of Chemistry: Cambridge, UK, 2002; pp 31–47.
26. Likos, C. N. Effective Interactions in Soft Condensed Matter Physics. *Phys. Rep.* **2001,** *348* (4–5), 267–439.
27. Liu, A. J.; Nagel, S. R. Jamming is Not Just Cool Any More. *Nature* **1998,** *396*, 21–22.
28. Lucas, P. W.; Prinz, J. F.; Agrawal, K. R.; Bruce, I. C. Food Physics and Oral Physiology. *Food Qual. Prefer.* **2002,** *13* (4), 203–213.

29. Luyet, B.; Rasmussen, D. Study by Differential Thermal Analysis of the Temperatures of Instability of Rapidly Cooled Solutions of Glycerol, Ethylene Glycol, Sucrose and Glucose. *Biodynamica* **1968,** *10* (210), 167–191.
30. Manski, J. M.; van der Goot, A. J.; Boom, R. M. Formation of Fibrous Materials from Dense Calcium Caseinate Dispersions. *Biomacromolecules* **2007,** *8* (4), 1271–1279.
31. Matveev, Y. I.; Grinberg, V. Y.; Tolstoguzov, V. B. The Plasticizing Effect of Water on *0*/Proteins, Polysaccharides and their Mixtures. Glassy State of Biopolymers, Food and Seeds. *Food Hydrocoll.* **2000,** *14* (5), 425–437.
32. McClements, D. J. Structural Design Principles for Improved Food Performance: Nano-laminated Biopolymer Structures in Foods. Chapter 1. In *Micro/Nanoencapsulation of Active Food Ingredients (ACS Symposium Series)*; Huang, Q.; Given, P., Qian, M., Eds.; American Chemical Society: Washington, DC, 2009; Vol. 1007; pp 3–34.
33. McClements, D. J.; Decker, E. A.; Park, Y.; Weiss, J. Structural Design Principles for Delivery of Bioactive Components in Nutraceuticals and Functional Foods. *Critic. Rev. Food Sci. Nutr.* **2009,** *49* (6), 577–606.
34. Michel, M.; Sagalowicz, L. Probing Food Structure. In *Food Materials Science: Principles and Practice*; Aguilera, J. M., Lillford, P. J., Eds.; Springer: Verlag, 2008; pp 203–226.
35. Mogensen, O. E., Ed. In *Positron Annihilation in Chemistry*, 1st ed.; Springer: Verlag, 1995; p 268.
36. Ougizawa, T.; Inoue, T. UCST and LCST Behavior in Polymer Blends and its Thermodynamic Interpretation. *Polym. J.* **1986,** *18* (7), 521–527.
37. Rahman, M. S.; Guizani, N.; Al-Khaseibi, M.; Al-Hinai, S. A.; Al-Maskri, S. S.; Al-Hamhami, K. Analysis of Cooling Curve to Determine the End Point of Freezing. *Food Hydrocoll.* **2002,** *16* (6), 653–659.
38. Rahman, M. S.; Kasapis, S.; Guizani, N.; Al-Amri, O. S. State Diagram of Tuna Meat: Freezing Curve and Glass Transition. *J. Food Eng.* **2003,** *57* (4), 321–326.
39. Rahman, M. S.; Sablani, S. S.; Al-Habsi, N.; Al-Maskri, S.; Al-Belushi, R. State Diagram of Freeze-Dried Garlic Powder by Differential Scanning Calorimetry and Cooling Curve Methods. *J. Food Eng.* **2005,** *70* (2), E135–E141.
40. Rahman, M. S. Food Stability Determination by Macro–Micro Region Concept in the State Diagram and by Defining a Critical Temperature. *J. Food Eng.* **2010,** *99* (4), 402–416.
41. Ren, G. Y.; Zeng, F. L.; Duan, X.; Liu, L. L.; Duan, B.; Wang, M. M.; Zhu, W. X. The Effect of Glass Transition Temperature on the Procedure of Microwave–Freeze Drying of Mushrooms (*Agaricus bisporus*). *Dry. Technol.* **2015,** *33* (2), 169–175.
42. Roos, Y. H.; Drusch, S. In *Phase Transitions in Foods,* 2nd ed.; Academic Press: Amsterdam, 2015; p 367.
43. Roos, Y. H. Phase and State Transitions and Transformations in Food Systems. Chapter 4. In *Handbook of Food Engineering*, 3rd ed.; Heldman, D. R.; Lund, D. B., Sabliov, C., Eds.; CRC Press: Boca Raton, 2018; pp 485–550.
44. Roudaut, G.; Dacremont, C.; Pàmies, B. V.; Colas, B.; Le Meste, M. Crispness: A Critical Review on Sensory and Material Science Approaches. *Trends Food Sci. Technol.*, **2002,** *13* (6–7), 217–227.
45. Ruiz-Cabrera, M. A.; Rivera-Bautista, C.; Grajales-Lagunes, A.; Gonzalez-Garcia, R.; Schmidt, S. J. State Diagrams for Mixtures of Low Molecular Weight Carbohydrates. *J. Food Eng.* **2016,** *171*, 185–193.
46. Sablani, S. S.; Rahman, M. S. Pore Formation in Selected Foods as a Function of Shelf Temperature during Freeze Drying. *Dry. Technol.* **2002,** *20* (7), 1379–1391.

47. Sablani, S. S.; Kasapis, S.; Rahman, M. S.; Al-Jabri, A.; Al-Habsi, N. Sorption Isotherms and the State Diagram for Evaluating Stability Criteria of Abalone. *Food Res. Int.* **2004,** *37* (10), 915–924.
48. Sablani, S. S.; Bruno, L.; Kasapis, S.; Symaladevi, R. M. Thermal Transitions of Rice: Development of a State Diagram. *J. Food Eng.* **2009,** *90* (1), 110–118.
49. Sablani, S. S.; Syamaladevi, R. M.; Swanson, B. G. Review of Methods, Data and Applications of State Diagrams of Food Systems. *Food Eng. Rev.* **2010,** *2* (3), 168–203.
50. Schellman, J. A. Protein Stability in Mixed Solvents: A Balance of Contact Interaction and Excluded Volume. *Biophys. J.* **2003,** *85*(1), 108–125.
51. Shi, Q.; Wang, X.; Zhao, Y.; Fang, Z. Glass Transition and State Diagram for Freeze-Dried *Agaricus bisporus*. *J. Food Eng.* **2012,** *111* (4), 667–674.
52. Strybulevych, A.; Leroy, V.; Scanlon, M. G.; Page, J. H. Characterizing a Model Food Gel containing Bubbles and Solid Inclusions using Ultrasound. *Soft Matter* **2007,** *3* (11), 1388–1394.
53. Syamaladevi, R. M.; Sablani, S. S.; Tang, J.; Powers, J.; Swanson, B. G. State Diagram and Water Adsorption Isotherm of Raspberry (*Rubusidaeus*). *J. Food Eng.* **2009,** *91* (3), 460–467.
54. Telis, V. N.; Sobral, P. D. A.; Telis-Romero, J. Sorption Isotherm, Glass Transitions and State Diagram for Freeze-Dried Plum Skin and Pulp. *Food Sci. Technol. Int.* **2006,** *12* (3), 181–187.
55. Tolstoguzov, V. Structure–Property Relationships in Foods. Chapter 1. In *Macromolecular Interactions in Food Technology (ACS Symposium Series)*; Parris, N.; Kato, A.; Creamer, L. K., Pearce, J., Eds.; American Chemical Society: Washington, DC, 1996; pp 2–14.
56. Tolstoguzov, V. B. The Importance of Glassy Biopolymer Components in Food. *Nahrung* **2000,** *44* (2), 76–84.
57. Tolstoguzov, V. Food Polymers. Chapter 3. In *Food Materials Science: Principles and Practice*; Aguilera, J. M., Lillford, P. J., Eds.; Springer: New York, USA, 2008; pp 21–44.
58. Ubbink, J.; Kruger, J. Physical Approaches for the Delivery of Active Ingredients in Foods. *Trends Food Sci. Technol.* **2006,** *17* (5), 244–254.
59. Ubbink, J.; Burbidge, A.; Mezzenga, R. Food Structure and Functionality: A Soft Matter Perspective. *Soft Matter* **2008,** *4* (8), 1569–1581.
60. Van der Sman, R. G. M.; Boer, E. Predicting the Initial Freezing Point and Water Activity of Meat Products from Composition Data. *J. Food Eng.* **2005,** *66* (4), 469–475.
61. Van der Sman, R. G. M.; Van der Graaf, S. Diffuse Interface Model of Surfactant Adsorption onto Flat and Droplet Interfaces. *Rheol. Acta* **2006,** *46* (1), 3–11.
62. Van der Sman, R. G. M. Soft Condensed Matter Perspective on Moisture Transport in Cooking Meat. *AIChE J.* **2007,** *53* (11), 2986–2995.
63. Van der Sman, R. G. M. Moisture Transport during Cooking of Meat: An Analysis Based on Flory–Rehner Theory. *Meat Sci.* **2007,** *76* (4), 730–738.
64. Van der Sman, R. G. M; Van der Goot, A. J. The Science of Food Structuring. *Soft Matter* **2009,** *5* (3), 501–510.
65. Van der Sman, R. G. M.; Meinders, M. B. J. Prediction of the State Diagram of Starch Water Mixtures Using Flory–Huggins Free Volume Theory. *Soft Matter* **2011,** *7* (2), 429–442.
66. Van der Sman, R. G. M. Soft Matter Approaches to Food Structuring. *Adv. Colloid Interf. Sci.* **2012,** *176*, 18–30.
67. Van der Sman, R. G. M. Thermodynamics of Meat Proteins. *Food Hydrocoll.* **2012,** *27* (2), 529–535.

68. Van der Sman, R. G. M.; Meinders, M. B. J. Moisture Diffusivity in Food Materials. *Food Chem.* **2013,** *138* (2–3), 1265–1274.
69. Van der Sman, R. G. M. Biopolymer Gel Swelling Analysed with Scaling Laws and Flory–Rehner Theory. *Food Hydrocoll.* **2015,** *48*, 94–101.
70. Vilgis, T. A. Soft Matter Food Physics - The physics of Food and Cooking. *Rep. Progr. Phys.* **2015,** *78* (12), E-article 124602.
71. Wan, J.; Ding, Y.; Zhou, G.; Luo, S.; Liu, C.; Liu, F. Sorption Isotherm and State Diagram for Indica Rice Starch With and Without Soluble Dietary Fiber. *J. Cereal Sci.* **2018,** *80*, 44–49.
72. Welti-Chanes, J.; Guerrero, J. A.; Barcenas, M. E.; Aguilera, J. M.; Vergara, F.; Barbosa-Canovas, G. V. Glass Transition Temperature (Tg) and Water Activity (a_w) of Dehydrated Apple Products. *J. Food Process Eng.* **1999,** *22* (2), 91–101.
73. Welti-Chanes, J.; Vergara-Balderas, F.; Ríos, L.; Mújica-Paz, H.; Valdez-Fragoso, A. Thermodynamics in Food Engineering. In *Food Engineering*; Barbosa-Cánovas, G. V., Ed.; EOLSS Publishers Co. Ltd.: Oxford, United Kingdom, 2009; pp 240–273.
74. Windhab, E. J.; Dressler, M.; Feigl, K.; Fischer, P.; Megias-Alguacil, D. Emulsion Processing—from Single-Drop Deformation to Design of Complex Processes and Products. *Chem. Eng. Sci.* **2005,** *60* (8–9), 2101–2113.
75. Witczak, T.; Witczak, M.; Stępień, A.; Bednarz, A.; Grzesik, M. State Diagrams of Candied Orange Peel Obtained using Different Hypertonic Solutions. *J. Food Eng.* **2017,** *212*, 234–241.
76. Zhao, J. H.; Liu, F.; Wen, X.; Xiao, H. W.; Ni, Y. Y. State Diagram for Freeze-Dried Mango: Freezing Curve, Glass Transition Line and Maximal-Freeze-Concentration Condition. *J. Food Eng.* **2015,** *157*, 49–56.

CHAPTER 8

APPLICATIONS OF BIOTECHNOLOGICAL APPROACHES IN THE FOOD INDUSTRY: REVIEW

SREEJIT VALIATHAN and PREETHA RADHAKRISHNAN

ABSTRACT

In the present scenario, biotechnology has been suggested as a tool to put an end to hunger and prevent famine. The purpose of this chapter is to provide an overview of the scope and significance of food biotechnology, the benefits, and the concerns of using biotechnology in food processing. Biotechnology applications in foods include the manufacture of fermented food products, biopreservatives, probiotics, and the development of genetically modified food crops with improved disease resistance/higher-yielding capacity. Biotechnological production of food additives, biosynthesis of enzymes required for making preferable modifications/manufacturing of food ingredients, biosurfactants, bio-based packaging materials, single-cell proteins, the fabrication of biosensors, and omics tools for ensuring food safety have also gained widespread attention nowadays. The above-mentioned biotechnological applications are also discussed in this chapter.

8.1 INTRODUCTION

Biotechnology deals with utilizing technologies that utilize the biological system, alive organisms, or derivatives to create or alter processes or products for definite applications.[33] Biotechnology has a prominent role in food production, medicine, beverages production, and textile/leather production

Novel and Alternative Methods in Food Processing: Biotechnological, Physicochemical, and Mathematical Approaches. N. Veena, Megh R. Goyal, & Ritesh B. Watharkar (Eds.)

and has an immense impact on the society. Biotechnology has the possibility to minimize the food shortage nowadays, and it can avoid the mass starvation in the near-term future.[33]

The Institute of Food Science and Technology has defined food biotechnology as the "*application of biological techniques to food crops, animals and microorganisms to improve the quality, quantity, safety, ease of processing and production, economics of food.*"[27] The food processing industries use the biotechnological methods for various applications, such as preserving food using microorganisms for manufacturing a variety of products with added value, such as flavoring substances, enzymes, vitamins, food ingredients, and microbial cultures. In the food processing industries, biotechnology helps select and improve the microorganisms to enhance process control, output, efficiency, and effectiveness of the bioprocess.[60]

Biotechnology is being used in food processing for various applications, such as developing genetically modified (GM) crops and organisms, manipulation of gene and its transfer, the utilization of molecular markers, in vitro vegetative propagation of plants, and DNA-based methods of disease identification. It also finds application in processing raw materials generated by fishery, crop, and livestock to assure food security in many developing countries.

Biotechnology has a crucial role in improving the flavor, taste, texture, color, aroma, and nutritional worth of food. The required changes are made in the food through the process of fermentation. Fermentation can take place naturally, or it can occur through deliberate inoculation. Fermentation is an anaerobic process in which bacteria, yeast, or both convert carbohydrates into alcohol, CO_2, or organic acids. Some of the examples are the production of beer, wine, cider, lactic acid, and leavening of bread.[33] Biotechnology also finds application in the synthesis of food ingredients production such as food flavors, food additives, and aroma compounds,

Biotechnology holds a pivotal role in the manufacturing of enzymes, and the application of certain enzymes induces the necessary alteration in food. Enzymes are used for food processing, as biopreservative and food industry waste management. Enzymes are commercially synthesized with the help of microorganisms. For example, *Bacillus megaterium* was reported for the commercial synthesis of keratinase using chicken feathers as the substrate. *Aspergillus niger* was used for the commercial synthesis of phytase using wheat bran as the substrate. Cellulase was commercially made using *Aspergillus awamori* with sugarcane bagasse as the substrate.[30] Advanced genetic technologies have led to an increase in the quality and cost-effectiveness of enzymes. Advanced genetic technologies, in turn, lead to a rise in the application of enzymes for the processing of food.

Nowadays, GM microorganisms are employed for the commercial manufacturing of enzymes. The microorganisms produce enzymes as secretions as a result of the fermentation. Then the enzymes are purified. The food manufacturing industries use these purified enzymes. Genetic engineering has enhanced the overall productivity of the enzymes. Carbohydrase and protease are examples of the enzymes that are manufactured using GM microorganisms.[86] The advancements in protein engineering have enabled the production of a specific enzyme with structural modifications. This structural modification improved the activity of the enzyme. It allowed the enzyme to act on a new substrate even at a higher pH. It also made the enzyme thermostable.

Food processing involves different unit processes to transform a raw, inedible product with limited shelf life into edible food products with enhanced storage life. The unit processes and the manufacturing procedure employed for food processing should be food grade and should not cause any health hazards. Biotechnology finds wide application as a diagnostic tool for examining food safety and hence prevents food-borne diseases.[33] The advanced biotechnology methods are utilized for testing of food and food ingredients.[86] Biotechnological testing methods give precise results within a reasonably shorter time than conventional testing methods.[33]

The demand for food-borne pathogen identification using biosensors is increasing because they rapidly and precisely detect pathogens. They have reasonable quality control for process control within the food industry.[76] Food products such as cheese, milk, chicken, meat, fruits, and uncooked vegetables are highly prone to pathogenic contamination. In the food industry, biosensors are used for various purposes; for example, for the identification of monosaccharides, glucose, organic acids, and alcohol. Biosensors are also used to determine the freshness and aroma of food products and find out antibiotic residues, pesticidal residues, harmful microorganisms, and toxins.[76]

The nutritional components present in the food can prevent the occurrence of diseases.[17] Hence, it is also essential to optimize the nutritional composition for a particular population, which is possible nowadays with the intervention of Omics technology.[13] The current advancements in Omics methods can also enabled a great extent of molecular level quantification of food-borne pathogens to assure food safety. The most important Omics methods used are genomics, proteomics, transcriptomics, and metabolomics.[75]

The biotechnological applications in foods have been discussed in detail in this chapter.

8.2 HISTORY OF BIOTECHNOLOGICAL APPLICATIONS IN FOOD INDUSTRIES

Biotechnology has been used to process food and beverages for the past 8000 years: for example, the processing of alcohol, vinegar, cheese, sourdough, and so on. Bioengineering plays an essential role in food processing and food production.[45] It has been observed that the biotechnological method influences food production, preservation of raw materials, and modification of the functional/nutritional properties of the raw materials. Besides that, the development of processing aids, production aids, and direct additives, for instance, flavors, enzymes, pigments, polysaccharides, and antioxidants with the help of biotechnological methods, can enhance the overall quality and acceptability of food. The preservation and enhancement of raw material quality using biological processes are crucial for the agriculture and food manufacturing industries.

Some of the examples are the fermentation of coffee beans and cocoa beans,[45] oxidation of tea leaves,[39] production of single-cell protein (SCP),[7] and so on. The extent of oil extraction from oilseeds has been enhanced using nonlipolytic enzymes.[31] The increase in the overall output of fruit juice processing using pectolytic enzymes.[45] Traditionally, biotechnological methods have been used to manufacture organic acids, fatty acids, flavors, and vitamins. Conventionally, microbial metabolism has been extensively used to preserve food, stabilization of food, particularly meat, dairy, fruit, fish, and vegetable products.[45]

8.3 BIOTECHNOLOGICAL METHODS IN THE FOOD INDUSTRY: CURRENT STATUS, SCOPE, AND SIGNIFICANCE

Genetic engineering, cell engineering, and proteomics tools are new developments in the food industry. Different genetic engineering methods (such as recombinant DNA technology, gene transfer, and cloning) are useful in enhancing the quality of raw materials for food processing. These techniques are also helpful in improving the performance of food microbes. Cell engineering methods (such as plant and animal cell culture and cell fusion technology) help synthesize new food and food additives. Protein engineering finds application for the modification of substances for the manufacture of food. Biotechnology has great potential for creating new types of food additives for the food industry application.[95]

The biotechnological methods are useful in manufacturing and processing food products, such as brewing and fermentation. Modern biotechnological processes help improve genes affecting the processing quality of raw materials.[95] Biotechnological tools help create superior quality agricultural products and food additives and are handy in the manufacturing/processing food.[95] Biotechnology can play a crucial role in the food sectors, and some examples are:

- Synthesize of biopreservatives and antioxidants with the aid of fermentation and genetic engineering,
- Enhanced production and yield of raw material for the food industry using genetic engineering,
- Provide cost-effective, safe/quality food with the help of enzyme engineering, fermentation, and genetic engineering,
- Reduction of food loss by the improvement of food processing methods using genetic engineering, enzyme engineering, bioprocess engineering, and protein engineering,
- Maximum use of resources via efficient food processing waste treatment using enzyme engineering and fermentation tools.[95]

8.4 BIOTECHNOLOGICAL APPLICATIONS IN FOOD INDUSTRIES

The various biotechnological applications in food industries include the production of fermented food, enzymes, biopreservatives, probiotics, prebiotics, symbiotic, SCPs, functional foods, food ingredients, biosurfactants, biodegradable packaging, edible films, GM food, use of biosensors, and foodomics for ensuring food safety. They are discussed in this section.

8.4.1 PRODUCTION OF FERMENTED FOODS

Fermented foods are produced due to the interaction of microbes and their enzymes on raw material under specific conditions. The essential factors required for fermentation to occur are the right substrate, the suitable strain of microorganism, and an environment with optimum pH, moisture content, and temperature.[54] More than thousands of fermented foods are being manufactured worldwide with various microorganisms, raw materials, and manufacturing techniques. A few examples of fermented foods are kombucha, kefir, sauerkraut, natto, tempeh, miso, kimchi, and sourdough bread.[54]

Fermented milk is manufactured by using *Bifidobacteria* and yeast. The starter culture of microorganisms influences the flavor and texture of fermented milk. It also has a vital role in forming bioactive compounds that provide antihypertensive, antioxidant, antiallergic, and antidiabetic property to the raw materials.[54] Similarly, fermentation of other raw food matrices such as grains, fruits, vegetables, meat, and fish results in the production of health-boosting compounds.[54] Some examples of health-boosting compounds formed as the result of fermentation of different types of raw food matrices are given in Table 8.1.

TABLE 8.1 Health Boosting Compounds in Fermented Food Products.

Food material	Microorganism involved in fermentation	Useful compounds formed	Health boosting activity	Refs.
Cashew apple juice	*Lactobacillus plantarum*	Phenolic compounds, folate	Antioxidant activity	48
Fermented sausages	*Bifidobacterium longum*	Phenolic compounds, folate	Antioxidant activity	87
Goat milk	*Pediococcus pentosaceus*	Phenolic compounds, folate	Antioxidant activity	54
Lupin tempeh	*Propionibacterium freudenreichii* and *Rhizopus oryzae*	Folate (vitamin B_9), vitamin K, riboflavin	Increase vitamin content	81
Milk	*Lactobacillus acidophilus*	Phenolic compounds, folate	Antioxidant activity	54
Whole meal wheat Sourdough	*Lactobacillus brevis*	Gama amino butyric acid	Prevents high blood pressure	64

8.4.2 ENZYMES

Generally, throughout the world, the food manufacturing industries are using enzymes that are made using GM organisms, for example, carbohydrases and proteases. The cloning of genes responsible for enzyme production ensures improved enzyme production within a short period. Such enzymes are useful in processing various food products such as curd, cheese, and flavors.[86] Enzymes prove to be helpful in food preservation. Some enzymes are bacteriolytic, and they kill the bacteria by lysis of the bacterial cell wall. Example: Enterolisin-A.[94] Enzymes obtained from milk and eggs are used as biopreservatives.[52]

Some of the antimicrobial peptides obtained from milk proteins such as Kappacin, k-casecidin, and lactoferricin help preserve food.[68] Some of the latest research showed that hen egg proteins, such as ovotransferrin, cystatin, and ovalbumin could produce bioactive egg-derived proteins. These are obtained through enzymatic hydrolysis and fermentation.[97] *Lactobacillus plantarum* was used to carry out the hen eggshell membrane's fermentation to synthesize functional egg-derived proteins that inhibit different food spoilage microorganisms.[39] The egg-derived proteins can substitute harmful nitrites used for food preservation.[97] Currently, the use of lipolytic enzymes obtained from microorganisms is used for bioremediation and bi-product manufacturing. Examples are synthesis of biofuel from waste cooking oil using lipase from *Candida antarctica* and synthesis of biofuel from coconut oil mill waste using lipase from *Staphylococcus pasteuri.*[47]

Enzymes find applications in different food industries, such as, beverage industries, baking industries, brewing industries, fruit juice processing, and dairy industries. They are used in various steps of food processing. They are used for preliminary treatment, manufacturing, purification, etc.[67] Lipases, amylases, proteases, cellulases, and pectinases are used for the clarification of fermented beverages. Transglutaminase is used to strengthen the texture of homogenized noodles and sausages.[9] α-Amylases break down starch by breaking down the α-1,4 glycosidic bond of starch. It results in the liberation of small chains of dextrin.[83]

The α-amylases help in starch liquefaction. It results in the conversion of starch into glucose and fructose syrups. This process is called saccharification. The major microbial sources of α-amylases are *Bacillus licheniformis*, *Bacillus amylolique faciencs*, and *Bacillus stearothermophilus.*[9] Glucoamylases help in the production of high fructose syrups and glucose syrups. Dextrins are metabolized by glucoamylase into alcohol-free beer and fermentable sugars.[11] Xylanases and cellulases help to improve bread quality by enhancing the elasticity and strength of the gluten network. Xylanases are also used to increase the crumb structure, bread volume, and reduce the stickiness.[9] Xylanases are synthesized intracellularly or extracellularly by fungi, bacteria, plant cells, and animal cells.[51] For example, xylanases are synthesized by *Trichoderma* and *Aspergillus*.

Tannases, pectinases, and cellulases find extensive application in the fruit processing industry.[9] Pectinases carry out the depolymerization of pectic materials. It helps in achieving clarification and stability.[46] Pectinase reduces the bitterness, viscosity, and cloudiness of fruit juices. Pectinases enhance the compactness of the pulp. Pectinase reduces the stability of pectin.[5] Tannase reduces the haze content and bitter taste of fruit juices.[9] *Lactobacillus*, *Paecilomyces*, *Aspergillus*, and *Bacillus* synthesize tannase to act over an extensive temperature and pH.[9] α-Amylases are used in the beer processing industries to enhance the number of carbohydrates produced during fermentation. This will lead to the production of beer with higher alcohol content.

Proteases reduce the viscosity of beer by decreasing the solubility of the protein. Proteases provide amount of amino acids required for yeast growth. Rennet finds application in the dairy industry for the manufacturing of cheese. It helps in the coagulation of milk protein. Lipases are used in the cheese processing industries to impart lipolytic taste.[9] Peptidases help to hydrolyze peptides that impart a bitter flavor to the cheese. It reduces the bitterness and helps maintain the conventional cheese aroma.[15] Lactases break down lactose into galactose and glucose. This will enhance the sweetness and solubility of different dairy products. Many people are deficient in lactase; hence, they are unable to use milk sugar.[65] Lactase is used in the dairy industry to make lactose-free dairy products. The hydrolysis of lactose will be of assistance to lactose-intolerant people.[24]

Improvements have been made in the enzymes and various strains of microorganisms through the implementation of genetic engineering to accomplish the requirements of the food manufacturing industry.[43] The use of genetically engineered microbes will considerably decrease the cost and labor needed to extract and refine the enzyme.[70] Protein engineering approaches give rise to developing and designing customized enzymes with diverse functionalities.[10] There are two approaches: (1) Rational design using site-directed mutagenesis consists of replacing a particular amino acid with the standard amino acid at a specific site. This approach requires prior details about the target protein[43]; and (2) The direct evolution approach consists of establishing genome mutation randomly using DNA shuffling or chemical intervention techniques. This approach does not require prior details of the protein.[9]

The immobilization of enzymes can improve the stability of enzymes. It enhances selectivity, activity, and the operational and thermal stability of enzymes. Hence the enzymes become competent enough to be used in the unfavorable industrial environment. Immobilization helps in making the downstream processing of the product easier.[9] Immobilized enzyme bioreactors are consistently employed in the food manufacturing industries.[70] Production of high-fructose corn syrup using glucose isomerase and inverted sugar syrup production using invertase was reported using immobilized enzymes. The manufacturer should carefully select the supporting matrix for the immobilization of enzymes. For example, lipases had undergone hyper-activation due to the immobilization of lipase upon hydrophobic matrix.[53]

8.4.3 BIOPRESERVATIVES

Biopreservation method is used to improve food storage by using natural or controlled microbiota or antimicrobials. The foods obtained from the animal source are extremely perishable owing to their neutral pH, high moisture content, and high nutritional content. Hence it is necessary to preserve these foods using appropriate methods for ensuring food quality and safety. In biopreservation, the antimicrobial capability of naturally occurring microorganisms and their metabolites is used. Fermentation results in the synthesis of several useful products that decrease food spoilage and keeps the food free from harmful microbes and their metabolites.[85]

Lactic acid bacteria (LAB) are one of the main microorganisms used in the fermentation process. It results in producing organic acids that act as an antimicrobial agent and provide unique texture and flavor to the food products. LAB comprises of *Lactococcus*, *Lactobacillus*, *Streptococcus*, *Leuconostoc*, *Pediococcus*, *Enterococcus*, *Aerococcus*, *Carnobacterium*, *Tetragenococcus*, *Oenococcus*, *Weisella*, and *Vagococcus* genera of microorganisms.[85] LAB possesses unique characteristics such as proteolytic activity, resistance to bacteriophage, citrate and lactose fermentation, high resistance to lyophilization and freezing, production of polysaccharides, synthesis of antimicrobial compounds, and the ability for adhesion and colonization of the digestive mucosa.

About 60% of food industries use fermentation to ensure quality, uniformity, and safety of food. They are mainly utilized in the processing of dairy, meat, and vegetable products. For example, Nisin synthesized by *Lactococcus lactis* is used in the dairy industry,[85] and sakacin synthesized by *Lactobacillus sakei* is used in the meat industry.[89] The biopreservative used in food processing should be generally recognized as safe and should not have any toxicity or pathogenicity.[85] Two types of biological agents are commonly used in the food industries, and they are starter cultures and protective cultures. Starter culture consists of a collection of microorganisms used to start the process of fermentation. It assists the production of certain compounds essential for providing texture and flavoring compounds in fermented products. The purpose of protective culture is to manage the antimicrobial activity and inhibit the growth of harmful disease-causing microorganisms on food.

Nevertheless, both the starter and protective cultures are typically used in the food processing industries.[85] Bacteriocins are low molecular weight peptides synthesized by bacteria. They possess antimicrobial activity.[52] They act as biopreservatives. Bacteriocins can be added directly to food products. The bacteriocin can be synthesized in situ by inoculating the bacteriocin synthesizing microbial strain in the food and providing conditions that support bacteriocin synthesis.[66] Some of the commonly used bacteriocins in the food industries and their uses are given in Table 8.2.[85]

TABLE 8.2 Bacteriocins Commonly Used in the Food Industries.

Type of food industry	Name of bacteriocin	Microbe used to produce bacteriocin	Uses
Dairy	Nisin	*Lactococcus lactis*	Inhibits the growth of *Clostridium botulinum* and *Listeria monocytogenes* in cheese.
	Pediocin AcH	*Pediococcus*	Active against *Bacillus cereus*, *Staphylococcus aureus*, and *Listeria monocytogenes* in milk and cheese.
Fish	Nisin	*Lactococcus lactis*	Effective against *Listeria monocytogenes* in ice-cold defrosted salmon
Meat	Enterocin and Sakacin	*E. faecalis* *Lactobacillus sakei*	Moderately effectual against *Listeria monocytogenes* and other disease causing microorganisms.
Vegetables	Enterocin	*E. faecalis*	Inhibits the growth of *Bacillus cereus* in rice and vegetables and effective against the bacterium *Alicyclo bacillus acidoterrestris* responsible for food spoilage.

8.4.4 PROBIOTICS, PREBIOTICS, AND SYNBIOTICS

Probiotics are described as "*live microorganisms which when administered in adequate amounts, confer a health benefit to the host.*" Some of the essential features of probiotic microorganisms are nonpathogenic nature, acid and bile tolerance, ability to adhere to the gut lining, short generation time, and genetically stable. Moreover, probiotic strain should have the capability to survive processing conditions.[63]

Some of the probiotic microorganisms are *Lactobacillus reuteri*, *Bifidobacteria*, *Lactobacillus rhamnosus*, *Lactobacillus casei*, *Lactobacillus acidophilus*, *Bacillus coagulans*, and so on. FAO/WHO defines prebiotics as a "nonviable food component that confers a health benefit to the host associated with modulation of the microbiota."[63] The essential features of a prebiotic are its resistance to the acids present in the gut and the intestinal enzymes, nonabsorbability of the prebiotic by the upper intestinal tract, and easy fermentability of the prebiotic by probiotic microorganisms. Inulin, pectin, and galactooligosaccharides are few examples of prebiotics. At present, the food processing industries need simple, cost-effective, sustainable, and highly efficient techniques for the large-scale manufacturing and utilization of prebiotics.

Enzymes from different sources (such as plants and microbes) are useful in the synthesizing of prebiotics. These enzymes can be engineered for improving the reaction yield, which in turn enhances the quality of the product formed. The utilization of genetically engineered microorganisms can further improve the synthesis of oligosaccharides during large-scale commercial manufacturing.[44]. Synbiotics refer to the combination of probiotic and prebiotic that results in enhancing the survival rate of probiotic microorganisms in the gastrointestinal (GI) tract.[63]

The commercial synthesis of probiotics incorporated products requires a sophisticated manufacturing process to guarantee high yield, good stability, and absence of specific food allergens. The probiotics can be successfully incorporated in an extensive variety of food products as per the consumer needs through the correct choice of the production process, the right selection of probiotic strain, and product formulation. The probiotics are added as frozen pellets and not as a lyophilized powder in the fermented milk products. It will ensure uniform mixing of the product and decreases mixing time.[28]

8.4.5 SINGLE-CELL PROTEINS (SCPS)

The increase in proteinaceous food requirement necessitated the development of new sources of protein for supplementing the traditional protein sources.[88] The increased production of dairy and meat products is not a sustainable solution to meet this increased protein-rich food requirement. Some of the plants are abundant sources of proteins, for example, beans.

However, available cultivable lands will become inadequate to meet the worldwide need for protein.[72] Instead of growing whole animals for obtaining meat proteins, in vitro technique may be used to synthesize meat protein in a relatively effective manner.[41] SCPs prove to be beneficial in meeting the increased requirement of protein food. The use of SCP is also another inventive solution for the global food problem.

SCP is a rich source of protein, and it also consists of carbohydrates, fats, minerals, vitamins, and nucleic acids.[88] SCP can be used directly, and it can also improve the properties of fermented food.[72] SCP manufacturing includes formulation of nutrient media probably out of wastes, growing suitable microbe using the fermentation process, processing products obtained from fermentation to separate and concentrate the SCP, drying, and transforming SCP into useful products. Usually, substances belonging to the food category are employed as substrates for SCP production meant for human intake.[72] Some of the examples are the production of SCP by *Aeromonas hydrophylla* using lactose as the substrate, *Aspergillus fumigatus* using maltose and glucose as the substrate, *Saccharomyces cerevisiae* and *Chlorella pyrenoidosa* using CO_2 as substrate.[57]

8.4.6 FUNCTIONAL FOODS

The Functional Food Center (Dallas, Texas, USA) defines functional food as "*natural or processed food that contains known biologically-active compounds which when in defined quantitative and qualitative amounts provides a clinically proven and documented health benefit, and thus, an important source in the prevention, management, and treatment of chronic diseases of the modern age.*"[22] There is an increase in the demand for functional food incorporated with probiotic over the last decade since it is proved that probiotic organisms are capable of providing health benefits to the consumers. If probiotic products are prepared with the required viable probiotic dose, it can assure the necessary health benefit to the consumer.

Maintenance of the viability of probiotic organisms, while incorporating in various food products is one of the most common problems faced by the food industries. The most appropriate solutions for this problem are microencapsulation and the formulation of synbiotic capsules incorporating both probiotics and prebiotic substances. This helps in retaining high levels of the viability of the probiotic. They can improve the rate of survival of the probiotic during the processing of food, storage of food, and within the GI tract.[29]

Functional foods that are treated with enzyme can have improved bioavailability and hence, it is having direct influence on the bioefficacy of the product. The capability of the extremozymes for catalyzing the reactions under severe environmental circumstances can be used for the economical production of functional foods more efficiently. The development of novel methods for the immobilization of the enzyme can offer the added benefit of easy separation, continual use, and enzymes stability in the reaction blend. Hence the utilization of enzymes in the industries will increase.[20]

8.4.7 BIOTECHNOLOGICAL PRODUCTION OF FOOD INGREDIENTS

Food ingredients are employed to enhance the dietary value, alter the consistency, and improve flavor. Microbial fermentation helps manufacture the flavors, amino acid supplements, flavor enhancers, and vitamins.[33] The food additives, such as organic acids, amino acids, and flavoring agents, are microbial metabolites formed due to fermentation. Hence the fermentation by microbes is used by the industries for the manufacture of food ingredients. The synthesis of these food ingredients can be enhanced by understanding and redirecting the metabolic pathway related to the fermentation. Some of the food ingredients are sucrose that acts as a sweetener, beta-carotene that acts as a color additive, and Monosodium glutamate that acts as a flavor enhancer.[33]

Plants and animals are good sources of commercially essential compounds, such as polysaccharides, flavors, and fatty acids. One of the main focuses of recent research was on the extraction of these compounds. These compounds have the benefit of being labeled as natural. These resources are not sufficient to meet the requirements of the food industry due to seasonality as well as low yield. Bioprocess has proved to be one of the better and safer methods to obtain these natural products. Consumers demand natural food ingredients, which has inspired the commercial food companies to use bioprocess to get useful and economically feasible products.[82]

8.4.8 ROLE OF BIOSURFACTANTS IN THE FOOD INDUSTRY

Biosurfactants are collection of surface-active substances that are synthesized by microorganisms. They minimize the interfacial and surface tension since they can gather in between the liquid–liquid interface or the air–liquid interface.[59] It has an amphiphilic nature. The main classes of biosurfactants are glycolipids, lipopeptides, and polymeric surfactants. Examples of glycolipids include rhamnolipids synthesized by *Pseudomonas aeruginosa* and sophoro lipids synthesized by *Torulopsis bombicola.* Examples of lipopeptides are surfactin synthesized by *Bacillus subtilis* and lichenysin synthesized by *B. licheniformis.* Examples of polymeric surfactants are liposan produced by *Candida lipolytica* and mannoproteins synthesized by *S. cerevisiae.*[59]

The biosurfactants are preferred over chemically manufactured surfactants due to the following reasons. The raw materials used are renewable, biodegradable, lesser toxic, and biocompatible. Biosurfactants have a diversity of biological activity and structure. Biosurfactants exhibit stability under extreme temperature, salinity, and pH. In the food industries, biosurfactants are used as a cleaning agent and as a food additive.

Generally, surfaces of glass, plastic, wood, metal, food products, and food processing equipment are susceptible to biofilm formation. Biofilm is the diverse collection of microorganisms ingrained within an extracellular polymeric matrix that can attach to abiotic and biotic surfaces. Biofilm protects the microorganisms from physical and chemical agents. Biofilm formation can worsen the quality of food products, and it can be hazardous to public health. The formation of biofilm will adversely affect the processing of food. The presence of disease-causing microbes such as *Salmonella spp, Escherichia coli,* and *Listeria monocytogenes* will lead to the spread of foodborne diseases.[59]

Chemical sanitizers are usually less potent in eliminating biofilms because they are protected by the extracellular polymeric matrix.[59] Biosurfactants prove to be helpful to the food processing industries for getting rid of biofilm-associated microorganisms. Two possible mechanisms are there for the removal of biofilms by biosurfactants.

- In one possible mechanism, the biosurfactant molecule crosses the biofilm matrix via water channels. The biosurfactant molecule will then get adsorbed to the interface, thereby decreasing the interfacial tension between biofilm and solid substrate. It will also result in the reduction of the attractive forces involve on the bacterial linkage and this results in the separation of the biofilm.
- According to the second mechanism, the biosurfactants will function as a wetting medium that supports the solvability of extracellular polymeric matrix in an aqueous medium.

As far as the antiadhesive properties of the biosurfactants are concerned, the treatment of surfaces with biosurfactants will modify the physicochemical characteristics, such as electrostatic forces, free energy, and hydrophobicity, which are directly responsible for the adhesion of microorganisms.[21] Food additives are the materials incorporated into the food to carry out definite technological tasks as per the European Food Safety Authority. Some of the examples of biosurfactants and their applications are given in Table 8.3.

Biosurfactants can also be manufactured from industrial food wastes, thus reducing waste treatment costs of food industrial waste residues. For example, *P. aeruginosa* 7815 was used to synthesize biosurfactant rhamnolipids using waste cooking oil as the source of carbon.[77]

TABLE 8.3 Applications of Biosurfactants in Foods.

Name of biosurfactant	Microbe producing the biosurfactant	Role	References
Biosurfactant	*Candida utilis*	Emulsifier	16
Exopolysaccharides	*Rhodobacter johrii*	Emulsion stabilizer in food industry	73
Glycolipids	*S. cerevisiae*	Replace yolk in cookie formulation	71
Rhamnolipids	*Pseudomonas aeruginosa*	Disruption of biofilm	35
Rhamnolipids	*Pseudomonas aeruginosa*	Cleaning mediator	55
Sophorolipids	*Torulopsis bombicola*	Antibacterial	19
Surfactin	*Bacillus subtilis*	Antiadhesive	21
Surfactin	*Bacillus subtilis*	Antimicrobial	38

8.4.9 BIODEGRADABLE PACKAGING FOR FOODS AND EDIBLE FILMS

Plastics are one of the most prominently used packaging materials. At present, the worldwide manufacture of plastics is around 320 million tons per year, with a growing demand that shows its extensive utilization in various fields.[61] There is an increasing demand for appropriate packaging material with good biodegradability and improved mechanical, thermal, and barrier characteristics to protect the food from contamination and avoid food materials loss. Compostability is also an important feature required for polymers since its reprocessing needs energy.[3] Bio-based polymers obtained from inexhaustible natural resources can be used for the manufacturing of biodegradable packaging materials.

The major types of bio-based packaging polymers used to synthesize smart packaging films are microbial polymeric materials, wood-based polymeric materials, and protein-based polymeric materials. Examples of microbial polymeric materials are polylactic acid, polyhydroxyalkanoates, and exopolysaccharides. Examples of wood-based polymeric materials are hemicellulose, starch, cellulose, and lignin. Examples of protein-based polymeric materials are keratin, gelatin, soy protein, whey protein isolate, and wheat gluten.[40,69] These bio-based packaging polymers can be used for manufacturing packaging films and coatings that resist the transfer of gases and the entry of pathogenic microorganisms. The barrier, antifungal, and antibacterial traits of the bioplastic packaging can be improved by incorporating bio-based strengthening substances such as essential oils extracted from plants.[80]

Bio-based packaging materials and nano-biocomposite packaging materials will substitute nonbiodegradable packaging materials in the food industry. These bio-based packaging materials are found to have features comparable to petroleum-based plastics. They have good biodegradability and compostability.[6] Edible films help in enhancing the storage life of food. It also helps in reducing the application of nonbiodegradable materials.[84]

Edible films can effortlessly be dissolved within the mouth and instantaneously liberate active components.[56] At present, several works are being reported on the application of edible films for delivering probiotics.[23] The entrapment of probiotic microorganisms within the edible film improves the survival rate of microorganisms.[56] A composite edible film fabricated using sodium alginate, pectin, and casein was used to carry the probiotic microorganism *Enterococcus faecium Rp1*.[56]

8.4.10 GM FOOD

Biofortification is one of the methods used wherein the nutritional characteristics of the food crops are enhanced by incorporating nutrients and other health-supporting properties with the help of conventional plant breeding, agronomic practices, or modern biotechnological methods such as genetic modification.[34] GM foods are those foods produced by organisms whose genetic composition was changed using genetic engineering tools.[34] Genetic composition changes help in increasing the level of micronutrients in the food. There are two categories of GM crops, i.e., first-generation and second-generation.

- The first-generation GM crops are farmer oriented, and its objective is to enhance the agricultural output of the food crops.

- The second-generation GM crops are cultivated to augment the nutritional quality or processing characteristics of the crops.[34]

Some of the examples of GM crops are given in Table 8.4. Some of the enzymes obtained from GM organisms are employed in the food industries. For instance, *Pichia pastoris* is GM by cloning and expressing yak active chymosin, and it is used for the manufacturing of rennet.[49] The synthesis of catalase in *Bacillus subtilis* was increased by 20-fold by overexpressing it with the *Kat* A gene responsible for catalase production.[93]

TABLE 8.4 Examples of Genetically Modified Crops.

Type of food	Characteristics of the product	Purpose	References
Banana	Supplemented with pro-vitamin A and Iron	To combat nutrient deficiency	34
Potato	Fortified with protein		
Rice	Fortified with β-carotene		
Sorghum	Fortified with Fe, Zn and protein		
Canola	High amount of omega-3 fatty acid		
Alfalfa	DNA silencing using Intragenesis	Decrease the amount of lignin	92
Apple	Expression of DNA using Intragenesis	Protect apple from scab disease	42
Barely	Over expression of DNA using Cisgenesis	Enhanced the activity of phytase	37
Durum Wheat	Expression of DNA using Cisgenesis	Enhancement of baking characteristics	32
Potato	DNA silencing using Intragenesis	Decrease the acrylamide content in french fries	18

8.4.11 BIOSENSORS

Campylobacter spp., *Salmonella* spp., *L. monocytogenes*, strains of *Staphylococcus aureus*, *Bacillus cereus*, and *Clostridium* spp.[76] are responsible for producing toxins and various cell metabolites capable of causing fatal diseases. The factors such as increased cost, analysis time, and skilled personnel requirements restrict the application of existing identification techniques such as culture-based conventional microbiological methods, enzyme-linked immunosorbent assay, microarray-based techniques, and polymerase chain reaction.[76]

Biosensor-based technology leads to rapid, portable, and sensitive detection of pathogenic microbes in the food industry. In addition to that, online biosensor technology can act as a tool for internal process control in the food industry to ensure quality control.[76] A biosensor is fundamentally an analytical appliance used for the quantification of the target molecule within a sample. It consists of a biorecognition component such as an antibody, aptamer, or enzyme specific to the target molecule.[58] The biosensors used to identify disease-causing microbes in food are based on either antigen-antibody or nucleic acid or bacteriophage.[76] A biological or physiochemical signal is formed due to the molecular detection of the target molecule by the biorecognition element. The transducer converts this signal into a measurable quantity. The transducers are either electrical (capacitance, impedance, voltammetry) or optical (fluorescence, colorimetric) or thermometric, piezoelectric, magnetic, or micromechanical.[58] Examples of different food-borne pathogen and the limit of detection using biosensors are given in Table 8.5.

TABLE 8.5 Examples of Biosensor for the Detection of Food-Borne Pathogens.

Pathogenic microorganism	Food sample	Type of biosensor	Limit of detection (CFU/mL)	References
Salmonella typhimurium	Apple juice	Aptosensors	10^2–10^8	78
Salmonella gallinarum	Eggs and chicken meat	Electrochemical immunosensor	3.0×10^3	26
Salmonella enteritidis	Milk	Surface plasmon resonance	1×10^2	25
Staphylococcus aureus	Raw milk	Colorimetric method	10^1–10^6	96

8.4.12 APPLICATIONS OF FOODOMICS IN FOOD SAFETY AND NUTRITION

Food-borne diseases are one of the major diseases of concern worldwide. Parasites, viruses, and bacteria are responsible for creating such illnesses. Omics technologies such as proteomics, transcriptomics, and genomics prove useful in changing the perspective for detecting, preventing, and treating food-borne pathogens. Foodomics is described as "*a discipline that studies the food and nutrition domains through the application of omics technologies.*"[75] The genome sequencing method helps improve the detection of the epidemic and its cause.

Moreover, genome sequencing also helps in creating large quantities of food-borne pathogen genome sequence data. This data is obtainable for data mining. This will consecutively assist enhanced source allocation and give novel insight about biology and transmission of food-borne pathogens.[8]

The application of molecular surveillance, for example, PulseNet, proves to be useful by enhancing the capability to identify food-borne ailment outburst at a faster rate, resulting in lesser outbreaks with a smaller number of cases.[12] It also helps recognize the epidemic sources, thus giving details about the routes of food-borne illness transmission. It will support the government endeavors to limit the transmission of food-borne pathogens. However, there are also shortcomings in the application of molecular tools. For instance, rapid identification techniques will mainly result in a reduction of pathogen isolation. It will consequently decrease the capability to subtype due to the shortage of pure isolates from food samples or human specimens. This shortcoming can be overcome to a certain extent by using omics tools, for instance, genomics, proteomics, metagenomics, and metabolomics, which can considerably enhance the ability to avoid food-borne disease outbursts.[8]

Whole genome sequencing can overcome the inadequacies of subtyping techniques such as pulsed-field gel electrophoresis and multiple-locus variable-number tandem repeat analysis to inspect and identify food-borne illness outbursts. For example, for extremely clonal pathogens, the use of pulsed-field gel electrophoresis and multiple-locus variable-number tandem repeat analysis will mostly have the only limited discriminatory ability.[4] However, whole genome sequencing can significantly increase discrimination, for example, the use of whole genome sequencing-based characterization of *Bacillus anthracis* isolates associated with the 2001 bioterrorism occurrence in the USA. Furthermore, DNA-based subtyping techniques such as multilocus sequence typing, mostly cannot distinguish between intimately related *Salmonella* serovars. At the same time, whole genome sequencing can make a distinction between intimately related *Salmonella* serovars. The advanced genome sequencing techniques can also facilitate the fast generation of data. Hence, it can be used to develop an identifying method for a novel epidemic strain.[8]

Metagenomics also acts as an important tool to diagnose disease and to test food safety. Metagenomics is the culture-independent examination of the genetic matter of microbes present in a specified environment. There are a lot of uses of metagenomics in the field of food safety. One such application is identifying new and nonculturable mediators in clinical specimens responsible for causing food-borne illness. Another application is the characterization of indicator and pathogenic organisms in foods and food products.

Healthy gut microbiomes will protect against infection with food-borne pathogenic microorganisms. Metagenomics can be used to study the microbiome of humans and animals. The metagenomics studies are also reported to identify microorganisms involved at different stages of fermentation.[8]

There are some difficulties faced in the application of metagenomics for the identification of food-borne pathogenic microorganisms. One of the challenges is that the metagenome sequencing detects the DNA from both dead and alive organisms. The food samples can be categorized as positive for dead cells of food-borne pathogenic microorganisms. For example, the food samples were evaluated following pasteurization. Metatranscriptomics can be used to overcome this difficulty. However, metagenomics and metatranscriptomics created a large amount of sequence data sets for a particular food. These data will probably include some of the information related to food safety, for example, the existence of genes with antimicrobial resistance.[8]

Simultaneously, transcriptomics, proteomics, and metabolomics offer prospects for creating advanced methods to control food-borne pathogenic microorganisms during production, storage, distribution, and food consumption. It is essential to find out the physiological condition of pathogenic microorganisms present in foods. The gene modulation and protein expression as a result of stress show either the activation or the suppression of a particular physiological reaction. This will be useful for finding out the physiological condition of the pathogenic microorganism under different conditions.[8] The transcriptional response of *E. coli* O157:H7 to the antimicrobial substance cinnamaldehyde was reported in the literature. The study showed that the primary reaction of the pathogenic microorganism was the activation of the oxidative stress reaction. Finally, the antimicrobial stress was overcome by the transformation of cinnamaldehyde into cinnamic alcohol.[90]

Transcriptomics and proteomics information will help design novel control approaches for food-borne pathogens. For instance, the information from these studies can be used to find out novel compounds that distinctively interfere with pathways critical for the survival of pathogenic microorganisms in foods. For example, fluoro-phenyl-styrene-sulfonamide immensely slows down the activation of the general stress response sigma factor in *L. monocytogenes*.[62] The transcriptomics information of pathogens under various environmental stresses are also used to find out biomarkers associated with the resistance feature of the pathogenic microorganism. Transcriptomics and proteomics information may be incorporated within mathematical models to estimate the performance of microorganisms to improve control measures.[1]

Metabolomics is applied for various purposes, and it includes the study of tiny molecules in a tissue.[75] Metabolomics studies are also used to discover potential biomarkers in nutritional interference.[14] They are used for predicting the risk during the shift from a healthy condition to an unhealthy condition.[91] They are employed as potential dietary biomarkers[36] and illustrate the detailed facts regarding food composition.[79] These biomarkers also help to analyzes pathological or physiological response of an individual to food components or diet. In short, this technique will help to give information of dietary deficiency, provide variations in dietary requirements among individuals and thereby help to design personalized diet for a particular person. The above tool can also use to recommend a diet for a particular population based on their common genetic factors and food habits.

New phages can be designed to either biocontrol or detect pathogenic microorganisms by using synthetic biology and genomics.[2] Genetically engineered phages are designed to detect pathogenic microorganisms, for example, identification of *B. anthracis.*[74] Bioinformatics tools that are less computationally rigorous, quicker, and effortless have to be developed to assist further application of omics tools. Furthermore, the fabrication of properly authorized frameworks for applying omics information is also essential to help the government and industries for the use of omics tools.[8]

8.5 BENEFITS AND CONCERNS OF APPLYING BIOTECHNOLOGICAL TECHNIQUES IN FOOD SECTOR

The application of biotechnological processes in food processing industries can provide the following benefits:

- The application of biotechnology has enabled the production of drought-resistant and disease-resistant crops, thereby resulting in a remarkable increase in the yield of food crops.
- The production of GM crops with biotechnological methods has enabled the cultivation of crops at a lower production cost without applying herbicides, pesticides, and chemicals. Hence these GM crops prove to be environment friendly.
- The GM crops produced using biotechnological methods have more nutritional value compared with ordinary crops. It makes the GM crops healthier for consumption.
- Biotechnology has also enabled to enhance the storage life of food crops, thereby helping the food to stay fresher for a longer time.[50]
- The microorganisms can use waste as a substrate for the synthesis of SCP. Therefore, it will help in making the environment clean.[88]

- A biosensor is portable, and it helps in quick and precise identification of microorganisms within the food processing industries.[76]
- Enzymes can be used as an eco-friendly additive in food manufacturing industries to enhance the uniqueness of the product. [9]

The concerns about applying biotechnological techniques in food sector especially GM crops are related to potential risk related to health of human beings. The risk factor related to unapproved GM food includes allergic reactions, antibiotic resistance, alteration in nutritional composition, and toxin formation.[50]. In additions to that, higher the cost of GM seeds is also a major concern for the producers.

8.6 SUMMARY

The application of biotechnological techniques in food industries has an enormous impact on society. Nowadays, customers demand good quality food, which is natural, appropriate, and safe. Biotechnology finds wide application as a diagnostic tool for examining food safety and hence prevent food-borne diseases. Currently, biotechnology finds many applications in the food industry, for instance, the use of genetic engineering, cell engineering, and protein engineering to enhance the quality of raw materials used in food processing. This chapter gives an overview of the various applications of biotechnology in food processing, the scope and significance of food biotechnology, the advantages, and limitations of using biotechnology in food processing industries. It is proved that biotechnological tools in food processing are vital in enhancing food quality and food security.

KEYWORDS

- **biosensors**
- **fermented foods**
- **food biotechnology**
- **food ingredients**
- **foodomics**
- **genetically modified**

REFERENCES

1. Abee, T.; Wels, M.; de Been, M.; den Besten, H. From Transcriptional Landscapes to the Identification of Biomarkers for Robustness. *Microb. Cell Factor.* **2011,** *10* (Suppl 1), S9–S13.
2. Agapakis, C. M. Designing Synthetic Biology. *ACS Synth. Biol.* **2014,** *3* (3), 121–128.
3. Al Hosni, A. S.; Pittman, J. K.; Robson, G. D. Microbial Degradation of Four Biodegradable Polymers in Soil and Compost Demonstrating Polycaprolactone as an Ideal Compostable Plastic. *Waste Manag.* **2019,** *97*, 105–114.
4. Allard, M. W.; Luo, Y.; Strain, E.; Pettengill, J.; Timme, R.; Wang, C.; Li, C.; Keys, C. E.; Zheng, J.; Stones, R.; Wilson, M. R.; Musser, S. M.; Brown, E. W. On the Evolutionary History, Population Genetics and Diversity among Isolates of *Salmonella enteritidis* PFGE Pattern JEGX01. 0004. *PLoS One* **2013,** *8* (1), E-article 55254.
5. Amin, F.; Bhatti, H. N.; Bilal, M.; Asgher, M. Purification, Kinetic, and Thermodynamic Characteristics of an Exo-polygalacturonase from *Penicillium notatum* with Industrial Perspective. *Appl. Biochem. Biotechnol.* **2017,** *183* (1), 426–443.
6. Asgher, M.; Qamar, S. A.; Bilal, M.; Iqbal, H. M. N. Bio-based Active Food Packaging Materials: Sustainable Alternative to Conventional Petrochemical-based Packaging Materials. *Food Res. Int.* **2020,** *137*, E-Article 109625.
7. Bajpai, P.; Bajpai, P. K. Repeated Fed Batch Fermentation for Single Cell Protein Production from Prehydrolysate of a Pulp Mill. *Enzyme Microb. Technol.* **1988,** *10* (5), 280–283.
8. Bergholz, T. M.; Switt, A. I. M.; Wiedmann, M. Omics Approaches in Food Safety: Fulfilling the Promise? *Trends Microbiol.* **2014,** *22* (5), 275–281.
9. Bilal, M.; Iqbal, H. M. N. State-of-the-Art Strategies and Applied Perspectives of Enzyme Biocatalysis in Food Sector—Current Status and Future Trends. *Critic. Rev. Food Sci. Nutr.* **2020,** *60* (12), 2052–2066.
10. Bilal, M.; Zhao, Y.; Rasheed, T.; Iqbal, H. M. N. Magnetic Nanoparticles as Versatile Carriers for Enzymes Immobilization: A Review. *Int. J. Biol. Macromol.* **2018,** *120* (Part B), 2530–2544.
11. Blanco, C. A. I.; Caballero, R.; Barrios; Rojas, A. Innovations in the Brewing Industry: Light Beer. *Int. J. Food Sci. Nutr.* **2014,** *65* (6), 655–660.
12. Boxrud, D.; Monson, T.; Stiles, T.; Besser, J. The Role, Challenges, and Support of Pulsenet Laboratories in Detecting Foodborne Disease Outbreaks. *Public Health Rep.* **2010,** *125* (2), 57–62.
13. Brennan, L. Metabolomics in Nutrition Research: Current Status and Perspectives. *Biochem. Soc. Trans.* **2013,** *41* (2), 670–673.
14. Brennan, L. *Metabotyping: Moving Towards Personalized Nutrition*; Woodhead Publishing: London, 2015; pp 137–144.
15. Budak, Ş. Ö.; Koçak, C.; Bron, P. A.; de Vries, R. P. Role of Microbial Cultures and Enymes during Cheese Production and Ripening. Chapter 10. In *Microbial Cultures and Enzymes in Dairy Technology*; Budak, Ş. Ö., Akal, H. C., Eds.; IGI Global: Pennsylvania, 2018; pp 182–203.
16. Campos, J. M.; Stamford, T. L. M.; Rufino, R. D.; Luna, J. M.; Stamford, T. C. M.; Sarubbo, L. A. Formulation of Mayonnaise with the Addition of a Bioemulsifier Isolated from *Candida utilis*. *Toxicol. Rep.* **2015,** *2*, 1164–1170.

17. Capozzi, F.; Bordoni, A. Foodomics: A Comprehensive Approach to Food and Nutrition. *Genes Nutr.* **2013,** *8* (1), 1–4.
18. Chawla, R.; Shakya, R.; Rommens, C. M. Tuber-Specific Silencing of Asparagine Synthetase-1 Reduces the Acrylamide-forming Potential of Potatoes Grown in the Field without Affecting Tuber Shape and Yield. *Plant Biotechnol. J.* **2012,** *10* (8), 913–924.
19. Chen, J.; Lü, Z.; An, Z.; Ji, P.; Liu, X. Antibacterial Activities of Sophorolipids and Nisin and Their Combination against Foodborne Pathogen *Staphylococcus aureus*. *Eur. J. Lipid Sci. Technol.* **2020,** *122* (3), E-article 1900333.
20. Chourasia, R.; Phukon, L. C.; Singh, S. P.; Rai, A. K.; Sahoo, D. Role of Enzymatic Bioprocesses for the Production of Functional Food and Nutraceuticals. Chapter 15. In *Biomass, Biofuels, Biochemicals: Advances in Enzyme Catalysis and Technologies*; Singh, S. P.; Pandey, A.; Singhania, R. R.; Larroche, C., Li, Z., Eds.; Elsevier: Amsterdam, Netherlands, 2020; pp 309–334.
21. de Araujo, L. V.; Guimarães, C. R.; Marquita, R. L. D. S.; Santiago, V. M. J.; de Souza, M. P.; Nitschke, M.; Freire, D. M. G. Rhamnolipid and Surfactin: Anti-adhesion/antibiofilm and antimicrobial effects. *Food Control* **2016,** *63*, 171–178.
22. de LeBlanc, A. D.; Luerce, T. D.; Miyoshi. A.; Azevedo, V.; LeBlanc, J. G. Functional Food Biotechnology: The Use of Native and Genetically Engineered Lactic Acid Bacteria. In *Omics Technologies and Bio-Engineering: Towards Improving Quality of Life*; Barh, D., Azevedo, V. Eds.; Academic Press: London, 2018; Vol. 2; pp 105–128.
23. Ebrahimi, B.; Mohammadi, R.; Rouhi, M.; Mortazavian, A. M.; Shojaee-Aliabadi, S.; Koushki, M. R. Survival of Probiotic Bacteria in Carboxymethyl Cellulose-based Edible Film and Assessment of Quality Parameters. *LWT* **2018,** *87*, 54–60.
24. Erich, S.; Kuschel, B.; Schwarz, T.; Ewert, J.; Böhmer, N.; Niehaus, F.; Eck, L.; Lutz-Wahl, S.; Stressler, T.; Fischer, L. Novel High-Performance Metagenome β-Galactosidases for Lactose Hydrolysis in the Dairy Industry. *J. Biotechnol.* **2015,** *210*, 27–37.
25. Eser, E.; Ekiz, O. Ö.; Çelik, H.; Sülek, S.; Dana, A.; Ekiz, H. İ. Rapid Detection of Foodborne Pathogens by Surface Plasmon Resonance Biosensors. *Int. J. Biosci. Biochem. Bioinform.* **2015,** *5* (6), 329–335.
26. Fei, J.; Dou, W.; Zhao, G. A Sandwich Electrochemical Immunosensor for *Salmonella pullorum* and *Salmonella gallinarum* based on a Screen-Printed Carbon Electrode Modified with an Ionic Liquid and Electrodeposited Gold Nanoparticles. *Microchim. Acta* **2015,** *182*(13–14), 2267–2275.
27. Fellows, P. J., Ed. In *Food Processing Technology: Principles and Practice*, 3rd ed.; Woodhead Publishing: Cambridge, 2009; p 913.
28. Fenster, K.; Freeburg, B.; Hollard, C.; Wong, C.; Rønhave Laursen, R.; Ouwehand, A. C. The Production and Delivery of Probiotics: A Review of a Practical Approach. *Microorganisms*, **2019**, *7* (3), 83–90.
29. Frakolaki, G.; Giannou, V.; Kekos, D.; Tzia, C. A Review of the Microencapsulation Techniques for the Incorporation of Probiotic Bacteria in Functional Foods. *Critic. Rev. Food Sci. Nutr.* **2021,** *61* (9), 1515–1536.
30. Francois, N. N. Production of Microbial Industrial Enzymes. *Acta Sci. Microbiol.* **2019,** *2* (12), 75–89.
31. Fullbrook, P. D. The Use of Enzymes in the Processing of Oilseeds. *J. Am. Oil Chemists' Soc.* **1983,** *60* (2), 476–478.
32. Gadaleta, A.; Giancaspro, A.; Blechl, A. E.; Blanco, A. A Transgenic Durum Wheat Line that is Free of Marker Genes and Expresses *1Dy10*. *J. Cereal Sci.* **2008,** *48* (2), 439–445.

33. Ghoshal, G. Biotechnology in Food Processing and Preservation: An Overview. Chapter 2. In *Advances in Biotechnology for Food Industry*; Holban, A. M., Grumezescu, A. M., Eds.; Academic Press: London, 2018; pp 27–54.
34. Glass, S.; Fanzo, J. Genetic Modification Technology for Nutrition and Improving Diets: An Ethical Perspective. *Curr. Opin. Biotechnol.* **2017,** *44*, 46–51.
35. Gomes, M. Z. V.; Nitschke, M. Evaluation of Rhamnolipid and Surfactin to Reduce the Adhesion and Remove Biofilms of Individual and Mixed Cultures of Food Pathogenic Bacteria. *Food Control* **2012,** *25*, 441–447
36. Hedrick, V. E.; Dietrich, A. M.; Estabrooks, P. A.; Savla, J.; Serrano, E.; Davy, B. M. Dietary Biomarkers: Advances, Limitations and Future Directions. *Nutr. J.* **2012,** *11*, 109–114.
37. Holme, I. B.; Dionisio, G.; Brinch-Pedersen, H.; Wendt, T.; Madsen, C. K.; Vincze, E.; Holm, P. B. Cisgenic Barley with Improved Phytase Activity. *Plant Biotechnol. J.* **2012,** *10* (2), 237–247.
38. Huang, X.; Suo, J.; Cui, Y. Optimization of Antimicrobial Activity of Surfactin and Polylysine against *Salmonella enteritidis* in Milk Evaluated by a Response Surface Methodology. *Foodbor. Pathog. Dis.* **2011,** *8* (3), 439–443.
39. Jain, J. C.; Takeo, T. A Review the Enzymes of Tea and their Role in Tea Making. *J. Food Biochem.* **1984,** *8* (4), 243–279.
40. Jancy, S.; Shruthy, R.; Preetha, R. Fabrication of Packaging Film Reinforced with Cellulose Nanoparticles Synthesized from Jack Fruit Non-edible Part using Response Surface Methodology. *Int. J. Biolog. Macromol.* **2020,** *142*, 63–72.
41. Kadim, I. T.; Mahgoub, O.; Baqir, S.; Faye, B.; Purchas, R. Cultured Meat from Muscle Stem Cells: A Review of Challenges and Prospects. *J. Integr. Agric.* **2015,** *14* (2), 222–233.
42. Kamthan, A.; Chaudhuri, A.; Kamthan, M.; Datta, A. Genetically Modified (GM) Crops: Milestones and New Advances in Crop Improvement. *Theor. Appl. Genet.* **2016,** *129* (9), 1639–1655.
43. Kapoor, S.; Rafiq, A.; Sharma, S. Protein Engineering and its Applications in Food Industry. *Critic. Rev. Food Sci. Nutr.* **2017,** *57* (11), 2321–2329.
44. Kerry, R. G.; Patra, J. K.; Gouda, S.; Park, Y.; Shin, H. S.; Das, G. Benefaction of Probiotics for Human Health: A Review. *J. Food Drug Anal.* **2018,** *26* (3), 927–939.
45. Knorr, D.; Sinskey, A. J. Biotechnology in Food Production and Processing. *Science* **1985,** *229* (4719), 1224–1229.
46. Kohli, P.; Gupta, R. Alkaline Pectinases: A Review. *Biocatal. Agric. Biotechnol.* **2015,** *4* (3), 279–285.
47. Kumar, A.; Gudiukaite, R.; Gricajeva, A.; Sadauskas, M.; Malunavicius, V.; Kamyab, H.; Sharma, S.; Sharma, T.; Pant, D. Microbial Lipolytic Enzymes–Promising Energy-Efficient Biocatalysts in Bioremediation. *Energy* **2020,** *192*, E-Article 116674.
48. Li, Z.; Teng, J.; Lyu, Y.; Hu, X.; Zhao, Y.; Wang, M. Enhanced Antioxidant Activity for Apple Juice Fermented with *Lactobacillus plantarum* ATCC14917. *Molecules*, **2019**, *24* (1), 51–58.
49. Luo, F.; Jiang, W. H.; Yang, Y. X.; Li, J.; Jiang, M. F. Cloning and Expression of Yak Active Chymosin in *Pichia pastoris*. *Asian-Austral. J. Anim. Sci.* **2016,** *29* (9), 1363–1370.
50. Maghari, B. M.; Ardekani, A. M. Genetically Modified Foods and Social Concerns. *Avicenna J. Medic. Biotechnol.* **2011,** *3* (3), 109–117.

51. Mandal, A. Review on Microbial Xylanases and their Applications. *Int. J. Life Sci.* **2015,** *4* (3), 178–187.
52. Mani-López, E.; Palou, E.; López-Malo, A. Biopreservatives as Agents to Prevent Food Spoilage. Chapter 8. In *Microbial Contamination and Food Degradation*; Holban, A. M., Grumezescu, A. M. Eds.; Academic Press: London, 2018; Vol. 10; pp 235–270.
53. Manoel, E. A.; dos Santos, J. C.; Freire, D. M.; Rueda, N.; Fernandez-Lafuente, R. Immobilization of Lipases on Hydrophobic Supports Involves the Open Form of the Enzyme. *Enzyme Microb. Technol.* **2015,** *71*, 53–57.
54. Melini, F.; Melini, V.; Luziatelli, F.; Ficca, A. G.; Ruzzi, M. Health-Promoting Components in Fermented Foods: An Up-to-Date Systematic Review. *Nutrients* **2019,** *11* (5), 1189.
55. Meng, Q.; Zhang, G. Application of Rhamnolipid as Biological Cleaning Agent. Chinese Patent No. CN102399644A, 2012; p 7.
56. Namratha, S.; Sreejit, V.; Preetha, R. Fabrication and Evaluation of Physicochemical Properties of Probiotic Edible Film based on Pectin–Alginate–Casein Composite. *Int. J. Food Sci. Technol.* **2020,** *55* (4), 1497–1505.
57. Nasseri, A. T.; Rasoul-Amini, S.; Morowvat, M. H.; Ghasemi, Y. Single Cell Protein: Production and Process. *Am. J. Food Technol.* **2011,** *6* (2), 103–116.
58. Neethirajan, S.; Ragavan, V.; Weng, X.; Chand, R. Biosensors for Sustainable Food Engineering: Challenges and Perspectives. *Biosensors* **2018,** *8* (1), 23–30.
59. Nitschke, M.; Silva, S. S. E. Recent Food Applications of Microbial Surfactants. *Critic. Rev. Food Sci. Nutr.* **2018,** *58* (4), 631–638.
60. Okechukwu, Q. N.; Javed, F.; Ivantsova, M. N. Advancement of Food Processing Biotechnology in Developing Countries. *AIP Conf. Proc.* **2019,** *2174*, E-Article 020236.
61. Paletta, A.; Leal Filho, W.; Balogun, A. L.; Foschi, E.; Bonoli, A. Barriers and Challenges to Plastics Valorization in the Context of a Circular Economy: Case Studies from Italy. *J. Clean. Prod.* **2019,** *241*, E-Article 118149.
62. Palmer, M. E.; Chaturongakul, S.; Wiedmann, M.; Boor, K. J. The *Listeria monocytogenes* σB Regulon and its Virulence-associated Functions are Inhibited by a Small Molecule. *MBio* **2011,** *2* (6), E-Article 00241-11.
63. Pandey, K. R.; Naik, S. R.; Vakil, B. V. Probiotics, Prebiotics and Synbiotics - A Review. *J. Food Sci. Technol.* **2015,** *52* (12), 7577–7587.
64. Peñas, E.; Diana, M.; Frías, J.; Quílez, J.; Martínez-Villaluenga, C. A Multistrategic Approach in the Development of Sourdough Bread Targeted Towards Blood Pressure Reduction. *Plant Foods Hum. Nutr.* **2015,** *70* (1), 97–103.
65. Pereira, R. N.; Teixeira, J. A.; Vicente, A. A.; Cappato, L. P.; da Silva Ferreira, M. V.; da Silva Rocha, R.; da Cruz, A. G. Ohmic Heating for the Dairy Industry: A Potential Technology to Develop Probiotic Dairy Foods in Association with Modifications of Whey Protein Structure. *Curr. Opin. Food Sci,* **2018,** *22*, 95–101.
66. Popa, M. E.; Mitelut, A. C.; Popa, E. E.; Matei, F. Creating Products and Services in Food Biotechnology. In *Introduction to Biotech Entrepreneurship: From Idea to Business*; Matei, F., Zirra, D., Eds.; Springer International Publishing: Switzerland, 2019; pp 141–178.
67. Porta, R.; Pandey A.; Rosell, C. M. Enzymes as Additives or Processing Aids in Food Biotechnology. *Enzyme Res.* **2010,** *2010*, E-Article 436859.
68. Rai, M.; Pandit, R.; Gaikwad, S.; Kövics, G. Antimicrobial Peptides as Natural Bio-Preservative to Enhance the Shelf-Life of Food. *J. Food Sci. Technol.* **2016,** *53* (9), 3381–3394.

69. Ramesh. S; Radhakrishnan, P. Cellulose Nanoparticles from Agro-Industrial Waste for the Development of Active Packaging. *Appl. Surf. Sci.* **2019,** *484*, 1274–1281.
70. Rastogi, H.; Bhatia, S. Future Prospectives for Enzyme Technologies in the Food Industry. Chapter 49. In *Enzymes in Food Biotechnology: Production, Applications, Future Prospects*; Kuddus, M., Ed.; Academic Press: London, UK, 2019; pp 845–860.
71. Ribeiro, B. G.; Guerra, J. M. C.; Sarubbo, L. A. Potential Food Application of a Biosurfactant Produced by *Saccharomyces cerevisiae* URM 6670. *Front. Bioeng. Biotechnol.* **2020,** *8*, 434–441.
72. Ritala, A.; Häkkinen, S. T.; Toivari, M.; Wiebe, M. G. Single Cell Protein—State-of-the-Art, Industrial Landscape and Patents 2001–2016. *Front. Microbiol.* **2017,** *8*, E-Article 2009.
73. Sałek, K.; Euston, S. R. Sustainable Microbial Biosurfactants and Bioemulsifiers for Commercial Exploitation. *Process Biochem.* **2019,** *85*, 143–155.
74. Schofield, D. A.; Sharp, N. J.; Vandamm, J.; Molineux, I. J.; Spreng, K. A.; Rajanna, C.; Stewart, G. C. *Bacillus anthracis* Diagnostic Detection and Rapid Antibiotic Susceptibility Determination using Bioluminescent Reporter Phage. *J. Microbiol. Methods* **2013,** *95* (2), 156–161.
75. Sébédio, J. L.; Malpuech-Brugère, C. Implementation of Foodomics in the Food Industry. Chapter 13. In *Innovation Strategies in the Food Industry: Tools for Implementation*; Galanakis, C. M., Ed.; Academic Press: London, UK, 2016, pp 251–269.
76. Senturk, E.; Aktop, S.; Sanlibaba, P.; Tezel, B. U. Biosensors: A Novel Approach to Detect Food-Borne Pathogens. *Appl. Microbiol. Open Access* **2018,** *4* (3), 151–157.
77. Sharma, S.; Datta, P.; Kumar, B.; Tiwari, P.; Pandey, L. M. Production of Novel Rhamnolipids via Biodegradation of Waste Cooking Oil using *Pseudomonas aeruginosa* MTCC7815. *Biodegradation* **2019,** *30* (4), 301–312.
78. Sheikhzadeh, E.; CHamsaz, M.; Turner, A. P. F.; Jager, E. W. H.; Beni, V. Label-Free Impedimetric Biosensor for *Salmonella typhimurium* Detection based on Poly [Pyrrole-Co-3-Carboxyl-Pyrrole] Copolymer Supported Aptamer. *Biosens. Bioelectron.* **2016,** *80*, 194–200.
79. Shepherd, L.V.T.; Dobson, G.; Verrall, S. R.; Conner, S.; Griffiths, D. W.; McNicol, J. W.; Davies, H.; Stewart, D. Potato Metabolomics by GC-MS: What are the Limiting Factors? *Metabolomics* **2007,** *3* (4), 475–488.
80. Shruthy, R.; Jancy, S.; Preetha, R. Cellulose Nanoparticles Synthesized from Potato Peel for the Development of Active Packaging Film for Enhancement of Shelf Life of Raw Prawns (*Penaeus monodon*) during Frozen Storage. *Int. J. Food Sci. Technol.* **2021,** *56* (8), 3991–3999.
81. Signorini, C.; Carpen, A.; Coletto, L.; Borgonovo, G.; Galanti, E.; Capraro, J.; Scarafoni, A. Enhanced Vitamin B_{12} Production in an Innovative Lupin Tempeh is due to Synergic Effects of Rhizopus and Propionibacterium in Co-fermentation. *Int. J. Food Sci. Nutr.* **2018,** *69* (4), 451–457.
82. Silveira, B. M. P.; Barcelos, M. C. S.; Vespermann, K. A. C.; Pelissari, F. M.; Molina, G. An Overview of Biotechnological Processes in the Food Industry. Chapter 1. In *Bioprocessing for Biomolecules Production*; Molina, G.; Gupta, V.; Singh B., Gathergood, N., Eds.; John Wiley & Sons, Ltd.: New York, USA, 2019; pp 1–19.
83. Sindhu, R.; Binod, P.; Madhavan, A.; Beevi, U. S.; Mathew, A. K.; Abraham, A.; Pandey, A.; Kumar, V. Molecular Improvements in Microbial α-Amylases for Enhanced Stability and Catalytic Efficiency. *Bioresour. Technol.* **2017,** *245* (Part B), 1740–1748.

84. Singh, P.; Magalhães, S.; Alves, L.; Antunes, F.; Miguel, M.; Lindman, B.; Medronho, B. Cellulose-based Edible Films for Probiotic Entrapment. *Food Hydrocoll.* **2019,** *88,* 68–74.
85. Singh, V. P. Recent Approaches in Food Bio-Preservation-A Review. *Open Vet. J.* **2018,** *8* (1), 104–111.
86. Snehal, S.; Dubey, A. Role of Biotechnology in Food Processing. *Acta Sci. Agric.* **2019,** *3* (5), 60–61.
87. Song, M. Y.; Van-Ba, H.; Park, W. S.; Yoo, J. Y.; Kang, H. B.; Kim, J. H.; Kang, S. M.; Kim, B. M.; Oh, M. H.; Ham, J. S. Quality Characteristics of Functional Fermented Sausages added with Encapsulated Probiotic *Bifidobacterium longum* KACC 91563. *Korean J. Food Sci. Anim. Resour.* **2018,** *38* (5), 981–994.
88. Ukaegbu-Obi, K. M. Single Cell Protein: A Resort to Global Protein Challenge and Waste Management. *J. Microbiol. Microb. Technol.* **2016,** *1* (1), 5–11.
89. Urso, R.; Rantsiou, K.; Cantoni, C.; Comi, G.; Cocolin, L. Sequencing and Expression Analysis of the Sakacin P Bacteriocin Produced by a *Lactobacillus sakei* Strain Isolated from Naturally Fermented Sausages. *Appl. Microbiol. Biotechnol.* **2006,** *71* (4), 480–485.
90. Visvalingam, J.; Hernandez-Doria, J. D.; Holley, R. A. Examination of the Genome-wide Transcriptional Response of *Escherichia coli* O157: H7 to Cinnamaldehyde Exposure. *Appl. Environ. Microbiol.* **2013,** *79* (3), 942–950.
91. Wang, T. J.; Larson, M. G.; Vasan, R. S.; Cheng, S.; Rhee, E. P.; McCabe, E.; Lewis, G. D.; Fox, C. S.; Jacques, P. F.; Fernandez, C.; O'Donnell, C. J.; Carr, S. A.; Mootha, V. K.; Florez, J. C.; Souza, A.; Melander, O. A.; Clish, C. B.; Gerszten, R. E. Metabolite Profiles and the Risk of Developing Diabetes. *Nat. Med.* **2011,** *17,* 448–453.
92. Weeks, J. T.; Ye, J.; Rommens, C. M. Development of an In Planta Method for Transformation of Alfalfa (*Medicago sativa*). *Transgen. Res.* **2008,** *17* (4), 587–597.
93. Xu, S.; Guo, Y.; Du, G.; Zhou, J.; Chen, J. Self-Cloning Significantly Enhances the Production of Catalase in *Bacillus subtilis* WSHDZ-01. *Appl. Biochem. Biotechnol.* **2014,** *173* (8), 2152–2162.
94. Yang, S. C.; Lin, C. H.; Sung, C. T.; Fang, J. Y. Antibacterial Activities of Bacteriocins: Application in Foods and Pharmaceuticals. *Front. Microbiol.* **2014,** *5,* 241–247.
95. Yu, P. Food Biotechnology-An Important and Promising Research Field. *J. Food Biotechnol. Res.* **2017,** *1* (1), 1–11.
96. Yuan, J.; Wu, S.; Duan, N.; Ma, X.; Xia, Y.; Chen, J.; Wang, Z. Sensitive Gold Nanoparticle-based Colorimetric Apta Sensor for *Staphylococcus aureus*. *Talanta* **2014,** *127,* 163–168.
97. Zhang, Y. H.; Bai, J.; Jiang, W. N.; Zhao, C. R.; Ji, J. J.; Wang, J. Z.; Liu, Y. W. Promising Hen Egg-derived Proteins/Peptides (EDPs) for Food Engineering, Natural Products and Precision Medicines. *Res. Vet. Sci.* **2020,** *128,* 153–161.

CHAPTER 9

VALORIZATION OF AGRO-INDUSTRIAL WASTES USING FERMENTATION TECHNOLOGY

MANDIRA KAPRI, PREM P. SRIVASTAV, and SATYAWATI SHARMA

ABSTRACT

Agro-industrial waste generated is a rich source of various nutrients, nutraceutical, and bioactive ingredients, such as phenols, flavonoids, polysaccharides, carotenoids, mangiferin, vitamins, minerals, etc. These bioactive components possess various therapeutic benefits like antidiabetic, antiproliferative, hepatoprotective, antihypertensive, antiangiogenic, antidepressive, etc. However, these agro-wastes posses certain antinutritional factors causing hindrance in digestion, absorption, and metabolism, limiting its application in various aspects. Various valorizing techniques are practiced to overcome the limitations, making the agro-waste suitable for human consumption. Fermentation is one of the superior valorizing techniques used for augmenting the agro-industrial wastes into valuable products, making them suitable for the pharma and functional food industries. The current chapter deals with the microbial valorizing approach as a novel technique and a boon source for nutrition and nutraceuticals.

9.1 INTRODUCTION

The present scenario of constantly increasing population and global food insecurity leads to the search for quality food which can secure sustainable food distribution, combat poverty, and malnutrition. "The state of food

Novel and Alternative Methods in Food Processing: Biotechnological, Physicochemical, and Mathematical Approaches. N. Veena, Megh R. Goyal, & Ritesh B. Watharkar (Eds.)

security and nutrition in the world, 2019"—a report by Food and Agriculture Organization (FAO) (2019) reported that about 820 million people in the world are still starving and suffering from food insufficiency.[40] According to this report, about 2 billion persons throughout the world are suffering from moderate to severe food insecurity. The report also reveals that among 20.5 million babies globally (one in seven lives) suffer from low birth weight and malnutrition. The 2030 agenda for sustainable food supply and Zero hunger challenge put forward the aim to reduce the wastage of food by the revalorization process and produce quality food products that can ensure safe and adequate nutritive access of food to the people.[40]

The postharvest processing industries (like juice, jam, pulp, ketchup, oil, sugar, rice, wheat, etc.) produce tons of agro-industrial waste every year in the form of peel, pomace, seeds, fibers, deoiled cakes, hulls, etc.[69] These agro-industrial wastes are considered to be the good source of proteins, vitamins, minerals, carbohydrates, sugars, fatty acids, fibers, bioactive components, micro- and macro-nutrients, and various other nutraceuticals.[100] Due to the lack of proper valorization process of the agro-wastes, these by-products either remain underutilized or are used as feed and fodder for animals, the substrate for industrial enzyme production, fertilizers, etc.[99] These agro-industrial wastes have been reported for their potential to provide quality nutrients. Times of India has reported the advantages of consuming banana peel for its potential to act as an antidepressant, lowering blood cholesterol, improving cardiovascular health, protecting red blood cells, reducing weight and detoxifying the body, improving eyesight and to act as prebiotics under the "Health and Fitness" column on June 1.[98]

This chapter explores the fermentation of agro-industrial wastes as a novel valorization technique and its potential environmental and health benefits.

9.2 CLASSIFICATION OF AGRO-INDUSTRIAL WASTE AND THEIR PROCESSING TECHNIQUES

Agro-industrial wastes rich in quality nutrients are discarded in tons worldwide from the postharvest processing industries which mainly contribute the landfills and increases environmental issues.[100] These generated wastes can be classified into two major types: edible and nonedible wastes. Nonedible wastes are those which are not suitable for consumption but are further processed into various valuable products, for example, coconut shell, paddy and wheat straw, bagasse, etc. The edible wastes comprised of both the processed and the unprocessed part, of which, the unprocessed ingredients consist of horticulture wastes, stems of banana, pineapple, shells of cashew, etc. The processed agro-wastes includes physically processed products (like deoiled cakes, bakery wastes, bran, pomace, etc.) and fermented products (like whey, malt, fermented peels, fruit residues, etc.). However, these agro-wastes remain underutilized due to a lack of proper processing. Promising valorization techniques can be a novel alternative in accessing the valuable nutrients by providing the society with the "zero waste industry" and will be easy to achieve the "zero hunger challenge."[2]

Processing of agro-wastes to generate valuable products needs proper attention in terms of toxicity, processing hindrances, final products, and can be classified into four types as depicted in Figure 9.1.[58]

a. **Recovery:** By this process, the generated wastes are further utilized for accessing the remained valuable products within it. For example, recovery of pectin from fruit peels; lactose, casein, minerals, organic acids, etc., from cheese whey wastewater (CWW).
b. **Product formation:** By this process, wastes are further utilized for the generation of various valorized and novel products. For example, utilization of apple wastes for the production of apple cider vinegar, utilization of CWW, and oilcake for the production of various organic acids, enzymes, etc.
c. **Energy production:** By this process, utilization of agro-industrial wastes for the production of new food products can be utilized as a source of energy. For example, production of single-cell protein, mushroom cultivation, mycelium production, etc.
d. **Direct utilization:** By this process, the industrial wastes produced are directly utilized for consumption, pharmaceutical purpose, animal feed, fertilizers, etc., without further processing, for example, utilization of oilcake (groundnut oil cake, mustard oil cake, etc.) for consumption, animal feed, etc.

9.3 BIOACTIVE INGREDIENTS PRESENT IN AGRO-INDUSTRIAL WASTES AND THEIR HEALTH BENEFITS

Agro-industrial wastes are rich in various bioactive ingredients having various health benefits.[56] The fruit and peel wastes are rich in dietary fiber, various organic acids, coloring pigments, proteins, vitamins, hetero-polysaccharides, antioxidants including phenols, flavonoids, etc. The leafy vegetable and their peels are a good source of vitamins, minerals, cellulose, hemicellulose, etc. The cereal wastes like wheat bran, rice bran, germs, etc., are rich sources of fatty acids, proteins, sugars, vitamins, minerals, and fibers. The defatted oilseed meals are rich sources of nitrogen, that is, protein, fats, various vitamins, tannins, etc.[104] These agro-industrial wastes also contain various bioactive agents possessing therapeutic benefits like antiproliferative activity, antidiabetes, antistress, antimicrobial, immune-boosting, antioxidant, antimutagenic, antiallergenic, neuro healing attributes, and are of great interest to functional food and pharmaceutical industries.[82] Few of the bioactive compounds present in various agro-industrial wastes have been enlisted in Table 9.1.

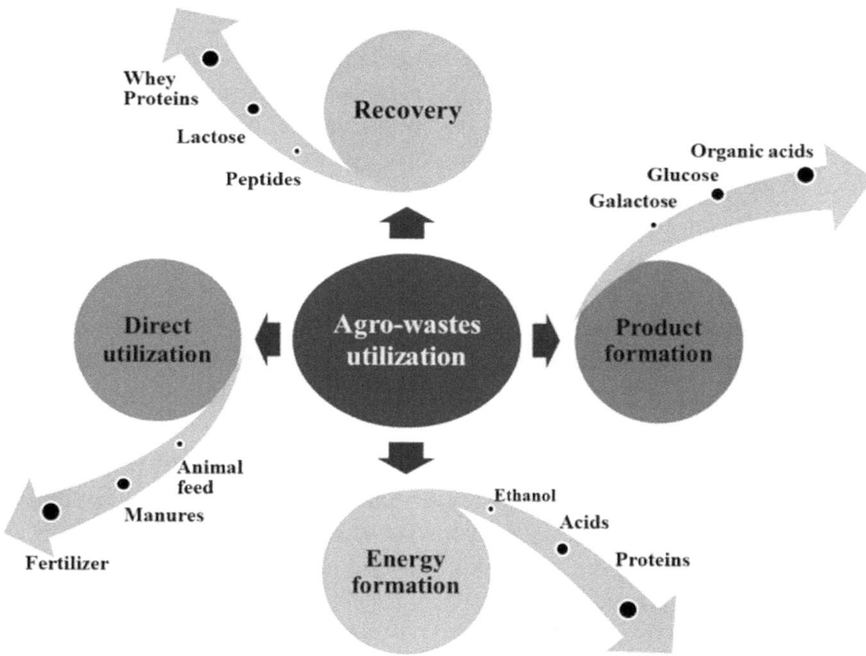

FIGURE 9.1 Various types of agro-waste processing techniques depending on the final products; adapted from Lin et al.[58]

9.4 FORTIFIED COMMERCIAL FOOD PRODUCTS: AGRO-INDUSTRIAL WASTE

Protein-energy-malnutrition is a severe problem seen in developed countries, where fortifying foods with micro- and macro-nutrients along with

TABLE 9.1 Bioactive Ingredients in Agro-Industrial Wastes With Potential Health Benefits.

Agro-industrial wastes		Bioactive ingredients	Potential health benefits	References
Fruits	Apple (pomace, peels, and residues)	Epicatechin, catechins, chlorogenic acid, hydroxycinnamates, phloretin glycosides, quercitin glycosides, procyanidins, anthocyanins	• Improves metabolism of lipid, antioxidant activity, and the function of gastro intestine. • Prevents metabolic disorders like hypergly-cemia, insulin resistance	93
	Avocado (peels, pomace, and seed shells)	Epicatechin, catechin, gallic acid, chlorogenic acid, cyanidin 3-glucoside, homogentisic acid	• Enhances antimicrobial, antioxidant, and anticancer attributes in the body • Cures dermatological problems	11
	Banana (peel, steam)	Gallocatechin, anthocyanins, delphinidin, cyaniding, catecholamine Tryptophan	• Prevents tumor, inflammation, proliferation, tuberculosis, HIV-1, hyperlipoproteinemic, blood cholesterol, diabetes, depression, etc.	65
	Berries (pomace, skin, and seeds)	Cyanidin, delphinidin, malvidin	• Prevents bacterium-induced peptic ulcer, urinary tract infections, coronary heart disease, and cardiac stroke	101
	Citrus (pomace and peels)	Hesperidin, naringin, eriocitrin, narirutin	• Prevents carcinogenicity, mutagenicity, allergenicity, and aging activity	60
	Lichi (peels and seeds)	Cyanidin-3-glucosides, cyanidin-3-rutonoside, malvidin-3-glucoside, gallic acid, epicatechin-3-gallate	• Hypo-glycaemic and hypolipidemic property • Prevents obese, cancer, and atherosclerosis • Hypotensive and immunomodulatory attributes	107
	Mango (seeds and peels)	Gallic acid, ellagic acid, gallates, gallo-tannins, condensed tannins, mangiferin	• Prevents inflammation, viral infection, diabetes, tumor, and cancer	47
	Olives	Cyanidin, delphinidin, malvidin, polyphe-nols such as oleuropein, hydroxytyrosol, and tyrosol, rutin, quercetin, caffeic acid, vanillic and o-, and p-coumaric acids	• Effective in combating coronary heart disease, atherosclerosis, and certain types (colorectal and breast) of cancer	37
	Palm (seeds and peels)	Tocopherols, tocotrienols, sterols, and squalene, phenolic antioxidants	• Antioxidant and anticancerous activity	55

TABLE 9.1 *(Continued)*

Agro-industrial wastes		Bioactive ingredients	Potential health benefits	References
	Pomegranate (peels and seeds)	Gallic acid, cyanidin-3,5-diglucoside, cyanidin-3-diglucoside, delphinidin-3,5-diglucoside, punicalagins	• Prevents bacterial and viral infection, hypo-lipidemia, and inflammation	94
Vegetables	Carrot	Phenols, β-carotene, ascorbic acid, and tocopherol	• Regulates the expression of the gene • Affects the function of a cell such as monocyte inhibition, adhesion, and platelets activation • Enhances immune system and decreases the threat of degenerative disease	88
	Onion	Quercetin, rutin	• Aggregates blood platelets • Cure cardiovascular health, apoptotic conditions, and inflammatory problems	17
	Tomato	Carotenoids	• Hinders abnormal cells growth • Blocks oxidation of LDL and thus prevents heart disease • Fulfils vitamin A deficiency	36
	Parsley	Apigenin, luteolin, quercetin	• Acts as antioxidant, antimicrobial, antiprotozoal, anti-inflammatory, antidiabetic, and anticancer properties	84
	Spinach	Apigenin, luteolin	• Improves eyesight, regulates blood pressure, strengthen muscles • Protects age-related macular degeneration, hemophilia, cataracts, atherosclerosis, heart attacks, neurological disorders, and bone mineralization	3

TABLE 9.1 *(Continued)*

Agro-industrial wastes		Bioactive ingredients	Potential health benefits	References
Cereals	Barley	β-Glucan	• Prevents hypercholesterolemia, obesity, antitumor, antidiabetes, anti-infection, and immune-modulating attributes	35
	Rice (bran and germ)	γ-Oryzanol, bran oil	• Prevents cardiovascular disease	48,51
	Wheat (bran and germ)	Sterols, mono- and di-glycerides, phospholipids, glycolipids, lipophilic antioxidants, tocopherols, and carotenoids	• Prevents cardiovascular disease and postprandial lipidemia	21
Beverages	Cocoa beans shell, pod husk	Epicatechin, phenolic compounds, dietary fibers, theobromine, and lipid	• Lowers blood cholesterol and glucose levels • Reduces bowling cancer, constipation, diverticulitis, and stimulates the immune system	70
	Green tea waste	(−)-epigallocatechin, (+)-gallocatechin, (−)epicatechin-3-*O*-gallate	• Prevents atherosclerosis, cellular oxidation, angiogenesis, and reduce blood cholesterol level	27
	Coffee beans shell	Caffeine, phenolic compounds, melanoidins, diterpenes, xanthines, and carotenoids	• Prevents cancer, chronic diseases, heart disease, and neurological disorders	54
Oil seeds	Groundnut cake	Fatty acids, resveratrol, phytoalexin, etc.	• Prevents cancer, inflammation, blood cholesterol, cardiovascular diseases, and neuroprotective agent	99
	Mustard cake	Sesamol, tocopherol, polyphenol, and β-carotene	• Antioxidant and anticancerous attributes	55
	Sesame cake	Lignan, sesamin, and sesamolin	• Antioxidant and anticancerous attributes	55
	Flaxseed cake	Lignan such as secoisolariciresinol, matairesinol, enterodiol, and enterolactone	• Antioxidant and anticancerous activities	55
	Sunflower cake	B-carotene, tocopherols, and tocotrienols	• Antioxidant and anticancerous activities	55

calorific ingredients can be a crucial strategy to serve the purpose. Various fortification programs have already been implemented with special attention toward mothers, children, and pregnant woman. One of these initiatives by Global Alliance for Improved Nutrition GAIN started from 2013 is National Food Fortification Program, which addresses the deficiency of vitamin and mineral by fortification of cereal flours, sugar, regular edible vegetable oil, various sauces with vitamins like A, B, D, and minerals like iron and zinc. GAIN also executed multinutrient supplementation program by fortifying regularly utilized staple products used for cooking (mainly wheat and oil) for addressing the micronutrient deficiency within the population. A community-based nutritional program has been implemented by the health minister of Indonesia between 2010 and 2014, which includes ionization of salt, fortification of flour, and oil with vitamin A. Joint effort of WHO and FAO to combat iron deficiency (anemia) by fortifying flour of wheat, maize, and salt with iron.[31,75]

For fulfilling the increasing demand and to come up with solutions for the malnourished population, food fortification using agro-industrial wastes (rich in protein, minerals, dietary fibers, and various bioactive ingredients) can be a good alternative compared to chemically fortified compounds. Few products fortified with agro-industrial wastes and are commercially available in the market that has been enlisted in Table 9.2.

The increased demand for the economic source for protein supplementation leads to the search of unconventional sources like deoiled cakes from oilseeds, which serves the quality protein required for combustion of malnutrition. One of the accessible sources for food fortification is defatted soybean flour used to fortify tortilla, pretzel, pasta, spaghetti, biscuits, and bread. The tortillas fortified with 3% and 6% defatted soybean flour showed maximum biological value, net protein utilization, and protein digestibility.[9]

Studies by Naik and Sekhon[67] revealed that an increased level of defatted soybean flour fortification in pretzel decreased the total calorific value and functional attributes with an increase in protein and mineral content. However, pretzel fortified with 5% defatted soybean flour revealed best organoleptic attributes in terms of texture, color, flavor, and consumer acceptance. Whereas pasta fortified with defatted soybean flour got deteriorated quality in terms of cooking loss, bulkiness and elasticity.[14] Spaghetti fortified with 35% defatted soybean flour inferences appreciative organoleptic attributes in terms of texture, flavor, and acceptance, showing no significant change in flavor. However, increased fortification level up to 50%, resulted in high protein and lysine content with increased bitter and beany flavor in the developed product.[89] Studies revealed that deoiled soybean flour fortified biscuits are

TABLE 9.2 Agro-Industrial Waste Fortified Commercial Food Products.

Agro-industrial waste	Fortified foodstuffs	Nutrients fortified	Reference
Banana peel	Baldy flatbread	Enhancing nutrients, fibers, and used as wheat supplementation	38
Defatted peanut flour (DPF) and peanut protein concentrate (PPC)	Meat analog	Potential to compete with commercial meat analogs and can substitute soya protein	79
Deoiled soybean cake	Tortilla chips	Protein	9
Deoiled soybean flour	Pretzel	Protein and minerals like Ca, K, Mg, P, and S	67
Deoiled soybean flour	Spaghetti	Protein and amino acids like lysine, threonine	89
Deoiled soybean flour	Biscuits	Protein, fat, minerals, crude fiber, and lysine	86
Deoiled soybean flour	Bread	Protein	62
Deseeded grape pomace	Cookies	Enhanced nutrition and dietary fiber	23
Maize germ flour (defatted)	Cookies	Protein and dietary fiber	68
Mango peel and kernel powder	Biscuit	Fiber and antioxidant	13
Partly Deoiled soybean flour	Pasta	Protein	14
Raspberry pomace	Shortcrust cookies	Enhanced dietary fiber	30
Rice bran (defatted)	Indian bread	Dietary fiber	106
Soybean and melon seeds flour (both defatted)	Fried balls	Protein and minerals	25
Wheat germ (defatted)	Cookies, noodles	Protein, fiber, and minerals like Ca, K, and Fe	12,44

rich in protein and lysine content with increased protein digestibility by 170%.[86] Whereas an increase in loaf volume and organoleptic attributes of bread was reported when soybean flour defatted was fortified at 3% and 7%.[62]

Apart from cereal-based products, fruit wastes can also be used for mineral, fiber, vitamin, and bioactives supplementation. Apple and grape pomace rich in fiber and antioxidants are used in cake and bread formulation, respectively, and were found to enhance the rheological properties.[49,95] Studies by Aslam et al.[13] reported that biscuits supplemented with banana peel and kernel powder showed an elevated amount of dietary fiber, antioxidants with enhanced functional, and organoleptic attributes. However, the substitution of the peel by 10% and kernel by 5% received the highest consumer acceptance.

9.5 LIMITATIONS ASSOCIATED WITH UTILIZATION OF RAW AGRO-INDUSTRIAL WASTE

Agro-industrial wastes though rich in various nutrients, nutraceuticals, and bioactive ingredients; their utilization in the raw form is very challenging due to the presence of various antinutritional factors present in the raw agro-wastes or is produced during the oxidative or enzymatic process.[41,80] Antinutritional compounds are naturally occurring ingredients or are produced after a certain reaction causing hindrance to various cellular functions or are harmful for the normal function of the body. For example, protein digestibility, absorption, and further utilization in the body gets interfered by the presence of trypsin and chymotrypsin inhibitors, lectins or hem-agglutinins, polyphenol compounds, and saponins, carbohydrate digestion gets compromised due to the presence of amylase inhibitors, polyphenolic compounds, NSP-s, flatulence factors, and mineral absorption gets hindered due to the presence of glucosinolates, oxalic acid, phytic acid, gossypol.[7,56]

Various defatted oilseed cakes, legumes, and other nutrient-enriched agro-industrial wastes remains underexploited due to the presence of these antinutritional factors or gets contaminated by certain toxins known as "Mycotoxin." These are secondary metabolites, present mainly in groundnut, soybean, sunflower, kernel of palm, corn, etc., in the form of aflatoxins, ochratoxin A, deoxynivalenol, and zearalenone.[18] Intake of the compounds creates a hindrance to nutrient digestion, utilization, and metabolism creating eventual sickness to the body and can be lethal in certain conditions.[15]

Various processing aids can reduce the content of these antinutritional factors namely: heating soaking, fermentation, etc.[10,71] However, heating and soaking causes loss and/or leaching of nutrients like vitamins, minerals, and other bioactive ingredients along with the antinutritional factors. Fermentation can be the possible alternative way for valorizing the agro-industrial waste by detoxifying these antinutritional components, making the nutrients and bioactive ingredients highly available to the human body. Studies reported that fermentation detoxified lupin flour by reducing the phytic acid content using *Aspergillus sojae, Aspergillus ficuum*, and their cocultures by 53%, 73%, and 65%, respectively.[72] However, contradictorily, a lower reduction level of phytic acid by 13.74%, 19.14%, and 18.21% was observed when canola meal was fermented using *A. sojae, A. ficuum*, and their cocultures.[72]

Studies revealed that detoxification of coffee husk was achieved by the fermentation treatment using *Rhizopus*, *Phanerochaete*, and *Aspergillus spp*.[20,74] Reduction of oligosaccharides, trypsin inhibitor, and phytic acid in soya meal along with degradation of amino acid and protein into simple peptides resulting into decreased allergic effects of soya protein with improved bioavailability by removing undesirable attributes were achieved by the fermentation process.[24] Also trypsin inhibitor, saponins, glucosinolates, and gossypols have been reported to get reduced by this valorization process.[56]

9.6 VALORIZATION TECHNIQUES OF AGRO-INDUSTRIAL WASTES

Valorization of agro-industrial wastes to yield value-added products can be achieved in various ways and has been depicted in Figure 9.2. The physical, chemical, and fermentation processing of agro-wastes are the potential way of transforming the waste into value-added products.

a. **Physical processing:** Agro-industrial wastes, being the rich source of nutritional and nutraceutical attributes, are utilized for various purposes like animal feeds, biofertilizers, pharmaceuticals, and commercial ingredients. Various deoiled cakes (like groundnut cake, mustard cake, mahua cake, etc.), vegetable and fruit peels are rich in macro- and micro-nutrients, making them suitable for their usage as animal feed, especially for ruminants and fish to get quality milk, meat, and other products.[1]

   However, nonedible deoiled cakes (like neem, karanj, and castor cake) though rich in various nutrients, remain underutilized due to the abundance of antinutritional factors like karanjin, tannin, trypsin inhibitor, etc. These agro-wastes being rich in nitrogen, phosphorus, potassium, and other micronutrients find their usage as biofertilizers by enhancing nutrient content of the soil and up-surging nitrogen uptake by the plant with reduced soil nitrification. Interestingly, the antinutritional factors present in these agro-wastes help in protecting the plant against nematodes, insects, parasites, and various infections.[77]

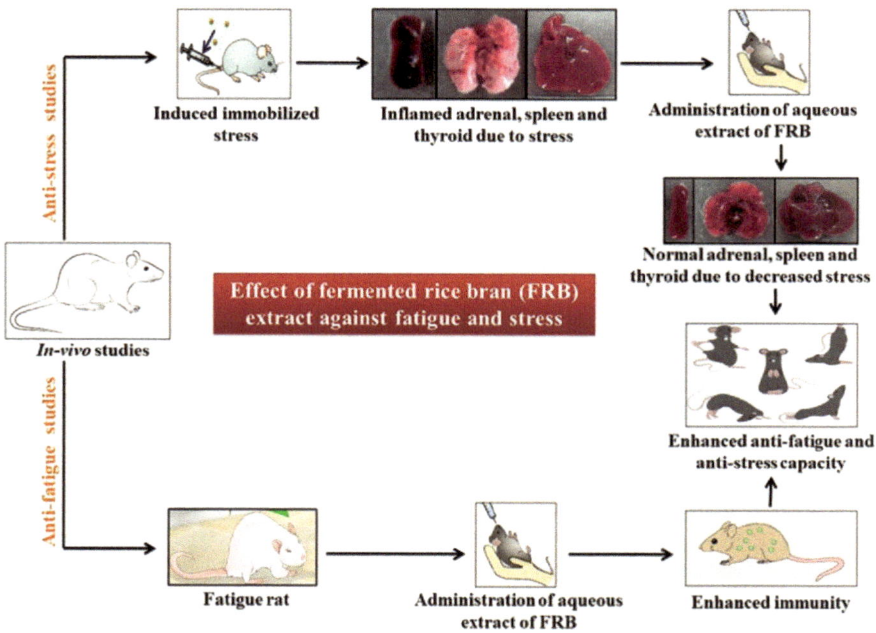

FIGURE 9.2 Possible mechanisms of antistress and antifatigue activity of induced fermented rice bran.[52]

Industrial wastewater, facilitating the potato processing industries is rich in insoluble potato starch, which can easily be separated by the centrifugation process.[32] According to Mironescu,[64] the recovered starch from industrial wastewater exhibited comparable attributes with the commercial starch in terms of physicochemical and functional attributes, having considerable water holding capacity (WHC) with respect to temperature (from 30 to 90°C).

b. **Chemical processing:** Studies have been reported on various chemical processing of agro-industrial wastes and the end products of which are further utilized for the welfare of the society. Agro-industrial wastes are treated for pyrolysis, catalysis, methanolysis, digestion for biochar, bioadsorbents, biofuel, and biogas formation.[4,85,105]

c. **Microbial processing:** Valorization using microbes is commonly termed as "fermentation," and is the most primogenital approach which can be achieved by various microbial cultures and is broadly classified into bacterial, fungal, and algal fermentation. Two types of fermentation process commonly practiced are solid-state fermentation (SSF) and submerged fermentation (SF). However, the process of fermentation depends on various regulating factors like microbial species, substrate, carbon–nitrogen ratio, temperature, pH, etc.[46,59]

9.7 TYPES OF FUNGAL VALORIZATION TECHNIQUES AND THEIR UTILIZATION FOR FOOD PRODUCT DEVELOPMENT

Fermentation is one of the most primogenital approaches for the valorization of agro-industrial wastes that are frequently used to obtain value-added byproducts. SSF and SF are the two commonly used for agro-waste valorization. However, the process of fermentation is a substrate-specific process, resulting in the production of various value-added nutraceuticals and bioactive compounds, which are of great interests for food and pharma industries.

a. **Solid-state fermentation (SSF) process:** The other name of SSF is koji fermentation, which is used worldwide and is considered to be the superior technique for degradation of lignocellulosic components, enzyme formation, single-cell protein formation (mushroom cultivation), etc.[34] This process leads to the growth of microorganism in lack of available liquid, promoting volumetric production of the fermented product.[81] SSF can be categorized into two different types: the first type deals with the utilization of the by-products of cereals, fruits, and vegetables to support the fermentation process by providing nutrition. Various fruit peel and pomace, oilseed wastes, cereal wastes have been reported for their potential to produce aromatic compounds, enzymes, edible mushroom, antioxidants.[5,28,34,66,78,91] Whereas, the second part deals with the soaking of by-products for volumetric productivity with decreased effluent producing sugars, lipids, organic acids, etc.[92]
b. **Submerged fermentation (SF) process:** By this technique, the process of fermentation is carried out within a liquid source providing economic products with high yield and is less prone to contamination. But this process has certain disadvantages like space, energy,

and requirement of water. Also, agitation, aeration, and foaming attributes are its limiting parameters.[53] However, the tranquil process of sterilization and control made the process user-friendly.[103] Studies revealed that this process is compatible in making an enzyme, exopolysaccharides, hetero-polysaccharide (i.e., xanthan, chitosan), proteins, volatile organic compounds.[16,19,22,42]

9.8 FERMENTED AGRO-INDUSTRIAL WASTE-INDUCED FOOD PRODUCTS AND THEIR HEALTH BENEFITS

The process of fermentation converts complex substances into simpler ones, by valorizing them with nutrients, nutraceuticals, and various other bioactive ingredients. By this process, digestibility and bioavailability of the process get increased providing healthier products to society.[72,73,81] Few functional food products developed from fermented agro-industrial waste have been highlighted in Table 9.3.

Vicenssuto and de Castro[102] reported an appreciable amount of antioxidant content at different fermentation stages with reduced growth period for lactic acid bacteria when the probiotic was supplemented with mango peel and has been suggested as beneficial for our health. However, the presence of mango peel showed no significant effect on acetic acid bacteria and yeasts. The study revealed 113% increase in 2,2-diphenyl-1-picrylhydrazyl (DPPH) radical scavenging at 10 h of the fermentation process, 153% increase in ferric reducing antioxidant power assay at 8 h of fermentation process, 21% increase in 2,2′-azino-bis(3-ethylbenzothiazoline-6-sulfonic acid) (ABTS)—radical scavenging activity at 6 h of fermentation process and 43% increase in total reducing capacity (Folin–Ciocalteu method) at 8 h of fermentation process compared to the controlled one which is not supplemented by mango peels.

Espírito Santo et al.[39] reported a decrease in fermentation time and rate of acidification value due to supplementation of passion fruit peel powder in whole milk yoghurt and also showed increased probiotic and *Streptococcus thermophilus* count during the fermentation process. In contrast, *Lactobacillus delbrueckii* showed similar or reduced growth compared to the control. This infers that supplementation of passion fruit peel powder enhances the growth of probiotics, which is healthier to our health. However, the presence of passion fruit peel powder was reported to show the reduced firmness and consistency to the final product.

TABLE 9.3 Functional Food Products Obtained From Valorized Agro-Industrial Wastes.

Functional food product	Agro-industrial waste used	Microorganism used	Enhanced attributes	Reference
A. blazei Murill fruiting bodies	Sunflower seed hulls and wheat bran	*A. blazei* Murill	Protein and ergosterol	63
Aroma compounds	Waste bread crumb	*Geotrichum candidum*	Pineapple like fruity aroma compounds	29
Fermented rice bran powder	Rice bran	*Saccharomyces cerevisiae*	Antistress and antifatigue attributes	51
Mango peel based probiotic milk product	Peels of ripped mango	A symbiotic culture of lactic acid bacteria, acetic acid bacteria, and yeast	Antioxidant activity and total reducing capacity	102
Milk-based passion fruit peel powder fortified probiotic yoghurt	Passion fruit peel powder	Yoghurt starter culture: *S. thermophilus* and *L. delbrueckii* Probiotic culture: *L.acidophilus* and *B. animalis*	Dietary fiber	39
Orange fiber fortified yoghurt	Orange fiber	*S. thermophilus* and *L. delbrueckii subsp. bulgaricus*	Dietary fiber and color	43
Pineapple peel based yoghurt	Pineapple peel	*S. thermophiles*, *L. bulgaricus*, *L. acidophilus*, *L. casei*, and *L. paracasei*	Dietary fiber	83
Sourdough bread	Whole wheat bread crumb waste	*Lactobacillus plantarum*	Organic acids	45
Yeast flavor	Orange peel wastes	*Saccharomyces cerevisiae*	Bioflavor (volatile aromatic esters)	61

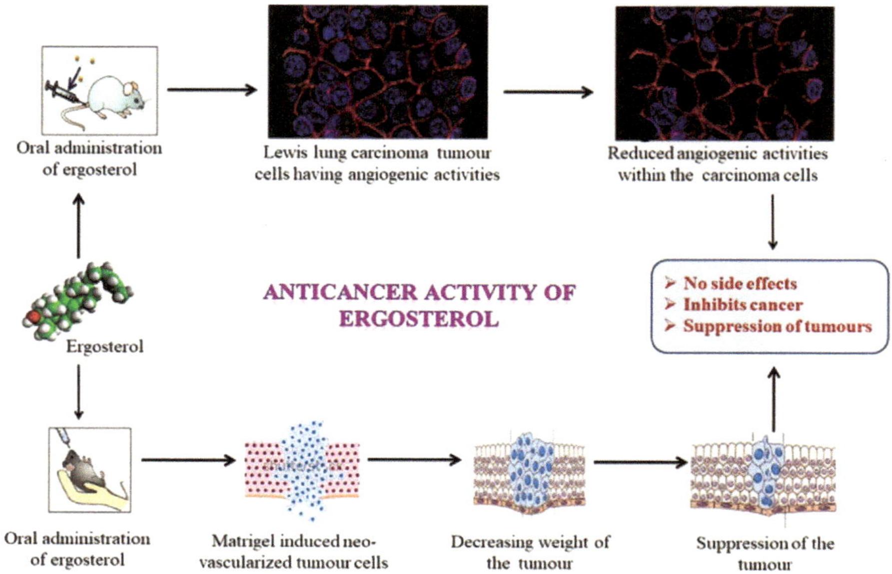

FIGURE 9.3 Possible mechanism for ergosterol-induced antitumor activity against cancer cells.[78]

Kim et al.[51] concluded that fermented rice bran (FRB) using *Saccharomyces cerevisiae* possess pharmacological attributes like antistress and antifatigue while administered in mice in a dosage of 1 gm/kg/day continued for 8 days. The possible mechanism of stress release by FRB was by affecting the weight of adrenal, spleen thymus, and thyroid. The possible mechanism of antifatigue might be resistant to induced stress and enhanced immunity and has been represented in Figure 9.2. However, rice bran is also used for various purposes other than food processing industries. Jafarpour et al.[50] reported rice bran was supplemented with soya cake and was used as a substrate for mushroom cultivation (*Pleurotus ostreatus*) with the highest yield of 794.3 g and biological efficiency of 158.9 g. Moreover, the heaviest fruiting body weight was obtained when palm fiber was supplemented with rice bran.

Matute et al.[63] reported that amendment of compost with sunflower seed hulls and wheat bran by 50% and 4.5% enhanced the yield and fresh weight of mushrooms by 11% and 6%, respectively. Studies by Takaku et al.[96] revealed antitumor effects while ergosterol extracted from *Agaricus blazei* Murill was administered orally to sarcoma 180–bearing mice. However, a direct inhibitory effect was observed on angiogenic activity and proliferation when the extract was intraperitoneally administered against a neovascularized tumor induced by Lewis lung carcinoma cells cell-packed chambers and matrigel (containing acidic fibroblast growth factor and heparin) induced neovascularized tumor in female C57BL/6 mice.[96] The potential mechanism of antitumor activity by ergosterol has been explained in Figure 9.3.

The agro-industrial wastes are also been used for the preparation of various aromatic flavors and fatty acids, which can help in avoiding the use of synthetic additives (like synthetic color and flavor) in the food processing industries.[43,61]

9.9 EFFECT OF FERMENTATION ON THE NUTRITIONAL AND FUNCTIONAL ATTRIBUTES OF AGRO-INDUSTRIAL WASTES

Research studies have reported that fermentation not only valorizes the agro-industrial wastes but also significantly affects their nutritional and nutraceutical attributes. Olukomaiya et al.[72] reported an increase in fiber fractions and minerals (especially calcium and phosphorous) when canola meal was fermented with *A. sojae*, *A. ficuum,* and in a combination of the two (*A. sojae* and *A. ficuum*). The study reported no significant increase in protein content during the fermentation process, which was in contrast with the findings of Aljuobori et al.[8] and Plaipetch and Yakupitiyage.[76]

However, Olukomaiya et al.[73] have reported an increase in fat and organic matter during the fermentation process using all the microorganisms. Different microorganisms and varying fermentation condition might be the possible reason for varied inferences. However, the maximum increment of mineral was observed when fermentation was achieved with the combination of both the microorganisms (*A. sojae* and *A. ficuum*). A significant decrease in in-vitro protein digestibility was observed in both *A. sojae* and with combined microorganism with conflicting inference with *A. ficuum.* The possible reason for decreased in-vitro protein digestibility might be due to reduced hydrolytic enzyme action or entangling of protein moieties into the fiber fractions. Whereas the decrease in phytic acid, glucosinolates, and phenol showed a diminishing effect in all the cases throughout the fermentation process. Fermentation with all organisms showed elevated water absorption capacity, solubility index swelling capacity, and bulk density with a decrease in oil absorption capacity and dispersibility while fermenting with *A. sojae* and combined microorganism.

Lin et al.[57] reported decreased glucose dialysis retardation index (GDRI) and decreased cholesterol absorbing capacity with increased inhibitory action of α-amylase and sodium cholate adsorption capacity. Increased GDRI was also observed by Ahmed et al.[6] and the possible reasons might be entrapment of glucose into the fiber matrix.

During fermentation, the enzymatic action may affect the fibers making modifying them to possess varying surface having varied responses. Increased inhibitory action of α-amylase was reported which might be because of the alteration of microstructure resulting due to the enzymatic action during microbial fermentation process resulting in elevated absorption of α-amylase by the fiber fractions. The decreased cholesterol absorption[33] by the elevated dietary fiber has been observed and this can be beneficial for health.[57] However, on the contrary, rice bran fermented by *Bacillus natto* revealed elevated cholesterol absorbing capacity by Chu et al.[26] The possible reason might be due to the presence of elevated soluble dietary fiber, which highly affects the serum cholesterol level.

Studies revealed an increase in total phenol content and total flavonoid content when peanut meal was fermented using *Aspergillus awamori* till 120 h which has a beneficial effect on health; however, a decrease content for both at 144 h was observed,[81] which was in contrast with the findings reported by Olukomaiya et al.[72] An increase in tannin content with respect to fermentation time was reported by Sadh et al.,[81] which was in contrary with findings of Shang et al.,[87] where tannin content was decreased due to lactic acid bacterial fermentation of Xuan Mugua. The increase in tannin content might due to the ability of the solvent to soluble available tannin in the substrate. DPPH was reported to get increased till 96 h of the fermentation process and later was found to get decreased, whereas ABTS was found to increase up to 120 h of the fermentation process.[81] This study also indicated 28.37% and 33.92% in WHC and OHC with significant increase in emulsion and foaming stability, which might be due to alteration of hydrophilic and lyophilic balance resulted due to the exposed hydrophilic groups by fungal enzymatic activity. However, a significant decrease in bulk density with respect to fermentation time which would be a potential advantage for baby food formulation.[81] An increase in mineral, namely, iron, calcium, and zinc was observed during 120 h of the fermentation process, after which an abrupt reduction of mineral content was observed in this study.

Valorization of rice and wheat bran by a fermentation process using *Rhizopus sp* and *Aspergillus oryzae* alternatively revealed an increase in protein content by 69% and 56% using *Rhizopus sp* fermentation, and 11% and 14% using *A. oryzae* fermentation, respectively, were reported by

Silveira and Badiale-Furlong.[90] The in-vitro protein digestibility of rice and wheat bran using *A. oryzae* was increased after the fermentation process, which was in contrast with the results for *Rhizopus sp* fermentation. The possible reason for the decrease of in-vitro protein digestibility might be due to the elevated amount of chitin content in the fungal biomass.[97]

Functional attributes of rice and wheat bran were found to get significantly affected by the fermentation process. The decrease in gel formation attributes was observed after the fermentation process in all cases except rice bran fermented with *A. oryzae.* The study also revealed that rice and wheat bran fermented using *Rhizopus sp* showed a significant decrease and increase in WHC, respectively, but the findings were in contrast with the findings after fermentation by *A. oryzae* where no significant change has been observed by Silveira and Badiale-Furlong.[90]

Supplementation of pineapple peel powder (PPP) in yoghurt fastened the fermentation process by 16% compared to yoghurt supplemented with inulin. Fastening of the fermentation process was also reported by Espírito et al.[39] and García-Pérez et al.,[43] when the yoghurt was supplemented with passion fruit and orange fiber, respectively. The presence of PPP also affected the textural properties of the yoghurt by elevating the syneresis process by 10% compared to the insulin substitution and was in agreement with the observation made by García-Pérez et al.[43] The process of syneresis in yoghurt occurs due to the weak water entrapping capacity resulting in fragile gel formation ability. The yoghurt supplemented with PPP also possesses reduced firmness capacity by 6.12% compared to the inulin-supplemented yoghurt.[83] Table 9.4 indicates effect of fermentation on nutritional and functional attributes of agro-industrial wastes.

9.10 SUMMARY

Tons of agro-industrial wastes generated by the processing industries are rich in nutrients, nutraceuticals, and bioactive components and can be used for nutrient supplementation worldwide. Apart from the traditional source of nutrition, fermented agro-industrial wastes can be a good source with enhanced digestibility and bioavailability. However, the quality and quantity of bioactive components depend on the raw materials and various processing aids. Fermentation is one of the superior processing techniques through which agro-industrial waste is valorized with enhanced nutritional and nutraceutical attributes at a low cost. This technique can be an alternate method of utilizing the wastes in the welfare of the society rather than contributing

TABLE 9.4 Effect of Fermentation on Nutritional and Functional Attributes of Agro-Industrial Wastes.

Substrate	Microorganism used	Nutritional attributes		• Functional attributes	Reference
		Protein	IVPD		
Canola meal	*Aspergillus sojae*	1.77% ↑	2.35% ↓	• 25% ↑ in water absorbing capacity (WAC) • 20% ↓ in OAC • 3.22% ↑ in bulk density • 0.13% ↓ in dispersibility • 24.81% ↑ in swelling index • 32.82% ↑ in swelling capacity	73
	Aspergillus ficuum	0.3% ↑	5.47% ↑	• 20% ↑ in WAC • 5% ↓ in OAC • 3.22% ↑ in bulk density • 0.26% ↓ in dispersibility • 15.03% ↑ in swelling index • 53% ↑ in swelling capacity	
	A. sojae and A ficuum	0.46% ↑	12.37% ↓	• 45% ↑ in WAC • 26.5% ↓ in OAC • 1.61% ↑ in bulk density • 0.26% ↓ in dispersibility • 24.81% ↑ in swelling index • 27.27% ↑ in swelling capacity	

TABLE 9.4 *(Continued)*

Substrate	Microorganism used	Nutritional attributes		• Functional attributes	Reference
		Protein	IVPD		
Okra	*N. crassa and L. bulgaricus*	–	–	• ↓ Glucose dialysis retardation index • 57.89% ↑ in α-amylase activity inhibition ratio • ↓ Cholesterol absorption capacity • ↑ Sodium cholate adsorption capacity	57
Peanut meal	*Aspergillus awamori*	–	–	• 28.37% ↑ in WHC • 33.92% ↑ in OHC • ↑ Emulsion ability and stability • ↑ Foaming ability and stability • 38.46% ↓ in bulk density	82
Rice bran (defatted)	*Rhizopus sp*	69% ↑	49% ↓	• 9.37% ↓ in WHC	90
Wheat bran	*Rhizopus sp*	56% ↑	64% ↓	• 3.75% ↓ in gelation • 12.85% ↑ in WHC	
Rice bran (defatted)	*A. oryzae*	11% ↑	18.75% ↑	• 7.35% ↑ in gelation	
Wheat bran	*A. oryzae*	14% ↑	6.25% ↑	• 3.75% ↓ in gelation	
Pineapple peel	*S. thermophiles, L. bulgaricus, L. acidophilus, L. casei,* and *L. paracasei*	–	–	• ↓ Storage modulus (G′) values • ↓ Shear rate • ↓ Yield stress	83
Milk-based passion fruit peel powder fortified probiotic yoghurt	Passion fruit peel powder	–	–	• ↑ Syneresis • ↓ Water entrapping capacity • ↓ Gel formation stability	[39]

(↑) indicates an increasing trend and (↓) indicates a decreasing trend.

to landfills by dumping them. Special attention and endorsement by the government agencies to promote this process of supplementation and further to commercialize the process are required. This novel and green approach not only open-up with innovative dimensions for bioeconomic projects but will also contribute to social health with reduced environmental hazards.

ACKNOWLEDGMENTS

This work was supported by DST INSPIRE [DST/INSPIRE/03/2015/005291], Indian Institute of Technology, Delhi, and Chaman Lal Setia Exports Ltd., Karnal, Haryana, which is greatly acknowledged.

KEYWORDS

- **ergosterol**
- **functional food**
- **nutraceuticals**
- **submerged fermentation**
- **therapeutic value**
- **waste management**

REFERENCES

1. Abarghuei, M. J.; Rouzbehan, Y.; Salem, A. Z. M.; Zamiri, M. J. Nutrient Digestion, Ruminal Fermentation and Performance of Dairy Cows Fed Pomegranate Peel Extract. *Livestock Sci.* **2013,** *157*, 452–461.
2. Abdel-Shafy, H. I.; Mansour, M. S. Solid Waste Issue: Sources, Composition, Disposal, Recycling, and Valorization. *Egypt. J. Pet.* **2018,** *27*(4), 1275–1290.
3. Abu Al-Qumboz, M. N.; Abu-Naser, S. S. Spinach Expert System: Diseases and Symptoms. *Int. J. Acad. Inf. Syst. Res.* **2019,** *3*(3), 16–22.
4. Abu-Jrai, A. M.; Jamil, F.; Al-Muhtaseb, A. H.; Baawain, M.; Al-Haj, L.; Al-Hinai, M.; Al-Abri, M.; Rafiq, S. Valorization of Waste Date-Pits Biomass for Biodiesel Production in Presence of Green Carbon Catalyst. *Energy Convers. Manage.* **2017,** *135*, 236–243.
5. Aggelopoulos, T.; Katsieris, K.; Bekatorou, A.; Pandey, A.; Banat, I. M.; Koutinas, A. A. Solid State Fermentation of Food Waste Mixtures for Single Cell Protein, Aroma Volatiles and Fat Production. *Food Chem.* **2014,** *145*, 710–716.

6. Ahmed, F.; Sairam, S.; Urooj, A. In Vitro Hypoglycemic Effects of Selected Dietary Fiber Sources. *J. Food Sci. Technol.* **2011,** *48*(3), 285–289.
7. Akogwu, R. D.; Aguoru, C. U.; Ikpa, F., Ogbonna, I.; Olasan, J. O. Relevance of Industrial Wastes from *Jatropha curcas L.* Seed in Agricultural Biotechnology. *Int. J. Environ. Agric. Biotechnol.* **2018,** *3*(4), 1547–1551.
8. Aljuobori, A.; Idrus, Z.; Soleimani, A. F.; Abdullah, N.; Juan Boo, L. Response of Broiler Chickens to Dietary Inclusion of Fermented Canola Meal under Heat Stress Condition. *Ital. J. Anim. Sci.* **2017,** *16*(4), 546–551.
9. Amaya-Guerra, M.; Alanis-Guzman, S. S.; Saldívar, S. O. S. Effects of Soybean Fortification on Protein Quality of Tortilla-based Diets Produced from Regular and Quality Protein Maize Plant Foods for Human. *Plant Foods Hum. Nutr.* **2004,** *59*(2), 45–50.
10. Arancon, R. A. D.; Lin, C. S. K.; Chan, K. M.; Kwan, T. H.; Luque, R. Advances on Waste Valorization: New Horizons for a More Sustainable Society. *Energy Sci. Eng.* **2013,** *1*(2), 53–71.
11. Araújo, R. G.; Rodriguez-Jasso, R. M.; Ruiz, H. A.; Pintado, M. M. E.; Aguilar, C. N. Avocado By-products: Nutritional and Functional Properties. *Trends Food Sci. Technol.* **2018,** *80*, 51–60.
12. Arshad, M. U.; Anjum, F. M.; Zahoor T. Nutritional Assessment of Cookies Supplemented with Defatted Wheat Germ. *Food Chem.* **2007,** *102,* 123–128.
13. Aslam, H. K. W.; Raheem, M. I. U.; Ramzan, R., Shakeel, A.; Shoaib, M.; Sakandar, H. A. Utilization of Mango Waste Material (Peel, Kernel) to Enhance Dietary Fiber Content and Antioxidant Properties of Biscuit. *J. Glob. Innov. Agric. Soc. Sci.* **2014,** *2*(2), 76–81.
14. Baiano, C.; Lamacchia, C.; Fares, C.; Terracone, C.; La Notte, E. Cooking Behaviour and Acceptability of Composite Pasta Made of Semolina and Toasted or Partially Defatted Soy Flour. *LWT Food Sci. Technol.* **2011,** *44*(4), 1226–1232.
15. Bankole, S. A.; Ogunsanwo, B. M.; Mabekoje, O. O. Natural Occurrence of Moulds and Aflatoxin B_1 in Melon Seeds from Markets in Nigeria. *Food Chem. Toxicol.* **2004,** *42*(8), 1309–1314.
16. Beg, Q. K.; Bhushan, B.; Kapoor, M.; Hoondal, G. S. Enhanced Production of a Thermostable Xylanase from *Streptomyces sp.* QG-11-3 and Its Application in Bio-bleaching of Eucalyptus Kraft Pulp. *Enzyme Microb. Technol.* **2000,** *27*(7), 459–466.
17. Benítez, V.; Mollá, E.; Martín-Cabrejas, M. A.; Aguilera, Y.; López-Andréu, F. J.; Cools, K.; Esteban, R. M. Characterization of Industrial Onion Waste (*Allium cepa*): Dietary Fiber and Bioactive Compounds. *Plant Foods Hum. Nutr.* **2011,** *66*(1), 48–57.
18. Bhat, R.; Sridhar, K. R.; Karim, A. A. Microbial Quality Evaluation and Effective Decontamination of Nutraceutically Valued Lotus Seeds by Electron Beams and Gamma Irradiation. *Radiat. Phys. Chem.* **2010,** *79*(9), 976–981.
19. Biz, A.; Finkler, A. T. J.; Pitol, L. O.; Medina, B. S.; Krieger, N.; Mitchell, D. A. Production of Pectinases by Solid-State Fermentation of a Mixture of Citrus Waste and Sugarcane Bagasse in a Pilot-Scale Packed-Bed Bioreactor. *Biochem. Eng. J.* **2016,** *111*, 54–62.
20. Brand, D.; Pandey, A.; Roussos, S.; Soccol, C. R. Biological Detoxification of Coffee Husk by Filamentous Fungi using a Solid State Fermentation System. *Enzyme Microb. Technol.* **2000,** *27*(1–2), 127–133.
21. Brandolini, A.; Hidalgo, A. Wheat Germ: Not Only a By-product. *Int. J. Food Sci. Nutr.* **2012,** *63*(Suppl 1), 71–74.

22. Budihal, S. R.; Agsar, D. Exploration of Agrowastes for the Production of Cellulase by *Streptomyces* DSK29 under Submerged and Solid State Systems. *Int. J. Curr. Microbiol. Appl. Sci.* **2015,** *4*(11), 681–689.
23. Canett, R. R.; Ledesma, O. A.; Robles, S. R.; Morales, C. R.; León-Martínez, L.; León-Gálvez, R. Characterization of Cookies made with Deseeded Grape Pomace. *Arch. Latinoam. Nutr.* **2004,** *54*(1), 93–99.
24. Chen, L.; Madl, R. L.; Vadlani, P. V.; Li, L.; Wang, W. Value-Added Products from Soybean: Removal of Anti-nutritional Factors via Bioprocessing. Chapter 8; In: *Soybean –Bio-Active Compounds*; El-Shemy, H., Ed.; IntechOpen: London, 2013; pp 161–179.
25. Chinma, C. E.; Ingbian, E. K.; Akpapunam, M. Processing and Acceptability of Fried Cassava Balls (Akara-Akpu) Supplemented with Melon and Soybean Flours. *J. Food Process. Preserv.* **2007,** *31*(2), 143–156.
26. Chu, J.; Zhao, H.; Lu, Z.; Lu, F.; Bie, X.; Zhang, C. Improved Physicochemical and Functional Properties of Dietary Fiber from Millet Bran Fermented by *Bacillus natto*. *Food Chem.* **2019,** *294*, 79–86.
27. Cooper, R.; Morré, D. J.; Morré, D. M. Medicinal Benefits of Green Tea: Part I. Review of Noncancer Health Benefits. *J. Altern. Complement. Med.* **2005,** *11*(3), 521–528.
28. Correia, R. T.; McCue, P.; Magalhães, M. M.; Macêdo, G. R.; Shetty, K. Production of Phenolic Antioxidants by the Solid-State Bioconversion of Pineapple Waste mixed with Soy Flour using *Rhizopus oligosporus*. *Process Biochem.* **2004,** *39*(12), 2167–2172.
29. Daigle, P.; Gélinas, P.; Leblanc, D.; Morin, A. Production of Aroma Compounds by *Geotrichum candidum* on Waste Bread Crumb. *Food Microbiol.* **1999,** *16*(5), 517–522.
30. Górecka, D.; Pachołek, B.; Dziedzic, K.; Górecka, M. Raspberry Pomace as a Potential Fiber Source for Cookies Enrichment. *Acta Sci. Pol. Technol. Aliment.* **2010,** *9*(4), 451–461.
31. Tablante, E. C.; Pachón, H.; Guetterman, H. M.; Finkelstein, J. L. Fortification of Wheat and Maize Flour with Folic Acid for Population Health Outcomes. *Cochrane Database Syst. Rev.* **2019,** *7*(7), E-Article CD012150.
32. Devereux, S.; Shuttleworth, P. S.; Macquarrie, D. J.; Paradisi, F. Isolation and Characterization of Recovered Starch from Industrial Wastewater. *J. Polym. Environ.* **2011**, *19*, 971–979.
33. Devi, P. B.; Vijayabharathi, R.; Sathyabama, S.; Malleshi, N. G.; Priyadarisini, V. B. Health Benefits of Finger Millet (*Eleusine coracana, L.*) Polyphenols and Dietary Fiber: A Review. *J. Food Sci. Technol.* **2014,** *51*(6), 1021–1040.
34. Dhillon, G. S.; Brar, S. K.; Kaur, S.; Verma, M. Screening of Agro-Industrial Wastes for Citric Acid Bioproduction by *Aspergillus niger* NRRL 2001 through Solid State Fermentation. *J. Sci. Food Agric.* **2013**, *93*(7), 1560–1567.
35. Du, B.; Bian, Z.; Xu, B. Skin Health Promotion Effects of Natural Beta-Glucan Derived from Cereals and Microorganisms: A Review. *Phytother. Res.* **2014,** *28*(2), 159–166.
36. Dutta, D.; Chaudhuri, U. R.; Chakraborty, R. Structure, Health Benefits, Antioxidant Property and Processing and Storage of Carotenoids. *Afr. J. Biotechnol.* **2005,** *4*(13), 1510–1520.
37. Ena, A.; Pintucci, C.; Carlozzi, P. The Recovery of Polyphenols from Olive Mill Waste using Two Adsorbing Vegetable Matrices. *J. Biotechnol.* **2012,** *157*(4), 573–577.
38. Eshak, N. S. Sensory Evaluation and Nutritional Value of Balady Flat Bread Supplemented with Banana Peels as a Natural Source of Dietary Fiber. *Ann. Agric. Sci.* **2016,** *61*(2), 229–235.

39. Espírito Santo, A. P.; Perego, P.; Converti, A.; Oliveira M. N. Influence of Milk Type and Addition of Passion Fruit Peel Powder on Fermentation Kinetics, Texture Profile and Bacterial Viability in Probiotic Yoghurts. *LWT Food Sci. Technol.* **2012,** *47*(2), 393–399.
40. FAO; IFAD; UNICEF; WFP; WHO. *The State of Food Security and Nutrition in the World (SOFI) 2019: Safeguarding against Economic Slowdowns and Downturns*; Food and Agriculture Organization (FAO): Rome, 2019, p 212.
41. Farias, D. F.; Rosa, M. F.; Melo, V. M. M.; Araújo, A. J.; Rocha-Bezerra, L. C. B.; Carvalho, A. F. U. Physicochemical and Biological Characterization of Agrowaste from Green Coconut Shell and its Potential Use in Laboratory Animal Breeding. *J. Solid Waste Technol. Manag.* **2012,** *38*(3), 194–201.
42. Favela-Torres, E.; Volke-Sepúlveda, T.; Viniegra-González, G. Production of Hydrolytic Depolymerising Pectinases. *Food Technol. Biotechnol.* **2006,** *44*(2), 221–227.
43. García-Pérez, F. J.; Lario, Y.; Fernández-López, J.; Sayas, E.; Pérez-Alvarez, J. A.; Sendra, E.; Effect of Orange Fiber Addition on Yogurt Color during Fermentation and Cold Storage. *Color Res. Appl.* **2005,** *30*(6), 457–463.
44. Ge, Y.; Sun, A.; Ni, Y.; Cai, T. Study and Development of a Defatted Wheat Germ Nutritive Noodle. *Eur. Food Res. Technol.* **2001,** *212*, 344–348.
45. Gelinas, P.; McKinnon, C. M.; Pelletier, M. Sourdough-type Bread from Waste Bread Crumb. *Food Microbiol.* **1999**, *16*(1), 37–43.
46. Ghosh, P.; Ghosh, U. Bioconversion of Agro-Waste to Value-Added Product Through Solid-State Fermentation by a Potent Fungal Strain *Aspergillus flavus* PUF5. In *Utilization and Management of Bioresources*; Ghosh, S., Ed.; Springer: Singapore, 2018; pp 291–299.
47. Gold-Smith, F.; Fernandez, A.; Bishop, K. Mangiferin and Cancer: Mechanisms of Action. *Nutrients* **2016,** *8*(7), 396–402.
48. Han, S. W.; Chee, K. M.; Cho, S. J. Nutritional Quality of Rice Bran Protein in Comparison to Animal and Vegetable Protein. *Food Chem.* **2014,** *172*, 766–769.
49. Hayta, M.; Özuğur, G.; Etgü, H.; Şeker, İ. T. Effect of Grape (*Vitis Vinifera L.*) Pomace on the Quality, Total Phenolic Content and Anti-Radical Activity of Bread. *J. Food Process. Preserv.* **2014**, *38*(3), 980–986.
50. Jafarpour, M.; Zand, A. J.; Dehdashtizadeh, B.; Eghbalsaied, S. Evaluation of Agricultural Wastes and Food Supplements Usage on Growth Characteristics of *Pleurotus ostreatus*. *Afr. J. Agric. Res.* **2010,** *5*(23), 3291–3296.
51. Kim, K. M.; Yu, K. W.; Kang, D. H.; Suh, H. J. Anti-Stress and Anti-Fatigue Effect of Fermented Rice Bran. *Phytother. Res.* **2002,** *16*(7), 700–702.
52. Kim, K. M., Yu, K. W.; Kang, D. H.; Koh, J. H.; Hong, B. S.; Suh, H. J. Anti-Stress and Anti-Fatigue Effects of Fermented Rice Bran. *Biosci. Biotechnol. Biochem.* **2001,** *65*(10), 2294–2296.
53. Knob, A.; Fortkamp, D.; Prolo, T.; Izidoro, S. C.; Almeida, J. M. Agro-residues as Alternative for Xylanase Production by Filamentous Fungi. *Bioresources* **2014,** *9*(3), 5738–5773.
54. Kourmentza, C.; Economou, C. N.; Tsafrakidou, P.; Kornaros, M. Spent Coffee Grounds Make Much More Than Waste: Exploring Recent Advances and Future Exploitation Strategies for the Valorization of an Emerging Food Waste Stream. *J. Clean. Prod.* **2018,** *172*, 980–992.
55. Kumar, P.; Shariff, R. Development of Nutraceutical Carriers for Functional Food Applications. *Nutr. Food Sci.* **2011,** *41*(1), 34–43.

56. Lai, W. T.; Khong, N. M.; Lim, S. S.; Hee, Y. Y.; Sim, B. I.; Lau, K. Y.; Lai, O. M. A Review: Modified Agricultural By-products for the Development and Fortification of Food Products and Nutraceuticals. *Trends Food Sci. Technol.* **2017,** *59*, 148–160.
57. Lin, D.; Long, X.; Huang, Y.; Yang, Y.; Wu, Z., Chen, H.; Tu, Z. Effects of Microbial Fermentation and Microwave Treatment on the Composition, Structural Characteristics, and Functional Properties of Modified Okara Dietary Fiber. *LWT Food Sci. Technol.* **2020,** *123*, E-Article 109059.
58. Lin, C. S. K.; Koutinas, A. A.; Stamatelatou, K.; Mubofu, E. B.; Matharu, A. S.; Kopsahelis, N.; Pfaltzgraff, L. A.; Clark, J. H.; Papanikolaou, S.; Kwan, T. H.; Luque, R. Current and Future Trends in Food Waste Valorization for the Production of Chemicals, Materials and Fuels: A Global Perspective. *Biofuels Bioprod. Biorefin.* **2014,** *8*(5), 686–715.
59. Madusari, S.; Firmanto, Z.; Sinuraya, R. Oil Palm Pressed Fiber Valorization: Improving Fiber Processing in Composting with *Azotobacter* for Use in Potting Media of *Brassica rapa*. *J. Appl. Sci. Adv. Technol.* **2020,** *2*(3), 65–72.
60. Mahato, N.; Sharma, K.; Sinha, M.; Cho, M. H. Citrus Waste Derived Nutra-/ Pharmaceuticals for Health Benefits: Current Trends and Future Perspectives. *J Funct. Foods* **2018,** *40*, 307–316.
61. Mantzouridou, F. T.; Paraskevopoulou, A.; Lalou, S. Yeast Flavor Production by Solid State Fermentation of Orange Peel Waste. *Biochem. Eng. J.* **2015,** *101*, 1–8.
62. Mashayekh, M.; Mahmoodi, M. R.; Entezari M. H. Effect of Fortification of Defatted Soy Flour on Sensory and Rheological Properties of Wheat Bread. *Int. J. Food Sci. Technol.* **2008,** *43*(9), 1693–1698.
63. Matute, R. G.; Figlas, D.; Curvetto, N. Sunflower Seed Hull Based Compost for *Agaricus blazei* Murrill Cultivation. *Int. Biodeterior. Biodegrad.* **2010,** *64*(8), 742–747.
64. Mironescu, M. Investigations on Wastewaters at Potato Processing and Starch Recovery and Characterisation. *J. Agroaliment. Processes Technol.* **2011,** *17*(2), 134–138.
65. Mordi, R. C.; Fadiaro, A. E.; Owoeye, T. F.; Olanrewaju, I. O.; Uzoamaka, G. C.; Olorunshola, S. J. Identification by GC-MS of the Components of Oils of Banana Peels Extract, Phytochemical and Antimicrobial Analyses. *Res. J. Phytochem.* **2016**, *10*(1), 39–44.
66. Murthy, P. S.; Naidu, M. M. Production and Application of Xylanase from *Penicillium sp.* Utilizing Coffee By-products. *Food Bioproc. Technol.* **2012,** *5*, 657–664.
67. Naik, H. R.; Sekhon K. S. Influence of Defatted Soy Flour Addition on the Quality and Stability of Pretzel Type Product. *J. Food Sci. Technol.* **2014,** *51*(3), 571–576.
68. Nasir, M.; Siddiq, M.; Ravi, R.; Harte, J. B.; Dolan, K. D.; Butt, M. S. Physical Characteristics and Sensory Evaluation of Cookies Made with Added Defatted Maize Germ Flour. *J. Food Qual.* **2008,** *33,* 72–84.
69. Nigam, P. S. N.; Pandey, A. Eds. Science & Business Media. In *Biotechnology for Agro-Industrial Residues Utilisation: Utilisation of Agro-Residues*, 1st ed.; Springer: Netherlands, 2009; p 466.
70. Okiyama, D. C.; Navarro, S. L.; Rodrigues, C. E. Cocoa Shell and its Compounds: Applications in the food industry. *Trends Food Sci. Technol.* **2017,** *63*, 103–112.
71. Olagunju, O. F.; Ezekiel, O. O.; Ogunshe, A. O.; Oyeyinka, S. A.; Ijabadeniyi, O. A. Effects of Fermentation on Proximate Composition, Mineral Profile and Anti-Nutrients of Tamarind (*Tamarindus indica L.*) Seed in the Production of Daddawa-type Condiment. *LWT* **2018**, *90*, 455–459.

72. Olukomaiya, O. O.; Adiamo, O. Q.; Fernando, W. C.; Mereddy, R.; Li, X.; Sultanbawa, Y. Effect of Solid-State Fermentation on Proximate Composition, Anti-nutritional Factor, Microbiological and Functional Properties of Lupin Flour. *Food Chem.* **2020**, *315*, E-Article 126238.
73. Olukomaiya, O. O.; Fernando, W. C.; Mereddy, R.; Li, X.; Sultanbawa, Y. Solid-State Fermentation of Canola Meal with *Aspergillus sojae, Aspergillus ficuum* and Their Co-cultures: Effects on Physicochemical, Microbiological and Functional Properties. *LWT* **2020**, *127*, E-Article 109362.
74. Pandey, R.; Daloul, R.; Coyne, D. W. Iron Treatment Strategies in Dialysis-dependent CKD. *Semin. Nephrol.* **2016,** *36*(2), 105–111.
75. Pasricha, S. R.; Drakesmith, H.; Black, J.; Hipgrave, D.; Biggs, B. A. Control of Iron Deficiency Anemia in Low-and Middle-Income Countries. *Blood* **2013,** *121*(14), 2607–2617.
76. Plaipetch, P.; Yakupitiyage, A. Effect of Replacing Soybean Meal with Yeast-Fermented Canola Meal on Growth and Nutrient Retention of Nile Tilapia, *Oreochromis niloticus* (Linnaeus 1758). *Aquac. Res.* **2014,** *45*(11), 1744–1753.
77. Ramachandran, S.; Singh, S. K.; Larroche, C.; Soccol, C. R.; Pandey, A. Oil Cakes and their Biotechnological Applications–A Review. *Bioresour. Technol.* **2007,** *98*(10), 2000–2009.
78. Rathore, H.; Prasad, S.; Kapri, M.; Tiwari, A.; Sharma, S. Medicinal Importance of Mushroom Mycelium: Mechanisms and Applications. *J. Funct. Foods* **2019,** *56*, 182–193.
79. Rehrah, D.; Ahmedna, M.; Goktepe, I.; Yu, J. Extrusion Parameters and Consumer Acceptability of a Peanut-based Meat Analogue. *Int. J. Food Sci. Technol.* **2009,** *44*(10), 2075–2084.
80. Rudra, S. G.; Nishad, J.; Jakhar, N.; Kaur, C. Food Industry Waste: Mine of Nutraceuticals. *Int. J. Sci. Environ. Technol.* **2015,** *4*(1), 205–229.
81. Sadh, P. K.; Chawla, P.; Duhan, J. S. Fermentation Approach on Phenolic, Antioxidants and Functional Properties of Peanut Press Cake. *Food Biosci.* **2018,** *22*, 113–120.
82. Sadh, P. K.; Kumar, S.; Chawla, P.; Duhan, J. S. Fermentation: A Boon for Production of Bioactive Compounds by Processing of Food Industries Wastes (By-products). *Molecules* **2018,** *23*(10), E-Article 2560.
83. Sah, B. N. P.; Vasiljevic, T.; McKechnie, S.; Donkor, O. N. Physicochemical, Textural and Rheological Properties of Probiotic Yogurt Fortified with Fiber-Rich Pineapple Peel Powder during Refrigerated Storage. *LWT Food Sci. Technol.* **2016,** *65*, 978–986.
84. Saleh, A. M.; Selim, S.; Al Jaouni, S.; AbdElgawad, H. CO_2 Enrichment can Enhance the Nutritional and Health Benefits of Parsley (*Petroselinum crispum L.)* and Dill (*Anethum graveolens L.*). *Food Chem.* **2018,** *269*, 519–526.
85. San Miguel, G.; Serrano, D. P.; Aguado, J. Valorization of Waste Agricultural Polyethylene Film by Sequential Pyrolysis and Catalytic Reforming. *Ind. Eng. Chem. Res.* **2009,** *48*(18), 8697–8703.
86. Serrem, C. A.; de Kock, H. L.; Taylor J. R. N. Nutritional Quality, Sensory Quality and Consumer Acceptability of Sorghum and Bread Wheat Biscuits Fortified with Defatted Soy Flour. *Int. J. Food Sci. Technol.* **2011,** *46*(1), 74–83.
87. Shang, Y. F.; Cao, H.; Ma, Y. L.; Zhang, C., Ma, F.; Wang, C. X.; Wei, Z. J. Effect of Lactic Acid Bacteria Fermentation on Tannins Removal in *Xuan Mugua* Fruits. *Food Chem.* **2019,** *274*, 118–122.
88. Sharma, K. D.; Karki, S.; Thakur, N. S.; Attri, S. Chemical Composition, Functional Properties and Processing of Carrot—A Review. *J. Food Sci. Technol.* **2012,** *49*(1), 22–32.

89. Shogren, R. L.; Hareland, G. A.; Wu Y. V. Sensory Evaluation and Composition of Spaghetti Fortified with Soy Flour. *J. Food Sci.* **2006,** *71*(6), S428–S432.
90. Silveira, C. M. D.; Badiale-Furlong, E. Sperathe Effects of Solid-State Fermentation in the Functional Properties of Defatted Rice Bran and Wheat Bran. *Braz. Arch. Biol. Technol.* **2009,** *52*(6), 1555–1562.
91. Singh, G.; Tiwari, A.; Rathore, H.; Prasad, S.; Hariprasad, P.; Sharma, S. Valorization of Paddy Straw Using De-oiled Cakes for *P. ostreatus* Cultivation and Utilization of Spent Mushroom Substrate for Biopesticide Development. *Waste Biomass Valorization* **2021,** *12*, 333–346.
92. Singhania, R. R.; Sukumaran, R. K.; Patel, A. K.; Larroche, C.; Pandey, A. Advancement and Comparative Profiles in the Production Technologies using Solid-State and Submerged Fermentation for Microbial Cellulases. *Enzyme Microb. Technol.* **2010,** *46*(7), 541–549.
93. Skinner, R. C.; Gigliotti, J. C.; Ku, K. M.; Tou, J. C. A Comprehensive Analysis of the Composition, Health Benefits, and Safety of Apple Pomace. *Nutr. Rev.* **2018,** *76*(12), 893–909.
94. Sorrenti, V.; Randazzo, C. L.; Caggia, C.; Ballistreri, G.; Romeo, F. V.; Fabroni, S.; Timpanaro, N.; Raffaele, M.; Vanella, L. Beneficial Effects of Pomegranate Peel Extract and Probiotics on Pre-adipocyte Differentiation. *Front. Microbiol.* **2019,** *10*, 660–667.
95. Sudha, M. L.; Baskaran, V.; Leelavathi, K. Apple Pomace as a Source of Dietary Fiber and Polyphenols and Its Effect on the Rheological Characteristics and Cake Making. *Food Chem.* **2007,** *104*(2), 686–692.
96. Takaku, T.; Kimura, Y.; Okuda H. Isolation of an Antitumor Compound from *Agaricus blazei* Murill and Its Mechanism of Action. *J. Nutr.* **2001,** *131*(5), 1409–1413.
97. Thanh, N. V.; Nout, M. J. R. *Rhizopus oligosporus* Biomass, Sporagiospore Yield and Viability as Influenced by Harvesting Age and Processing Conditions. *Food Microbiol.* **2002,** *19*(1), 91–96.
98. TOI (Times of India). Why You Should Have Banana Peel? [Online] 2017. https://timesofindia.indiatimes.com/life-style/health-fitness/diet/why-you-should-have-banana-peel/articleshow/58909498.cms (accessed Apr 20, 2021).
99. Torres-León, C.; Ramírez-Guzman, N.; Londoño-Hernandez, L.; Martinez-Medina, G. A.; Díaz-Herrera, R.; Navarro-Macias, V.; Alvarez-Pérez, O. B.; Picazo, B.; Villarreal-Vázquez, M.; Ascacio-Valdes, J.; Aguilar, C. N. Food Waste and Byproducts: An Opportunity to Minimize Malnutrition and Hunger in Developing Countries. *Front. Sustain. Food Syst.* **2018,** *2*, 52–59.
100. Varzakas, T.; Zakynthinos, G.; Verpoort, F. Plant Food Residues as a Source of Nutraceuticals and Functional Foods. *Foods* **2016,** *5*(4), 88–97.
101. Vattem, D. A.; Ghaedian, R.; Shetty, K. Enhancing Health Benefits of Berries Through Phenolic Antioxidant Enrichment: Focus on Cranberry. *Asia Pac. J. Clin. Nutr.* **2005,** *14*(2), 120–130.
102. Vicenssuto, G. M.; de Castro, R. J. S. Development of a Novel Probiotic Milk Product with Enhanced Antioxidant Properties using Mango Peel as a Fermentation Substrate. *Biocatal. Agric. Biotechnol.* **2020,** *24*, E-Article 101564.
103. Vidyalakshmi, R.; Paranthaman, R.; Indhumathi, J. Amylase Production on Submerged Fermentation by *Bacillus spp. World J. Chem.* **2009,** *4*(1), 89–91.
104. Wadhwa, M.; Bakshi, M. P. S.; Makkar, H. P. S. Wastes to Worth: Value Added Products from Fruit and Vegetable Wastes. *CAB Rev.* **2015,** *10*, 43–49.

105. Woon, K. S.; Lo, I. M. C.; Chiu, S. L. H.; Yan, D. Y. S. Environmental Assessment of Food Waste Valorization in Producing Biogas for Various Types of Energy Use based on LCA Approach. *Waste Manage.* **2016,** *50*, 290–299.
106. Yadav, K.; Singh,J.; Rehal. Studies on Fortification of Wheat Flour with Defatted Rice Bran for Chapatti Making. *J. Food Sci. Technol.* **2012,** *49*(1), 96–102.
107. Zhao, L.; Wang, K.; Wang, K.; Zhu, J.; Hu, Z. Nutrient Components, Health Benefits, and Safety of Litchi (*Litchi chinensis* Sonn.): A Review. *Compr. Rev. Food Sci. Food Saf.* **2020,** *19*(4), 2139–2163.

PART III

NUTRACEUTICALS AND ENCAPSULATION METHODS IN THE FOOD INDUSTRY

CHAPTER 10

PROSPECTS OF CITRUS AS A NUTRACEUTICAL FOOD

UPASANA DEB and SHEENA HAORONGBAM

ABSTRACT

The Nokrek biosphere reserve of West Garo Hills in Meghalaya, India, houses the genetic reserve of the ancestor of oranges (*Citrus indica* or the Indian Wild Orange). In addition, Meghalaya also hosts some other species of *Citrus* that are endemic to the region, such as *C. latipes, Citrus macroptera* Montrouz, *Citrus limetta, Citrus medica, Citrus limon*, and *Citrus reticulata*, out of which *C. indica, C. latipes,* and *C. reticulata* are found specifically only in Meghalaya. The locals of Meghalaya have been using these fruits as a food and also in their traditional medicine. In this chapter, an attempt was made to study some local *Citrus* species of Meghalaya with discussions on the beneficial impact of biocompounds on our health. The potential for utilization of the *Citrus* fruits from Meghalaya as a nutraceutical to provide both nutritional as well as health benefits has been discussed. The addition of *Citrus* as a part of daily diet could help boost the immune system in humans during the prevailing global health conditions.

10.1 INTRODUCTION

The world is facing a severe crisis as the pandemic Sars-Cov 2 that infected 79 million persons all around the world and claimed the life of 1,730,411 people.[23] The advent of coronavirus has taught us the lesson that we should be prepared for any outbreak of disease. However, the only preparation that

Novel and Alternative Methods in Food Processing: Biotechnological, Physicochemical, and Mathematical Approaches. N. Veena, Megh R. Goyal, & Ritesh B. Watharkar (Eds.)

we can have is enhancing our immunity power. Therefore, the need of the moment is to concentrate on the consumption of foods with functionality as a nutraceutical.

The term "nutraceutical" was first used by Stephen Defelice in 1989, that was framed from two words "nutrition" and "pharmaceuticals," which accordingly means, either the food or a part of the food giving some benefits to health and also providing some medicinal properties like prevention and treatment of a diseases.[20] Nutraceuticals consist of dietary supplements, genetically engineered foods, herbal products, isolated nutrients, processed foods like beverages, soups, cereals, etc.[7]

If we look back to our history, when recent medicines were not yet discovered, people relied on the use of fruits and vegetables as a remedy for their diseases. One such fruit with nutritional benefits is the *Citrus* fruit. The Genus *Citrus* is placed under the subfamily Aurantioideae of Family Rutaceae. This Genus is composed of the fruits of the orange family. The fruits of the *Citrus* family have cosmetic, dietary, medicinal, and nutritional properties. In addition to being a source of vitamin C or ascorbic acid, it serves as a source of citric acid, flavonoids, limonoids, pectins, and phenolics.[9] Citrus fruits are also known to fight against cancer, AIDS, cough and also help to reduce blood pressure.[11]

This chapter focuses on the potential for utilization of fruits from the *Citrus* family as a nutraceutical with the main potentiality for research to be made from the untapped resources found in Meghalaya, India.

10.2 IMPORTANT CITRUS SPECIES IN MEGHALAYA—INDIA

The citrus roots originated from *Citrus indica* Yu Tanaka or the Indian wild orange in the Nokrek region of the West Garo Hills of Meghalaya—India (which was entitled by UNESCO as a Biosphere Reserve in May, 2009).[11] Meghalaya also hosts other varieties of *Citrus* species, like *Citrus reticulata, Citrus latipes, Citrus medica*, etc. Among them, *C. reticulata* or Khasi Mandarin is the most popular one. Table 10.1 shows the species belonging to the *Citrus* family reported from Meghalaya, India. Seven species have been reported namely those of *C. indica, C. latipes, Citrus macroptera* Montrouz, *Citrus limetta, C. medica, Citrus limon*, and *C. reticulata,* out of which *C. indica* and *C. reticulata* are genomes endemic to the region of Meghalaya. As per their taxonomical details and properties which give them their nutraceutical potential, the species are further studied as given further.

TABLE 10.1 Different Species of *Citrus* Found in Meghalaya.

Species of citrus	Regions of Meghalaya from which species have been reported	Reference
Citrus indica	Nokrek biosphere reserve	11
Citrus latipes	Laitjem and Sadew, Khasi Hills	26
Citrus limetta	West Garo Hills	25
Citrus limon	No specific region; found throughout the state	26
Citrus macroptera Montrouz	Mawlong, Wahlong, Tyrna, Khasi Hills; Nokrek biosphere reserve, West Garo Hills	11 26
Citrus medica	Garo Hills	6
Citrus reticulata	Khasi Hills	24

10.2.1 CITRUS INDICA

The species of *C. indica* is found only in the Nokrek biosphere reserve of Meghalaya (Table 10.1).[11] The plant type of the species varies from small bush to climber. The plants are usually strong and thorny in nature. The average height of the plant is 90–180 cm (3–6 ft). The branches rise almost at right angles and the thorn length may be 2–3 cm or above.[12]

The fruits are small in size with a diameter ranging from 3.1 to 4.0 cm and the height measuring 3.0–3.7 cm. The shapes can be spherical or depressed globose and sometimes subpyriform.[12] The outer surfaces of the fruits have a smooth texture and shine due to the presence of oil glands. Different types of colors are observed during the maturation stages starting from immature dark green to partially mature greenish yellow and fully matured deep orange. The spongy mesocarp and thin rind are characteristics of a matured fruit. The segment of a fruit varies from 9 to 12 cm, while 7–12 seeds can be present in a fruit. The seeds are spheroid, creamish colored having green cotyledons and brownish spots in the chalazal area.[12] One characteristic of the fruit is that it will not fall from the branches even if ripe, so they are harvested by plucking the fruits directly from the branches.[12]

The fruits have a pleasant fruity smell but are very sour in taste. It is usually not consumed as fruit but eaten for medicinal purposes along with its skin. The fruit after being plucked may stay fresh for 2 weeks.[8]

Since the fruiting happens annually, hence to preserve the fruit, the people of the Garo tribe of Meghalaya apply two traditional methods. In the first method, they dry the whole fruit by placing it under the sun, while, in the other method, they apply the traditional smoking method where they pack the fruit inside bamboo racks, "gamchang" placed over the traditional fireplace.[18]

The study and local opinions of the Garo people on the plant had described *C. indica* as a strong and hardy plant, which can withstand abiotic and biotic stress. It hardly gets infected by pests and is resistant to diseases such as greening disease. Thus, the study of its genes becomes very important. The gene coding for its hardy traits could be incorporated into related *Citrus* groups of commercial importance, to make them strong.[9,12]

10.2.1.1 MEDICINAL PROPERTIES OF C. INDICA

The consumption of whole fresh and raw fruits and dried fruit powder can cure smallpox.[8,11]

- The fruit is also a remedy for jaundice.[8,11]
- The juice of the fruit taken in a small amount in a cup of water along with a pinch of salt can cure stomach ailments of human beings and domesticated animals.[8,11]
- The oral intake of the whole fruit or its juice can act as an antidote for food poisoning.[8]
- It is also reported to reduce hypertension and cure snake-bites.[11]
- The fruit juice also serves as an energy drink for people suffering from dehydration and fatigue.[11]

Apart from the above health benefits (Table 10.2), the Indian Wild Orange is also used by *Garos* in their final rites to ward off spirits from the body of the deceased. Thus, they locally call the fruit "Memang narang," which means fruit of the ghosts.[18]

10.2.2 CITRUS LATIPES

The plant *C. latipes* is reported from the Laitjem and Sadew, Khasi Hills region in Meghalaya (Table 10.1) and is commonly known as "Khasi papeda."[26] The Khasi tribes of Meghalaya call *C. latipes* as "Soh Kymphor." It is a small and thorny tree with fruit edible but rarely consumed. The fruit tastes bitter sour and eaten raw.[26]

10.2.2.1 APPLICATIONS OF C. LATIPES AS TRADITIONAL MEDICINE

- The leaves of *C. latipes* are put in water and boiled till the appearance of green color develops in the water. The green water can be used as bathing water, which relives aches of body, common cold, fever, and headaches,[26] as shown in Table 10.2.

- The peels of the fruit can be boiled, cooled, and filtered by muslin cloth. The filtrate can then be stored. The addition of water can dilute the filtrate, and it can be taken orally as a treatment for stomach disorders, problems of skin, and constipation. The diluted filtrate can also be used to treat dry and chapped skin.[26]
- The fruit juice can be combined with mustard oil and applied as a balm in the nose and forehead during cold or fever, thereby lowering body temperature.[26]
- The fruit juice can be applied in wounds and cuts as it has antiseptic properties.[26]

10.2.3 CITRUS MACROPTERA MONTROUZ

The plant of *C. macroptera* is reported from the regions of Mawlong, Wahlong, Tyrna, Nokrek biosphere reserve, West Garo hills of Meghalaya (Table 10.1).[11,26] The Khasi tribes call the fruit Soh Kwit while the Garo tribes call it Chambal.[26] Nongalleima et al.[14] confirmed the existence of phytochemical with medicinal properties. Alkaloid, flavonoid, phenolics, glycosides, etc., are also present in the fruit.[14]

10.2.3.1 USE OF C. MACROPTERA AS TRADITIONAL MEDICINE (Table 10.2)

- The fruit juice of *C. macroptera* is reported to give relief from ailments such as fever and stomach disorders. The juice's dilution in water and salt gives relief from constipation and flatulence.[26]
- The Garo tribes of Meghalaya consume the fruit juice as a medicine for food poisoning.[26]
- The fruit peels help to neutralize the meat fat.[26]
- The Khasi tribes simply store the food while the Garos boil the juice, cool it and keep it in bottles made of glass.[26]

10.2.4 CITRUS LIMETTA

The variety of *C. limetta* (Figures 10.1 and 10.2) grows in the regions of West Garo hills of Meghalaya (Table 10.1).[25] It is known as sweet lime or mosambi. Botanically, *C. limetta* is a hybrid of Citron (*C. medica*) and bitter orange (*Citrus aurantium*). It is rich in phytochemicals such as alkaloids, flavonoids, hesperidins, and limonoids.[5] D-limomene is a safe agent used for flavoring and can easily dissolve into the organs, such as intestines, liver, lungs, mammary glands, and kidneys, thereby speeding up the metabolism process. Other beneficial compounds present in *C. limetta* are carotenoids, essential oils, glycosides, phenols, saponins, and steroids.[5]

FIGURE 10.1 *Citrus limetta* fruit.

FIGURE 10.2 Peeled *Citrus limetta* fruit.

TABLE 10.2 Nutritional Aspects of *Citrus* Species From Meghalaya, India.

Citrus sp.	Compounds	Nutraceutical aspects	References
Citrus indica	Not yet isolated or studied	• Traditional medicine: for smallpox, jaundice, stomach ailment; antidote for food poisoning, hypertension, and snake bite	9,12
Citrus latipes	Not yet isolated or studied	• Traditional medicine: for body ache, common cold, fever, dry and chapped skin, and antiseptic	26
Citrus limetta	Ascorbic acid, calcium, potassium, phytochemicals, such as alkaloids, flavonoids, hesperidins, and limonoids; Carotenoids, essential oils, glycosides, phenols, saponins, and steroids	• Flavonoids: secretes bile, acid and digestive juices; • D-limonene: fasten the metabolism process; antioxidant, antibacterial; flushes out toxins; regulates blood circulation, good for skin	5,25
Citrus limon	Carbohydrates (10%) include fructose, glucose, sucrose, pectin, and fiber; Water (88–89%); vitamin C, vitamin B6, potassium, citric acid, hespiridin, diosmin, eriocitrin, D-limonene	• Vitamin C: enhances immune system and skin health • Vitamin B6: generated energy • Potassium: controls hypertension • Citric acid: prevents the formation of kidney stones • Hesperidin: antioxidant, strengthen blood vessels, and prevents atherosclerosis • Diosmin: antioxidant, regulates circulatory system, improves muscle tones • Eriocitrin: antioxidant • D-limonene: stomach reflex, and treat heartburn	3
Citrus macroptera Montrouz	Phenolic compounds: flavonoids, phenolic acid, stilbenes, coumarins, tannins; Flavonoids: Flavanones, flavones, flavonols, hesperidine, narirutum, naringin, eriocitrin	• Flavonoids: reduce the risk of coronary heart diseases, anticarcinogenic, anti-inflammatory; • Traditional medicine: fever, stomach disorder like constipation, antidote for food poisoning	14,26

TABLE 10.2 *(Continued)*

Citrus sp.	Compounds	Nutraceutical aspects	References
Citrus medica	Flavonoids: hespiridin, 3,5,6-trihydroxyl-4,7-dimethoxy flavones, and 3,5,6-trihydroxy-3′,4′,7-trimethoxy flavones. Peel contains coumarins, umbelliferone, limettin, scopoletin and scoparone, Seeds contains nomilinic acid, limonol, and limonin. Essential oils: isolimonene and erucylamide	• Phenols and flavonoids: analgesic, Hypoglycaemic, anticholinesterase, antihelminthic, antimicrobial, anticancer	15
Citrus reticulata	37 compounds in the essential oils: neryl acetate (1.1%), nerol (2.3%), β-caryophyllene (2.6%), geraniol (3.5%), geranyl acetate (3.9%), neral (14.5%), and limonene (46.7%); Maltol Glycerol	• Traditional medicine: for constipation, pain, better vision, good skin • Maltol: an antioxidant, anticonvulsion, antitumor, and antifatigue • Glycerol: antineuralgic and anticataract • Essential oil: antibacterial, anti-inflammatory, and antiviral	4,24

10.2.4.1 EFFECT OF SWEET LIME ON HUMAN HEALTH

- Sweet lime is a stress buster and good detoxifier.[25]
- The daily intake of mosambi (sweet lime) can neutralize the effects of pollution as it helps in flushing out toxins from the body.[25]
- Damian et al.[5] found that the concentration of ascorbic acid in sweet lime is 3.36 ± 0.25 mg/g dextrose water.
- The presence of phytocompounds like flavonoids helps in the secretion of bile, acids, and digestive juices. The potassium helps to prevent dysentery and diarrhea. It also prevents constipation and indigestion. It also has a great potential in the therapy of gastro-esophageal reflux disorder.[25]
- The juice of sweet lime plays a role in the regulation of blood circulation.[25]
- Sweet lime has antibacterial and antioxidant properties.[25]
- The sweet lime juice bleaches the dark spots and blemishes on the skin, thereby giving the skin a natural shine.[25]
- The high calcium and phosphorus content are very good for the unborn babies of the pregnant ladies.[25]

10.2.5 CITRUS MEDICA

C. medica is also called citron and is grown in Garo hills region of Meghalaya (Table 10.1).[6] The exterior of the fruit is rough and lemony in color. The pulp is dry so it is not as juicy as other species of *Citrus*. It is composed of alkaloids, carbohydrates, flavonoids, and phenols. Some of the flavonoids known in the fruit are hespiridin, 3,5,6-trihydroxyl-4, 7-dimethoxy flavones, and 3,5,6-trihydroxy-3',4',7-trimethoxy flavones.[15] The peel contains coumarins, umbellic-ferone, limettin, scoparone, and scopoletin while the seeds contain limonin, limonol, and nomilinic acid. The essential oils like erucylamide and isolimonene are also reported.[15]

10.2.5.1 PHARMALOGICAL PROPERTIES OF C. MEDICA

- Analgesic: The doses of 1, 2, and 4 mL/kg of the fruit of *C. medica* is reported to be effective for the pain stimulated by the method of hot-plate while the doses of 2 and 4 mL/kg were effective in mice for the pain induced by the method of tail immersion.[15] The dose of 4 mL/kg was effective for the pain induced by both the methods of hot plate and tail immersion. The potency of the dose 4 mL/kg was equivalent to diclofenac sodium injection. The analgesic property could be due to the presence of phenols and flavonoids.[15]

- Anhelminthic activity: The exact reason for the anhelminthic activity is reported for *C. medica* and has not been proved significantly. It may be due to its property of stopping glucose inhibition, whereby, the parasites get depleted of glucose, and hence cause death of the parasites. The fruit may even activate the worm's receptor of nicotinic cholinergic, thereby causing hyperpolarization or depolarization.[15]
- Anticancer activity: The ripe and half-ripe fruit juices exhibited anti-cancer properties.[15]
- Antimicrobial activity: The agar cup method proved the antimicrobial property of the fruit. The antibacterial properties were functional against *Bacillus subtilis, Escherichia coli, Proteus vulgaris, Klebsiella pneumonia, Pseudomonas aeruginosa,* and *Staphylococcus aureus.*[15]
- Hypoglycemic and anticholinesterase activity: The peel extract of citron has anticholinesterase and hypoglycemic activity.[15]

10.2.6 CITRUS LIMON

C. limon is known in simple English as lemon and grows throughout the region of Meghalaya (Table 10.1).[3] The fruit of lemon is mainly composed of carbohydrates (10%) and water (88–89%) and the concentration of proteins and fats are very low.[3] The carbohydrates present in lemon are simple sugars like fructose, glucose, and sucrose.[3] The main fiber in lemon is pectin. Pectin may reduce sugar levels in blood by decreasing the breakdown of starch and sugar.[3]

Lemon is rich in Vitamin C, which enhances the immune system and health of the skin.[3] The presence of vitamin B_6 in lemon helps in generating energy by the conversion of food.[3] Potassium is also present in high amount which helps in reducing hypertension.[3]

10.2.6.1 SOME IMPORTANT COMPOUNDS IN LEMON (C. LIMON)

- Citric acid: It is one of the most common organic acid in lemons. Citric acid may help in the prevention of the formation of stones in kidneys.[3]
- D-limonene: It is responsible for the unique aroma of lemon. If isolated, it can treat stomach reflux and heartburn.[3]

- Diosmin: It is an antioxidant that may affect the functioning of the circulatory system. It works in the improvement of muscle tone. Also helps in the reduction of chronic inflammation in the blood vessels.[3]
- Eriocitrin: It is an antioxidant.[3]
- Hesperidin: It is an antioxidant that will give strength to blood vessels and helps in the prevention of atherosclerosis.[3]

10.2.6.2 HEALTH BENEFITS OF LEMONS

- Anemia prevention: Citric acid and vitamin C help in the absorption of iron from the other foods. Therefore, the chance of occurrence of anemia diminishes.[3]
- Cancer: The compounds like D-limonene and hesperidin have anti-cancer properties.[3]
- Digestion: The lemon juice is taken with warm water and it acts like a laxative and thereby treats constipation.[3]
- Heart health: The regular intake of vitamin C can reduce the chances of stroke. The fibers have a role in reducing cholesterol levels in blood. The essential oils in lemons can also protect the oxidation of LDL cholesterol. Diosmin and hesperidin may also have positive effects on the heart.[3]
- Prevention of kidney stones: The citric acid in lemons may have the property to reduce the risk of formation of kidney stones.[3]
- The essential oil of lemons had inhibitory effect on *Enterobacter amnigenus, Enterobacter gergoviae,* and *Staphylococcus carnosus.*[21]

10.2.7 CITRUS RETICULATA

C. reticulata is called Khasi Mandarin or the King of Oranges in common language, while the Khasis call it Soh Niamtra or Soh Myntra and is found endemic to the Khasi hills of Meghalaya (Table 10.1).[24] It is one of the most aromatic and sweetest oranges and has a rich nutritional value, unique taste and quality, bright golden color, and a sweet tanginess. It was granted the geographical identification tag in 2014.[24]

The fruit of *C. reticulata* is distinguished from other *Citrus* fruits as it has tight and smooth skin, which makes the peeling process difficult. The thickness of the peel is 3–3.5 mm. The fruit is slightly larger than a tennis ball, with a weight of 134–150 g. The size is 60–70 mm in diameter. The fruit is segmented into 8 to 10 parts having 18–20 per seed.[24]

10.2.7.1 PHYSICOCHEMICAL COMPOSITION OF C. RETICULATA

Chutia et al.[4] isolated and analyzed the oil of the peels of ripe *C. reticulata* by hydro-distillation and GC–MS, respectively, and identified 37 compounds in the oils. The compounds isolated were neryl acetate (1.1%), nerol (2.3%), β-caryophyllene (2.6%), geraniol (3.5%), geranyl acetate (3.9%), neral (14.5%), and limonene (46.7%).[4]

10.2.7.2 MEDICINAL PROPERTIES OF KHASI MANDARIN

- As traditional Khasi medicine: The Khasi tribes of Meghalaya believe that the fruit could ward off anger. In the region, if any misunderstanding occurs among the tribes, the fruits are gifted to make peace.[24] They also believe that it may protect from constipation and pains. The fibers present in the fruit could be a reason for smooth bowel movements. The ancient Khasis had a belief that the oils present in the peels could provide better vision, good for skin, and could ward off pimples.[24]
- The bioactive compounds present in the fruit such as Maltol is an antioxidant, anticonvulsion, antitumor, and antifatigue. Glycerol present in the fruit has antineuralgic and anticataract properties.[24] There are many other compounds that are antibacterial, anti-inflammatory, antiviral, and hyperglycemic.[24]
- The Khasi Mandarin has antimicrobial properties. The essential oil of the fruit is effective in decreasing the propagation of *Aspergillus flavus*.[22] Chutia et al.[4] found that the minimum inhibitory concentration of 0.2 mL per 100 mL is potent for *Alternaria alternata, Curvularia lunata,* and *Rhizoctonia solani*, while it was more than 0.2 mL per 100 mL for *Helminthosporium oryzae* and *Fusarium oxysporum.*

10.3 SUMMARY

The *Citrus* family is the gift of India to the world. It comes in many varieties, each having their own significance. The beneficial compounds that the *Citrus* fruits have to make them a great nutraceutical. Apart from that, it makes our cuisines more delicious. Seven species of *Citrus,* namely *C. indica, C. latipes, C. macroptera* Montrouz, *C. limetta, C. medica, C. limon, and C. reticulata* are reported to grow throughout the vast area of the state of Meghalaya in

India, out of which the species of *C. indica, C. latipes,* and *C. reticulata* are yet to be tapped into. The indigenous people of Meghalaya have been using these plants of *Citrus* since generations. Due to the geographical region, while medical facilities are limited in some isolated villages, *Citrus* functions as a part of their ethnic medicines. Further study on the *Citrus* plant endemic to the region of Meghalaya may lead to the isolation of compounds useful in the pharmaceutical fields. Therefore, portrayal of the *Citrus* from the region in the field of nutraceutical could help boost the health of the people in the region as well as provide nutritional benefits. As goes the proverb "prevention is better than cure," adding citrus to one's diet, not only makes the diet tastier but also healthy. Therefore, the recommendation of at least a variety of *Citrus* as a part of the daily diet could provide immense health and nutritional benefits, further proving that *Citrus* is a fruit, which in itself is packed with various nutraceutical properties.

KEYWORDS

- **antioxidant**
- **citrus**
- **flavonoid**
- **Meghalaya**
- **nutraceutical**
- **vitamin C**

REFERENCES

1. Azman, N. F. I. N.; Azlan, A.; Khoo, H. E.; Razman, M. R. Antioxidant Properties of Fresh and Frozen Peels of Citrus Species. *Curr. Res. Nutr. Food Sci. J.* **2019,** 7(2), 331–339.
2. Bhutani, K. K.; Goyal, A.; Singh, S. *Herbal Wealth of Northeast India: A Pictorial and Herbaria Guide,* 1st ed.; *S.A.S.*: Nagar, Punjab, India, 2008; p 400.
3. Bjarnadottir, A. Lemons 101: Nutrition Facts and Health Benefits. [Online] 2019, p 1. https://www.healthline.com/nutrition/foods/lemons#benefits (accessed Sept 20, 2021).
4. Chutia, M.; Bhuyan, P. D.; Pathak, M. G.; Sarma, T. C.; Boruah, P. Antifungal Activity and Chemical Composition of *Citrus reticulata* Blanco Essential Oil against Phytopathogens from North East India. *LWT Food Sci. Technol.* **2009,** *42*(3), 777–780.

5. Damián-Reyna, A. A.; González-Hernández, J. C.; Maya-Yescas, R.; de Jesús Cortés-Penagos, C.; del Carmen Chávez-Parga, M. Polyphenolic Content and Bactericidal Effect of Mexican *Citrus limetta* and *Citrus reticulata*. *J. Food Sci. Technol.* **2017,** *54*(2), 531–537.
6. Department of Agriculture and farmer's Welfare, Government of Meghalaya (India); Horticulture Crops. [Online] 2020, p 1. http://megagriculture.gov.in/PUBLIC/crops_fruits.aspx (accessed Dec 21, 2021).
7. Dudeja, P.; Gupta, R. K. Nutraceuticals. Chapter 40. In *Food Safety in the 21st Century*, 1st ed.; Academic Press: Cambridge, 2017; pp 491–496.
8. Government of India. Geographical Indications Journal No. 65. [Online] 2014, pp 7–8. http://www.ipindia.nic.in/writereaddata/Portal/IPOJournal/1_71_1/gi-journal-65.pdf (accessed May 20, 2021).
9. Hazarika, T. K. Citrus Genetic Diversity of North-East India, their Distribution, Ecogeography and Ecobiology. *Genet. Resour. Crop Evol.* **2012,** *59*(6), 1267–1280.
10. Kumar, S.; Nair, K. N.; Jena, S. N. Molecular Differentiation in Indian *Citrus* L. inferred from nrDNA ITS Sequence Analysis. *Genet. Resour. Crop Evol.* **2013,** *60*(1), 59–75.
11. Malik, S. K.; Chaudhury, R.; Dhariwal, O. P.; Kalia, R. K. Collection and Characterization of *Citrus indica* Tanaka and *C. macroptera* Montr.: Wild Endangered Species of Northeastern India. *Genet. Resour. Crop Evol.* **2006,** *53*(7), 1485–1493.
12. Malik, S. K.; Kumar, S.; Singh, I. P.; Dhariwal, O. P.; Chaudhury, R. Socio-Economic Importance, Domestication Trends and *In Situ* Conservation of Wild *Citrus* Species of Northeast India. *Genet. Resour. Crop Evol.* **2013,** *60*(5), 1655–1671.
13. Montanari, A.; Widmer, W.; Nagy, S. Health Promoting Phytochemicals in Citrus Fruit and Juice Products. Chapter 2, In *Functionality of Food Phytochemicals*; Johns, T., Romeo, J. T., Eds.; Springer: Boston 1997; vol 31, pp 31–52.
14. Nongalleima, K.; Ajungla, T.; Singh, C. B. Phytochemical, Total Phenolic, Total Flavonoid and Total Flavonol Content Estimation in *Citrus macroptera*. *J. Med. Plants Stud.* **2017,** *5*(3), 114–118.
15. Panara, K.; Joshi, K.; Nishteswar, K. Review on Phytochemical and Pharmacological Properties of *Citrus medica*. *Int. J. Pharm. Biol. Arch.* **2012,** *3*(6), 1292–1297.
16. Rafiq, S.; Kaul, R.; Sofi, S. A.; Bashir, N.; Nazir, F.; Nayik, G. A. Citrus Peel as a Source of Functional Ingredient: A Review. *J. Saudi Soc. Agric. Sci.* **2018,** *17*(4), 351–358.
17. Sanabam, R.; Singh, N. S.; Sahoo, D.; Devi, H. S. Genetic Relationship of Rough Lemon Landraces and Under-Utilized *Citrus* Genotypes from North-East India Revealed by SSR and RAPD Markers. *Trees* **2018,** *32*(4), 1043–1059.
18. Slow Food. Memang Narang. [Online] 2020, p 1. https://www.fondazioneslowfood.com/en/ark-of-taste-slow-food/memang-narang/ (accessed May 20, 2021).
19. Upadhaya, A.; Chaturvedi, S. S.; Tiwari, B. K. Utilization of Wild Citrus by Khasi and Garo Tribes of Meghalaya. *Indian J. Tradit. Knowl.* **2016,** *15*(1), 121–127.
20. Verma, B.; Popli, H. Regulations of Nutraceuticals in India and US. *Pharma Innov.* **2018,** *7*(7), 811–816.
21. Viuda-Martos, M.; Ruiz-Navajas, Y.; Fernández-López, J.; Pérez-Álvarez, J. Antifungal Activity of Lemon (*Citrus lemon* L.), Mandarin (*Citrus reticulata* L.), Grapefruit (*Citrus paradisi* L.) and Orange (*Citrus sinensis* L.) Essential Oils. *Food Control* **2008,** *19*(12), 1130–1138.
22. Viuda-Martos, M.; Ruiz-Navajas, Y.; Fernandez-Lopez, J.; Perez-Álvarez, J. Antibacterial Activity of Lemon (*Citrus lemon* L.), Mandarin (*Citrus reticulata* L.), Grapefruit (*Citrus

paradisi L.) and Orange (*Citrus sinensis* L.) Essential Oils. *J. Food Saf.* **2008,** *28*(4), 567–576.
23. World-meter. Covid-19 Coronavirus Pandemic. [Online] 2020, p 1. https://www.worldometers.info/coronavirus/ (accessed May 23, 2021).
24. Zizira. King of Oranges–The Khasi Mandarin. [Online] 2018, p 1. https://www.zizira.com/blogs/plants/king-oranges-khasi-mandarin (accessed May 20, 2021).
25. Zizira. Meghalaya's Mosambi–Sweet Lime for Refreshment and Nutrition. [Online] 2018, p 1. https://www.zizira.com/blogs/plants/meghalayas-mosambi-sweet-lime-refreshment-nutrition#:~:text=Well%2C%20it%20is%20a%20cross,west%20Garo%20hills%20of%20Meghalaya (Accessed May 20, 2021).
26. Zizira. Wild Citrus Fruits of Meghalaya and Their Uses You Never Knew About. [Online] 2017, p 1. https://www.zizira.com/blogs/plants/wild-citrus-fruits-meghalaya-uses (accessed May 20, 2021).

CHAPTER 11

POTENTIAL OF FUNCTIONAL LIPIDS: PRODUCTION, PROPERTIES, AND APPLICATIONS

ANKITA KATARIA, BOBADE HANUMAN, and RAJAN SHARMA

ABSTRACT

Dietary sources of functional lipids include plants, meat, and bovine milk from which functional lipids can be obtained for the improvement of our health. For fortification in foods, these may be extracted from these and from other sources like seaweeds and microorganisms by solvent or supercritical fluid extraction. Enzymes can also be used to produce structured lipids mainly by the process of interesterification. Major functional lipids present and obtained from the sources include medium chain fatty acids, dietary phospholipids, and ω−6 and ω−3 fatty acids. The most important are conjugated linoleic acid, docosahexaenoic acid, and eicosapentaenoic acid. The functional lipids play a significant role in preventing chronic risks like cancer, cardiovascular diseases, and atherosclerosis apart from brain development functions. For incorporation in food, these lipids may be so designed to attain certain physical properties like a specific melting point or particular solid fat content. The functional lipids can be incorporated in emulsions mainly beverages, infant formulas, and parenteral nutrition, or used in frying oils to exhibit better thermostability and flavor. These may also be utilized for the production of low-calories foods and biopackaging materials. Thus, functional lipids exhibit a high potential for incorporation in foods with respect to both health and food functionality.

Novel and Alternative Methods in Food Processing: Biotechnological, Physicochemical, and Mathematical Approaches. N. Veena, Megh R. Goyal, & Ritesh B. Watharkar (Eds.)

11.1 INTRODUCTION

Lipids are macromolecules constituted by fatty acids (FAs) and their esters. Mostly, dietary lipids are constituted by triglycerides with microquantities of certain complex lipids like phospholipids, which exist in the cell membranes. They supply 9 kcal/g of energy, which is more than double that supplied by the other two macromolecules, that is, proteins and carbohydrates. Generally, lipids are negatively associated with human health since they contribute to cardiovascular diseases, cancer, type-2 diabetes, and obesity.

Functional foods are essentially considered as conventional foods, which provide physiological and nutritional health benefits on consumption. Functional lipids form a significant part, which are gradually evolving with the increase in industrialization and awareness. Major classes of functional lipids are medium chain fatty acids (MCFAs), polyunsaturated fatty acids (PUFA, ω−6 and ω−3 fatty acids), conjugated linoleic acid (CLA), and phospholipids. These are available from the conventional dietary sources, for example, milk, eggs, plants, and meat. Microbially and enzymatically produced lipids are also gaining attention since dietary sources are limited in nature and may also present hazardous risks. Structured lipid (SL) is another class of functional lipids that are not found naturally but are produced enzymatically or by genetic modification. Novel sources (such as, microalgae and insects) are particularly attractive due to higher lipid content and quality. Functional lipids essentially play biological and clinical roles since they act as regulators and mediators of factors responsible for chronic disorders like inflammation, cancer, cardiovascular diseases, and obesity. The recommended dietary intakes specified for various functional lipids have been specified for achieving the positive outcomes. Thus, these may be supplemented to prepare tailor-made foods for patients with these chronic disorders.

In addition to impressive biofunctionality of lipids, they exhibit certain techno-functional properties making them potential ingredients in value-added food products. They are known to impart physical, mechanical, rheological, and textural attributes to many food preparations such as bakery products and beverages. Novel functional lipids have been given critical attention for their emulsifying properties to replace conventional saturated fats. SLs are designed to favor the desirable properties without compromising the mechanical potential. They are majorly intended to reduce the caloric value for health benefits. Their functionality on the quality attributes of fried products is a novel concept of investigation. Furthermore, lipid-based biodegradable and edible films/coatings induce higher moisture barrier and flexibility owing to hydrophobicity.

This chapter focuses on the major classes of functional lipids, their dietary sources as well as processes for the production (mainly microbial and enzymatic). The clinical roles of the functional lipids are elaborated along with their techno-functional attributes.

11.2 MAJOR CLASSES OF FUNCTIONAL LIPIDS

11.2.1 MEDIUM CHAIN FATTY ACIDS (MCFAS)

A FA mixture constituted by 6–10 carbons is MCFA, which includes caproic acid ($C_{6:0}$), caprylic acid ($C_{8:0}$), and capric acid ($C_{10:0}$). These are esterified to glycerol backbone in medium chain triglycerides (MCTGs), which are majorly present in palm-kernel oil and coconut oil. The metabolization process of MCTGs is similar to carbohydrates since they are rapidly absorbed and utilized like glucose rather than being stored in adipose tissues or impacting the cholesterol levels. Pancreatic lipases can easily hydrolyze MCTGs compared to long chain triglycerides (LCTGs) leading to rapid production of MCFAs. Due to higher water solubility, MCFAs do not require bile salt emulsification to enter the circulatory system. Rapid transportation to liver followed by quick oxidation leads to release of energy.[44] On the contrary, the pancreatic lipases hydrolyze LCTGs to LCFAs and *sn*-2 monoglyceride. Upon bile salt emulsification, absorption into intestinal cells occurs followed by reformation of LCTGs due to reattachment of LCFAs to the *sn*-2 monoglyceride. These form chylomicrons, on combining with apolipoproteins and phospholipids, which enter the lymphatic system and get stored in adipose tissues.[71]

11.2.2 POLYUNSATURATED FATTY ACIDS (PUFAS)

FAs containing 18–20 carbon atoms or more with two or more double bonds are categorized as long-chain PUFAs. The position of the first double bond from the methyl end group categorizes the PUFAs into $\omega-6$ and $\omega-3$ FAs.[121] Major $\omega-6$ FAs are linoleic acid (LA) (C18:2 $\Delta^{9,12}$) and arachidonic acid (AA) (C20:4 $\Delta^{5,8,11,14}$). Major $\omega-3$ FAs are α-linolenic acid (ALA) (C18:3 $\Delta^{9,12,15}$), docosahexaenoic acid (DHA) (C22:6 $\Delta^{4,7,10,13,16,19}$), and eicosapentaenoic acid (EPA) (C20:5 $\Delta^{5,8,11,14,17}$). These cannot be efficiently produced in the human body making them essential FAs. Other PUFAs of nutritional significance such as γ-linolenic acid (GLA) (C18:3 $\Delta^{6,9,12}$) and stearidonic acid (SDA) (C18:4 $\Delta^{6,9,12,15}$) are intermediates formed during the production of ω−6 and ω−3 FAs.[1]

Figure 11.1 shows the steps for the synthesis of certain PUFAs like SDA and EPA in eukaryotes and humans. The formation of DHA requires further steps via Sprecher pathway in mammals. Two successive elongation steps of EPA via Δ6 desaturation to generate docosa-pentaenoic acid ($C_{22:5}$ ω−3 FA) and tetra-cosanol-pentaenoic acid ($C_{24:5}$ ω−3 FA). β-oxidation of $C_{24:5}$ shortens the carbon chain to produce DHA.[101]

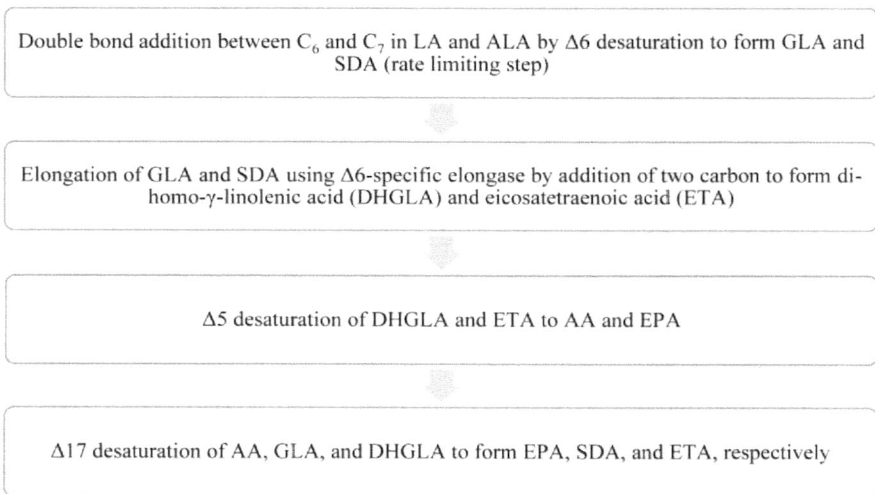

FIGURE 11.1 Synthesis of certain PUFAs in eukaryotes and humans.[1]

The ratio of polyunsaturated to saturated FAs is important to signify the lipid quality of a diet. The optimum value is close to 1. A lower ratio of 0.33 or less promotes atherogenicity and higher ratio of 3 or more promotes tumor.[117] The intake ratio of ω−6 and ω−3 FAs decides the bioconversion processes since both compete for their respective elongase and desaturase enzymes. The formation rate of DHA and EPA is highest if the ratio of ALA to LA is 1:1 in human hepatoma cells.[50] Recommended daily intakes of EPA plus DHA are[73]:

- Healthy adults—250 mg (World Health Organization); 500 mg (American Heart Association).
- Individuals with coronary heart diseases (CHD): 1–2 g.
- Hypertriglyceridemic adults: 4 g.

11.2.3 CONJUGATED LINOLEIC ACID

One of the isomers of LA is CLA, which is a mixture of geometrical and positional isomers of LA typically containing *cis*-9, *trans*-11 CLA and *trans*-10, *cis*-12 CLA isomers in equal concentrations.[127] Its metabolic precursor is *trans*-vaccenic acid (TVA) or *trans*-11 C18:1.[56] Different positions contain the conjugated double-bond configurations wherein the double bonds have either a *trans* or a *cis* orientation.[13] *Cis*-9, *trans*-11 CLA, also known as rumenic acid, is the predominant isomer formed during the biohydrogenation process by ruminant organisms like *Butyrivibrio fibrisolvens* to convert LA to stearic acid (C18:0).[127] Ruminant meat and milk are superior sources of CLA since it is primarily produced in ruminant animals. The action of Δ9-desaturase in the mammary gland leads to the formation of CLA from TVA.[13,46]

11.2.4 DIETARY PHOSPHOLIPIDS

Phospholipids, the structural and functional elements of cell membranes, have a glycerol backbone, two FAs esterified at *sn*-1 and *sn*-2 (conferring hydrophobicity) and phosphate group at *sn*-3 conferring hydrophilicity due to attachment of a polar head group mainly ethanolamine, choline, serine, and inositol.[21] With respect to these polar groups, phospholipids are named as phosphatidylethanolamine, phosphatidylcholine, phosphatidylserine, and phosphatidic acid. Generally, *sn*-1 and *sn*-2 positions are occupied by saturated and unsaturated FAs, respectively.[112]

In the intestinal lumen, phospholipids are cleaved at *sn*-2 by pancreatic lipases to yield lysophospholipids and FFAs. Upon entering the enterocyte, new PLs are formed due to re-esterification of these FFAs which become a part of the outer layer of chylomicrons. These FFAs may also be re-esterfied to form TAG, which move into the nuclei of these chylomicrons which then circulate after entering into the lymph. Some phospholipids may be incorporated into HDL (when unhydrolyzed) or very low-density lipoprotein (VLDL) (when re-esterified to novel phospholipids and TAGs).[21]

11.2.5 STRUCTURED LIPIDS

SLs are defined as chemically or enzymatically modified TAGs containing a different composition of FAs and/or distribution according to position in a glycerol backbone. These are tailor-made lipids with enhanced functional and nutritional attributes.[84] Precise modifications are made in the structure of TAGs for obtaining desirable physiochemical characteristics and incorporating and increasing the essential FAs, such as SCFAs, MCFAs, and PUFAs for improved nutrition.[57] Generally, a liquid oil and a high-melting point fat (like palm oil) are used in the interesterification reaction to form a unique TAG with novel attributes.[78] Certain commercially available SLs are[109]:

- Human milk fat substitutes—Betapol™ and InFat™.
- Low-calorie lipid—Resetta™.
- Fat for baked goods and margarine—NovaLipid™ and SansTrans™.

Medium- and long-chain TAGs wherein *sn*-1, 3 position are esterified with MCFAs and *sn*-2 position with LCFAs are desirable since these can efficiently carry the LCFAs and are easily absorbed in the intestine. In human milk fat substitutes, unsaturated FAs are esterified at *sn*-1,3 and palmitic acid is esterified at *sn*-2 positions. Such a unique composition exhibits multiple benefits such as enhanced absorption of fat and minerals, less constipation, and soft stools. These infant formulas are made using plant oils, such as sunflower, rapeseed, hazelnut, and olive oils.[60]

Apart from interesterification, traditional breeding techniques are used for modifying the crop plant oil FA composition to a limited extent for the production of SLs more recently. Approximately 16–20% oleates are present in sunflower oil which can be increased to 90% with less than 7% saturates using mutagenesis. Moreover, no catalytic hydrogenation is required for its stabilization. Similarly, low saturates but high linolenic acid (flaxseed oil) and high LA (sunflower oil) lipids can be produced using genetic engineering.[84]

11.3 PRODUCTION OF FUNCTIONAL LIPIDS

11.3.1 DIETARY SOURCES

11.3.1.1 MARINE SOURCES

The human body cannot produce ω−3 FAs and thus, dietetic interference is necessary. The most significant source of DHA and EPA is oil of fishes like mackerel, tone, and code. The ratio of ω−3 to ω−6 FAs in the fish tissues varies from 1–10 with variations in the composition of FAs in different tissues.[1] PUFAs like EPA and DHA are produced by phytoplankton which is transferred via the animal food chain. The flesh lipids of fishes are mostly phospholipids (choline phosphatide, ethanolamine phosphatide, and little serine phosphatide). They constitute 0.6–0.8% of the fish muscle but 30–40% of PUFAs are present in these phospholipids. Among ω−3 FAs, the ratio of DHA to EPA is 6:1, and EPA plus DHA may constitute 12–30% of FAs.[2]

11.3.1.2 ANIMAL SOURCES

Milk contains TVA and CLA but in minor quantities, that is, 3% and 1% of the total milk fat, respectively. Thus, it is considered as a poor delivery agent of these functional lipids since this is lower than the required limit for anticarcinogenic response.[72] In bovine milk fat, the most significant isomer of CLA is *cis*-9, *trans*-11 CLA constituting more than 82%. For improved responses, the diet of dairy cattle can be manipulated to obtain a certain profile of FAs in the milk for instance the administration of diet fortified with safflower oil and monensin can increase the production of *cis*-9, *trans*-11 CLA by more than 10-fold. Moreover, safflower and flaxseed oil addition also reduce the levels of short- to medium-chain FAs ($C_{4:0}$–$C_{15:0}$) especially myristic and palmitic acids which are considered as hypercholesterolemic.[13] Moreover, among the dairy and meat products, TVA is the chief isomer of all the *trans*-FAs.[56] Dairy products like yoghurt, milk, butter, and cheese are also significant sources of MCFAs[82] and PUFAs.[1] Milk fat globule membrane is also a rich source of phospholipids containing 46% saturated and 20% unsaturated FAs with negligible PUFA.[132]

The tissues of meat contain ω−6 and ω−3 FAs but red meat is a significant source of docosapentaenoic acid ($C_{22:5}$ $\Delta^{7,10,13,16,19}$).[47] The content of PUFAs depends mainly on the species for instance, the TAG content of beef is 2–3%, whereas that of pork is 7–15%. These PUFA are present in phospholipids of the muscle whose composition is not affected by the diet to a greater extent.[1] One of the most significant sources of phospholipids is egg yolk. Lecithin and cephalin constitute 78% and 18%, respectively, of the total phospholipids which account of 10% of the egg yolk. Their oxidative stability is high due to the presence of saturated FAs at higher levels.[112]

11.3.1.3 PLANT SOURCES

Plants are major sources of ALA which upon consumption are converted to the essential ω−3 FAs like EPA and DHA by desaturases and elongases.[1] Grains (such as chia, soybean, lupin, walnuts, canola, basil, and hemp) are rich sources of PUFAs. Hemp seed oils are also a rich source of GLA (precursor of AA) and SDA (intermediate product during *de novo* conversion of ALA to DHA and EPA).[58] In chia seed oil, the PUFA level of 83% is the maximum in edible oils and in basil seeds, the ω−6 to ω−3 FA ratio ranges from 1:1.6 to 1:3.6.[1]

Fruit seeds like those of blueberry, cranberry, raspberry, rose fruit, black currant, and primrose are also rich in PUFA. More specifically, the ω−6 to ω−3 FA ratio of raspberry is very low, that is, 1.6:1.[87] In total, 12–25% of the FA content is constituted by GLA in black currant seed oil. Evening primrose seed oil contains 8–14% GLA and 65–80% LA among total FAs.[1] Soybean and sunflower oils are also rich sources of phospholipids majorly lecithin or phosphatidylcholine and cephalin or phosphatidylethanolamine and are devoid of cholesterol wherein LA is the major PUFA.[112]

11.3.2 MICROBIAL PRODUCTION

Microbial production can overcome barriers related to production capacity, nutritive requirements, climatic conditions, and space requirements along with higher productivity. Certain oleaginous microorganisms can accumulate a high amount of diverse lipids known as single cell oils with lipid content ranging from 19% (*Cunninghamella echinulata*) to 83% (*Streptomyces coelicolor* TR0958),[7,123] on certain low-cost substrates such as molasses,[96] tomato waste hydrolysate,[41] glycerol,[40] rice hull hydrolysate,[37] and food waste.[92]

There are two stages in the cultivation process: biomass growth and accumulation of lipids. Biomass growth occurs in a nutrient-rich culture medium to enhance cell reproduction. Accumulation of lipids results under stressful conditions wherein the organism faces lack of nitrogen and thus stores energy in the form of lipids.[42] During a stage of lipid turnover, microorganisms may consume short chain lipids consequently leading to higher DHA contents.[25,99] Figure 11.2 shows the process for the production of functional lipids microbially. Cell disruption step is necessary for extracting the lipids efficiently, since they are accumulated in the intracellular vacuoles.[42]

The mechanism for the production of lipids from oleaginous fungi and yeasts involves two steps. Acetyl-CoA is produced in the first step followed by its conversion via lipogenesis pathway into palmitic acid in the subsequent step. The net equation for the process is[12]:

$$\text{7 Malonyl CoA} + \text{2 Acetyl CoA} + 14\ H^+ + \text{14 NADPH} \rightarrow$$

$$\text{Palmitic acid} + \text{8 CoA} + \text{14 NADP}^+ + 7\ CO_2 + 6\ H_2O$$

Further, unsaturated FAs and PUFAs are formed by the elongation and double bond(s) introduction using elongase and desaturase, respectively.[12]

High unsaturated lipids are accumulated by fungal species compared to those by yeasts. Single-cell oils containing higher quantities of γ-linolenic acid are produced by fungal strains of Zygomycetes which possess anticancer activities.[4] *Mortierella* fungal species produce a biomass wherein 50% of total FAs is constituted by AA.[12]

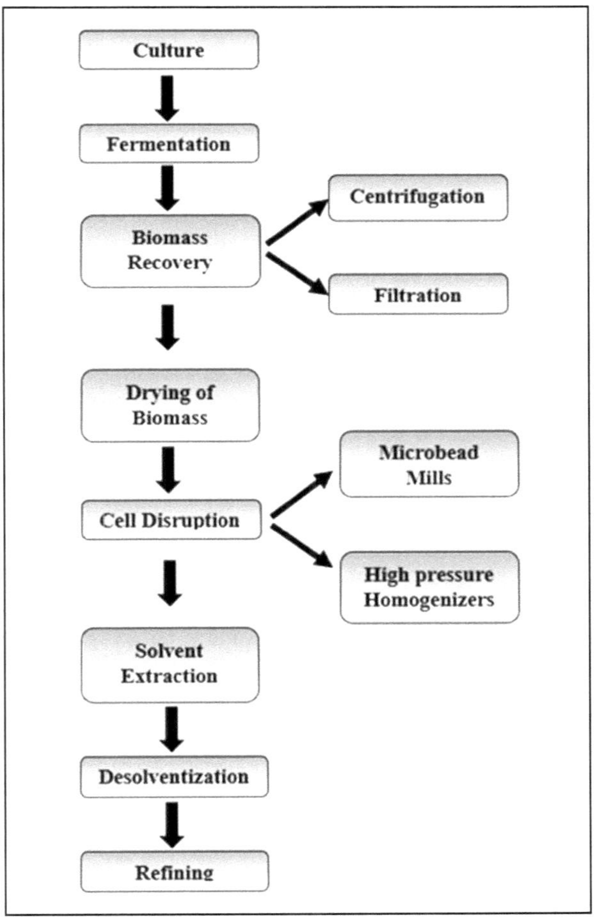

FIGURE 11.2 Schematic diagram for microbial production of functional lipids.

Oleaginous yeast strains such as *Rhodotorula*, *Candida*, *Cryptococcus*, and *Trichosporon* produce mainly oleic (70% w/w), linoleic, palmitic, and stearic acids. For conversion to functional lipids (PUFAs), various approaches used are genetic engineering, desaturase catalysis, and fermentation under reduced oxygen conditions.[9]

11.3.3 EXTRACTION METHODS

The choice of extraction method varies with the raw material, food applications, and toxicity of the solvents. The main objective of lipid extraction is efficient separation and isolation of the desirable lipids from the various components like polysaccharides, proteins, fibers, and miscellaneous molecules.

11.3.3.1 CLASSICAL METHODS

The classical method includes Soxhlet extraction method which is semi-continuous extraction of lipids in solvents followed by desolventization. Extraction efficiency can be enhanced by the application of strategies such as microwave to reduce the consumption of solvent and time of extraction.[45] The major disadvantages are inability to extract single class of lipids, over-estimation of total FA yield due to presence of coextracted compounds, and instability of unsaturated FAs at high temperatures.[83]

Another method for extraction of total lipids is the Folch-method in which the solvent system of chloroform: methanol (2:1) and addition of salts to the crude extract to form biphasic system wherein lower phase contains the lipid fraction and upper phase contains the nonlipid fraction.[83] Other solvents like hexane can be applied to prevent the usage of a toxic mixture of methanol and chloroform but the extraction efficiency decreases with higher moisture sensitivity.[107]

11.3.3.2 PRESSURE LIQUID EXTRACTION

This method is an advancement of the classical method of Soxhlet extraction wherein high pressure and temperature conditions are employed to improve the extraction. Such conditions facilitate improved mass transfer and solubility leading to reduced extraction times and consumption of solvents. Commonly employed conditions are 10–20 MPa and 80–200°C. Desolventization is carried out using pressurized nitrogen.[83] Total lipids have been extracted from *Rhodotorula glutinis*, an oleaginous yeast.[22] Extraction of different lipid classes is also possible by application of sequential pressure liquid extraction.[93]

11.3.3.3 SUPERCRITICAL FLUID EXTRACTION (SFE)

SFE is based on improved penetration and effusion of a sample by increasing the temperature and pressure above the critical point of a supercritical fluid to achieve higher solubility of a particular analyte.[83] CO_2 is commonly used solvent due to gentle supercritical conditions, that is, low critical temperature of 31.1°C. Moreover, only slightly polar or nonpolar substances can dissolve, and oxidation is also prevented which makes it highly suitable to extract PUFAs.[15,100] DHA can be selectively extracted from heterotrophic microalgae *Crypthecodinium cohnii* with a purity of 72% using supercritical fluid extraction (SFE).[31]

11.3.4 ENZYMATIC PRODUCTION

Interesterification process is necessary to modify the natural lipids for certain applications. These may be used in food and pharmaceutical industries for specific applications such as infant foods and patients with fat malabsorption disorders.[109] Interesterification may include:

- Acidolysis—FA and TAG.
- Alcoholysis—Alcohol and TAG; avoided due to formation of undesirable products.
- Glycerolysis—Glycerol and TAG.
- Transesterification—Splitting of ester bonds between the glycerol backbone and FA followed by random shuffling in the pool of FAs; re-esterification at a new position of the same (intraesterification) or another glycerol (inter-esterification): most widely used process.

It may be carried out chemically or enzymatically. Enzymatic interesterification is advantageous due to better selectivity, less by-products, feasible reaction conditions, and ease of recovery of products along with minimum environmental damage.[126] Lipases utilized for the reaction may be FA specific, stereospecific, or regiospecific. Stereospecific lipases act at different rates on positions *sn*-1 and *sn*-3 of TAG. FA lipases act on a specific FA class. Regiospecific lipases (from *Rhizopus arrhizus* and *Aspergillus niger*) are specific to FA ester bonds positioned at *sn*-1 and *sn*-3 of TAG.

Due to steric hindrance, *sn*-2 position is not acted upon by the lipases.[57] Immobilized lipases are commercially utilized for different applications. These include Lipozyme® TL IM from *Thermomyces lanuginosus* (low-calorie SL, SL with ω−3 FAs), Lipozyme® RM IM from *Rhizomucor miehei* (human milk fat substitute, trans-free margarine), and Novozyme® 435 from *Candida antarctica* (low-calorie SL).[109]

Single-step transesterification using *sn*-1, 3 lipases shows lower efficiency of FA insertion and can produce partial acylglycerols as byproducs along with the required TAGs making it difficult to purify them. Two-step transesterification is thus used wherein, alcoholysis of TAG produces *sn*-2 monoacylglycerol which is recovered by crystallization. This is esterified with the desired FA to yield a pure TAG.[109] Short-path distillation can be used for purification of ω−3 PUFA (31%) rich monoacylglycerols of higher purity (91%) prepared from sardine oil by the application of immobilized lipase from *C. antarctica* after enzymatic glycerolysis.[110]

11.3.5 MISCELLANEOUS SOURCES

Although fish oil is a significant source of DHA, consumers refrain from consuming it due to certain undesirable flavors and odors. Moreover, it is also reported to possess hazardous levels of antibiotics, heavy metals, and methyl mercury apart from environmental contaminants like dioxins, dichlorodiphenyltrichloroethane, and polychlorinated biphenyls. Additionally, it is expensive and a declining resource making it necessary to search for safe and sustainable sources. Dietary ALA is a precursor for DHA production by the body but the efficiency is low due to deficiency in cofactors like proteins, B vitamins, vitamin C, magnesium, and zinc necessary for the conversion. ALA alone cannot meet the demand during the development of the fetus and neonates.[26]

Microalgae is a source for DHA and the cofactors necessary for its synthesis from ALA. DHA cannot be synthesized de novo even in the fishes but they obtain it from microalgae.[26] DHA is produced by species of Euglenophyceae, Prymnesiophyceae, and Dinophyceae. EPA is also produced by certain species of Rhodophyceae, Glaucophyceae, Eustigmatophyceae, Cryptophyceae, Prasinophyceae, and Xanthophyceae. Their biomass may contain upto 50% lipids constituting 30–70% of DHA.[73] Since they usually possess thick cell walls, microalgae cells need to be processed using methods, such as microwaves, sonication, thermolysis followed by conventional extraction of lipids. PUFAs can be extracted using gas chromatography–mass spectrometry after derivatization into methyl or ethyl esters.[19,66]

Orthopteran insects [such as house cricket (*Acheta domesticus*), wild meadow grasshoppers (*Chorthippus parallelus* Zetterstedt), and long-winged conehead (*Conocephalus discolor* Thunberg)] are commonly consumed for their high levels of proteins.[23] But, high levels of lipids (10–24%) are also present in these insects making them sustainable sources of food lipids.[90] ALA is the major lipid in *Ch. parallelus* constituting 38.83% of its total lipid content whereas for *Ac. domesticus* and *Co. discolor*, oleic and LAs are the major constituents. Thus, ALA can be supplemented using *Ch. parallelus*. Moreover, the polyunsaturated to saturated FAs ratio for lipids *Ac. domesticus* nearing one and low ratio of ω−6 to ω−3 FAs imply superior lipid quality.[88]

11.4 BIOLOGICAL AND CLINICAL ROLES OF FUNCTIONAL LIPIDS

11.4.1 BRAIN DEVELOPMENT

ω−3 FAs are essential for processes involved in brain development especially cell differentiation, neurogenesis, and neuronal cell membrane integrity apart from the optimum neurologic function such as synaptic transmission and brain plasticity.[10] They act as precursors to neurotransmitters and thus deficiency of these PUFAs is linked to mental illnesses.[26] The mechanism of DHA to modulate neurochemical activities includes modulation of neurotransmitter reuptake, degradation, synthesis, and receptor binding; anti-inflammatory and antiapoptotic effects; and the enhancement of cell membrane fluidity and neurogenesis via upregulation of brain-derived neurotrophic factor.[111] It is facilitated by certain cofactors like N-acetyl cysteine, S-adenosyl methionine, folate, B-vitamins, vitamin C, vitamin D, magnesium, and zinc.[26]

11.4.2 PREVENTION OF CHRONIC RISKS

Patients of diseases related to lifestyle like diabaters, atherosclerosis, obesity, and hypertension are increasing due to industrialization. Dietary lipids are significant in modulating risks related to such disorders. ω−3 FAs like ALA, EPA, and DHA are the precursors of molecules known to aid in resolution of inflammation (proresolving mediators), whereas ω−6 PUFAs like LA and AA have notable proinflammatory attributes.[68] Certain chronic diseases, which can result in significant morbidity and mortality, are regulated by potent lipid mediators for which DHA acts as a precursor.[43] According to WHO and Food and Drug Administration (FDA), lipid peroxidation increases can be prevented by the consumption of 2 and 3 g per day, respectively, of EPA and DHA. For patients of prior myocardial infarction or CHD, mortality can be reduced by daily consumption of 0.85–4 g EPA plus DHA for 12–42 months with treatment.[73] Since MCTGs are rapidly oxidized, they can be supplemented in diets for patients with Alzheimer's disease due to improvement in neuronal metabolism.

Certain bioactive lipids like TVA and CLA promote health and disease prevention like bone formation, weight control, and decrease of risk of atherosclerosis and cancer.[124] The level of CLA required to exhibit anticarcinogenic response is 0.055–3.5 g per day.[72] Various functional lipids reduce the risks of these chronic disorders by playing significant biological roles as shown in Table 11.1.

11.4.3 HYPOLIPIDEMIC EFFECTS

TVA exhibits lipid reduction characteristics causing reduced hepatic lipogenesis and secretion of chylomicrons consequently inducing hypotriglyceridemic effects during dyslipidemia. This can be explained by two possible reasons[56]:

- CLA is a natural peroxisome proliferator-activated receptor (PPAR) ligand. Adipogenesis and metabolism of FFAs in the adipose tissue can be induced for treating type 2 diabetes by PPAR-agonists. Thus, TAG synthesis and FA oxidation are promoted by CLA which leads to the lowering of circulating concentrations of nonesterified FAs (NEFA) and TAG.
- Two lipogenic enzymes namely acetyl-CoA carboxylase-1 (ACC-1) and FA synthase (FAS) promote FA and consequent TAG synthesis which is then stored in the hepatocyte in the form of lipid droplets and secreted as VLDL in the blood or it may undergo oxidative hydrolysis. ACC-1 and FAS pathways may be acted upon or differently regulated by TVA and CLA leading to post-translational modifications which then cause reduced VLDL secretion and TAG.

Therefore, metabolism of body lipids is enhanced due to lipid-lowering attributes of TVA and CLA upon chronic dietary supplementation.[56]

TABLE 11.1 Roles of Various Functional Lipids in Certain Chronic Disorders.

Functional lipid	Biological roles	Reference
Medium chain fatty acids	Obesity and lipid metabolism:	
	• Suppression of lipogenesis and enhancement of lipolysis	108
	• Higher expenditure of postprandial energy	104
	Diabetes and hypertension:	
	• Increase in insulin-mediated glucose metabolism	36
	• Increased adiponectin levels which reduce obesity	114
	• No effect on blood pressure with capacity to overcome adrenergic stress	54
Polyunsaturated fatty acids	Cardiovascular diseases and obesity:	
	• Regulation of cholesterol, inflammation, and lipogenesis	30
	• Activation of brown adipose tissues and increase in thermogenesis for expenditure of energy	85
	Cancer:	
	• Induction of tumor cell apoptosis	16
	• Influence membrane fluidity by altering the plasma membrane lipid composition thereby affecting interaction between antigen-presenting cells and T cells	131
	• ω−3 FAs suppress and ω−6 FAs promote development of cancer	101
	Bone health:	
	• Mediators of hormonal changes, calcium absorption, lipid peroxidation • Improvement of bone mineral density; therefore, useful for osteoporosis patients	38
Conjugated linoleic acid	Cancer:	
	• Reduced expression of 5-lipoxygenase and cyclooxygenase by *cis*-9, *trans*-11 CLA	127
	• Apoptosis and cell cycle modulation by *trans*-10, *cis*-12 CLA	127

TABLE 11.1 *(Continued)*

Functional lipid	Biological roles	Reference
	Obesity:	
	• Transcription of the enzymes required for synthesis of fatty acids and TAGs is inhibited	69
	• Enhancement of lipolysis and fatty acid oxidation thereby impairing storage of adipocyte TAG	34
Phospholipids	Immunology:	
	• Enhancement of phagocytic potential	79
	• Restoration of lymphocytic potential	70
	Inflammation:	
	• Reduction of rheumatoid arthritis	51
	• Reduced levels of tumor necrosis factors	116

TABLE 11.2 Significance of Balance of ω–6 to ω–3 Fatty Acids in the Diet.

Functional deficit	Effects of lower ratio of ω–6 to ω–3 fatty acids	Reference
Cognitive impairment in children with epilepsy	• Improvement in seizure controls on appropriate supplementation of ω–6 and ω–3 FAs • ω–3 FA levels are positively correlated with cognitive function scores • ω–6 to ω–3 should not exceed 4:1	10
Increased maternal adiposity	• ω–3 FAs can reduce the gestational inflammation during maternal obesity • Placental functioning and fetal growth and brain development are optimum with ω–3 FAs • Less infant negative affect (crying, fear reactivity, and fearfulness) on higher EPA levels	49
Chronic obstructive pulmonary disease (COPD)	• Higher lung function with higher ω–3 Fas • Higher COPD morbidity with increased ω–6 FA intake (lesser quality of life, improper respiratory health, worse dyspnea)	68
Cardiovascular diseases (CVDs)	• Reduction in serum cholesterol, non-HDL-cholesterol, and LDL-cholesterol • Lower contents of serum ApoC3 protein which has proatherogenic characteristics, that is, inflammation and dysfunction of endothelial cells • Reduction in inflammation by reduction in proinflammatory cytokines tumor necrosis factor α (TNF α) and interleukin 6 (IL 6)	128

11.4.4 OPTIMAL BALANCE OF Ω–6 AND Ω–3 FOR HUMAN NUTRITION

Certain therapeutic foods are administered for the management of nutrition, especially for children suffering from severe malnutrition. Conventional ready-to-eat therapeutic foods constitute dietary sources like peanuts and vegetable oils with ω−6 LA in higher levels than its requirements but negligible amounts of ω−3 α-linolenic acid (ALA). Both LA and ALA are the chief PUFAs that act as substrates for the formation of long-chain PUFAs especially ω−6 FAs (AA) and ω−3 FAs (EPA and DHA) during the endogenous metabolism. DHA is significantly present in neural tissue membranes and, immune functions and inflammation modulation are influenced by EPA and DHA. But, EPA and DHA are present in limiting levels in the diet which can be improved by consumption of marine foods that present the challenge of spoilage.[18]

Apart from availability, the optimum balance of the FAs is necessary. ω−6 FAs should constitute only 20–40% of total long-chain PUFAs in the diet.[73] Modifications in diet over the years have led to a marked reduction in the consumption of ω−3 FAs causing higher ratios of ω−6 to ω−3 FAs, as high as 20:1 compared to ancient diets wherein this ratio was 1:1.[128] The average human diet contains ω−6 and ω−3 FAs in the ratio of 10:1 and the recommended ratio is 5:1.[128] Poor balance of various PUFAs shows ample evidence of functional deficits (Table 11.2).

11.5 FOOD FUNCTIONALITY

11.5.1 PHYSICAL PROPERTIES

Lipids strongly influence the physical, textural, and rheological properties of food systems and their mechanical performance is the key element dictating the quality of finished products. Since melting is associated with techno-functional characteristics of food, biochemical composition, and structural orientation of fats are vital components to induce required functionality in the food matrix. The functioning of fats in different food products is dependent on the source, FA composition, structural features of lipids, and characteristics of the end products, for instance, higher fat content enhances the tenderness of cookies otherwise they exhibit higher breaking strength.[64,86]

Moreover, they also affect the properties of complementing constituents as lipids at a 30% level in dough have been reported to inhibit the retrogradation properties of starch and further extended the plasticity of the gluten matrix.[89] They are also responsible to impart lubricity, mouthfeel, spreading, flavor, cracking, and shelf life to many bakery and related products. Shortening, an important ingredient and process parameter in bakery, is characteristic to fewer of the lipids especially smooth and plastic fats.[35] Polymorphism attributes fundamental functionality to laminated doughs and other bakery products characterized by the presence of β-crystals.[74] Melting point representing the saturation of FAs and globule size of fat may influence the thermo-mechanical properties of the cheese including flowability, oiling off, stretchability, and meltability.[105]

Introduction of foreign fat restructured the milk fat globules and acted as a potential lubricant in protein layers and contributed as active fillers in the formulated cheese.[105] Lipids also enhance the water holding capacity and emulsion stability of the meat products at elevated levels due to higher linking with protein molecules or water induced by enhanced availability of free radicals.[103] The authors also reviewed that lower fat content may increment the cooling losses and alter the harness of meat emulsions. Fats destroy the protein structure of the meat and facilitate the protein–gel matrix.[74] Certainly, FAs composition governs the consistency of the fats and thus modifies the textural and physical attributes of emulsion-based products.

11.5.2 EMULSIONS

In food systems, when oil and water being two immiscible liquids are dispersed in another phase generate emulsion; however, they are generally unstable under given thermodynamic conditions. There are numerous physicochemical processes that facilitate the instability of the emulsions such as flocculation, foaming, creaming, partial or complete coalescence, phase inversion, or Ostwald ripening.[76] Lipid-based emulsifiers include monoglycerides, diglycerides, phospholipids, spans, and tweens but they must be amphiphilic in nature. Most of them are characterized by single or multiple aliphatic chains and may be nonionic, anionic, or zwitterionic. The generation of association colloids is the main mechanism by which thermodynamically stable system is formed as these emulsifiers minimize the contact area between water and nonpolar ends of emulsifiers.[76] Phospholipids are one of the most exploited agents as lipid-based emulsifiers. Phosphoric acid and related groups make up the hydrophilic fraction while glycerol backbone chain comprising of FAs form the lipophilic tail.[77]

Soya, milk, and rice phospholipids have been reported to induce high-fat destabilization ratio and resistance to melt in addition to improvement in the desirable attributes of ice cream.[97] Phospholipids transform their physical structure into bilayers majorly responsible to influence the particle size; however, in certain cases, they may construct several bilayers, piled head to head disturbing the properties of resultant emulsions.[77] Also, the balance of hydrophilic and hydrophobic fractions governs the particle size distribution.[55] Sunflower phospholipids have been reported as natural emulsifiers to deliver encapsulated omega-3 FAs with low pH and high electrostatic screening to alter electrostatic repulsion favoring the physical stability of generated nanoemulsions.[62]

Lecithin is the commercial phospholipid emulsifier comprising of varied concentrations of phosphatidylinositol, phosphatidic acid, phosphotidylethanolamine, and phosphatidylcholine in different biological ingredients, such as eggs, milk, canola, sunflower, and cottonseed.[39,77,125] Other than this, avocado phospholipids when used as an emulsifier at 2% level exhibited gel-like and pseudoplastic behavior, resulting in the formation of stable oil-in-water emulsions; however, at 1% avocado phospholipid, inversion interval was noted between 60% and 70% oil phase.[133]

11.5.3 FRYING OILS

Frying is the among the best conventional techniques for cooking imparting palatability and favorable characteristics to food such as golden color, desirable texture, and savory flavor.[129] Functionality and quality of lipids during frying operations are dependent on physicochemical properties including viscosity, color, degree of unsaturation, acid value, p-anisidine value, and polarity.[75] Various biochemical changes such as oxidation, hydrolysis, and polymerization occur during frying due to high temperatures depending upon their composition and antioxidant potential. Generally, higher degree of unsaturation causes thermal degradation of oils during frying process and volatile compounds also dictate the quality of oils.[29,115]

Studies have indicated the presence of more than 200 volatile compounds in different frying oils responsible for the distinguished flavor of fried foods including aldehydes, ketones, alcohols, hydrocarbons, and furans developed by two major biochemical pathways of oxidation and Maillard reaction.[24] Flavoring compounds are also synthesized as a result of several secondary processes such as degradation, rearrangement, and fragmentation due to the interference of oxidation products in Maillard reaction.[24] Egg yolk phospholipids have been referred to as potential functional agents for frying being rich in arachidonic and LAs for the synthesis of odorant compounds during frying.[27]

Certainly, the texture is an important parameter associated with fried foods defined in terms of crispiness, crunchiness, and crumbliness; however, sporadic reports are available underlying the absolute mechanism and impact of operational parameters on these responses. Bahram-Parvar et al.[11] stated that hardness, modulus of elasticity, and fracture force are inversely related to the severity of the frying treatment in terms of time and temperature. Golden yellow color is often a desirable attribute of fried products; however, treatment temperature, oil type, and thickness of the food product may negatively influence the color of the end product. Since the color profile is governed by polyphenolic composition of oils, Salehi[102] suggested that relatively lower temperature and use of hydrogenated oils may reduce color degradation during processing. However, coated fried food has been reported to exhibit brighter red and yellow colors.[63]

11.5.4 LOW-CALORIE PRODUCTS

A great deal of efforts has been witnessed in the recent past to develop low-calorie fats which impart desirable functionality to food without increment in the calorie content. Bakery products have observed replacement of shortenings which are 100% refined vegetable oils or blended fats comprising of few hydrogenated lipids by low-calorie substitutes such as oatrim and oleogels of canola oil-carnauba wax.[28,61,67]

Pintado and Cofrades[91] found that emulsion gels of chia and olive oils and beeswax oleogels when used in place of animal fats maintained the technological properties of fermented meat product. Despite of the use of plant-based fats, oxidative stability was not lowered. Gelled oil in water emulsions (linseed oil in κ-carrageenan) also have been successfully used to modify the FA composition and calorific value of meat sausages with no compromise to taste, texture, color, and sensory profile; however, minor concentration of volatile compounds as a result of oxidation was observed.[5]

Berasategi et al.[14] also used linseed gelled emulsion to produce low-energy bologna sausage with reduced saturated fats having PUFAs and stated no deterioration in the functionality. Amylose-lipid complex (ALC) is also a key ingredient to develop low-calorie products as supported by Lee et al.[65] who found that substitution of ALC in bakery retained the quality attributes of bread while reducing the energy intake. Replacement of conventional fats with quality lipids rich in bioactive compounds are sometimes better choice than calorie deficient diet due to necessity of metabolic processes. Olein, a fraction of chia oil rich in omega 3 FAs, when used in ice cream to replace milk fat did not significantly varied the energy value but improved the antioxidant potential without altering organoleptic profile.[118]

Structural lipids are modified fats obtained by restructuring the FA composition with the aim of refining functional and nutritional characteristics of existing fats. Mota et al.[81] utilized crude oil of olive pomace and spent coffee grounds to develop low-calorie structured fats which may be investigated for their potential in food products to target calorie reduction and nutritive enhancement.

11.5.5 BIOPACKAGING

Concerning the environmental impact and toxicity of synthetic polymers, biodegradable materials including polysaccharides, proteins, and fats have been assessed for containment and barrier properties. Lipid-based biopackages are generally made of glycerides and waxes to induce high moisture barrier and hydrophobicity; however, they possess brittle nature. Rice bran oil, beeswax, carnauba, and paraffin are some of the important candidates explored for functionality in biodegradable and edible films.[106]

Several authors have investigated lipids as biopackaging materials such as edible coating of sunflower oil on pork hamburgers,[120] rice bran oil to maintain freshness, and extend shelf stability of kiwi[53] and palm fruit oil[98] which resulted in promising results.[80] Substitution of functional lipids to other biopolymers inhibits moisture penetration and enhances the flexibility, cohesiveness, and hydrophobic nature in addition to the retention of aroma, tenderness, freshness, color, and quality of food products.[33] Acetylated monoglycerides exhibit improved moisture barrier attributed to elimination of hydrophilic hydroxyl groups and this property augments with increment in degree of acetylation.[32]

Wax-based coatings and films have highest moisture barrier and improve the visual appeal and surface appearance of food products; however, if dense layers are used, they must be discarded while thin coatings are fit for consumption.[17] Shellac resins that are aliphatic alicyclic hydroxyl acid polymers have been permitted to use as tributary chemicals mainly in pharmaceutical and edible films but their effectiveness and potential in food products remain to be explored.[52] Since lipids are prone to hydrolytic rancidity and oxidation, incorporation of fats in films at higher level may impart undesirable taste and appearance to contained products. Lipid-based coatings and films can be applied in various forms including dispersions, emulsion, monolayer, and composite layers.[95]

Although different in composition from lipids, essential oils also have remarkable hydrophobicity, excellent polyphenol composition to induce bioactivity. Bioactive compounds in lipids and essential oils may incorporate antioxidant and antimicrobial effects making them biofunctional additives to valorize the product and extend its shelf life; however, they may influence the tensile strength and transparency of the films.[8] Table 11.3 indicates recent applications of lipids in novel food functionalities and their influence on product quality.

11.6 SUMMARY

With the evolution of the functional food category, functional lipids have gained significance due to their beneficial attributes both physiologically and technically. Dominant classes are MCFAs, PUFAs, CLA, and phospholipids. Although they are present in conventional ingredients like seeds, eggs, milk, and meat, the natural forms pose challenges in their absorption, processing, and storage. Supercritical extraction or microbial and enzymatic production of the required lipids are thus necessary for ease of supplementation in foods and pharmaceuticals. Lifestyle diseases or chronic disorders can be cured or prevented by regular consumption of the functional lipids as recommended. $\omega-3$ PUFAs and CLA are essential specifically for brain development and hypolipidemic effects, respectively. Physical and techno-functionality of lipids in food systems have been mainly determined in terms of physical and textural attributes of several bakery preparations. Mechanical and rheological characteristics induced by lipids include lubricity, spreading, cracking, flowability, oiling off, stretchability, and meltability. They are potential emulsifiers reducing the surface contact area between water and lipid fractions thus augmenting the stability of emulsion and transforming the system into pseudoplastic fluids. Frying is another inseparable property of oils which is a function of FA composition, degree of unsaturation, and other physicochemical properties. Several novel fats have been explored and utilized for the development of low-calorie food products comprising of SLs and emulsion gels. Several researchers have investigated the lipid-based biofilms and coating and found improved flexibility, cohesiveness,

TABLE 11.3 Recent Applications of Lipids in Novel Food Functionalities and Their Influence on Product Quality.

Functionality in food system	Food product	Functional lipid	Influence on food properties	Reference
Deep frying	Wheat/rice flakes	Coconut oil	Higher oil absorption and better textural properties when compared to palm oil; lower density of the flakes; healthier frying medium	113
Edible coating	Guava fruit	Date pit oil	Considerable increment in the shelf life; retention of quality attributes indicated by pH, moisture loss, vitamin C content, firmness, and titratable acidity	3
Edible nanocoating	Strawberry	Beeswax	Decline in the decay rate; inhibition of fungal growth; limited respiration, reduced water loss, and oxygen migration	130
Emulsification	Meat products	Olive oil	Successful replacement of animal fat (pork back fat); no significant variation in product profile; improved bioactive composition due to healthy lipid profile	20
Emulsification	Chicken sausage	Canola oil	Improved emulsion stability, replacement of animal fat; reduction in cooking losses during preparation; increased level of ALA	119
Frying	Banana chips	Sunflower oil	Higher retention of flavor compared to palm oil; shelf stability for 2 weeks; acceptable sensory characteristics	59
Low-calorie food	Cake	Beeswax–sunflower oil (75:25)	Structured low-calorie fat caused reduction in weight gain; modification in plasma cholesterol level; reduction in low-density lipoproteins and triglycerides	94
Low-fat food for cancer patients	Ice cream	Olive oil	Lower fat content; acceptable sensory parameters; improved nutritive potential; higher PUFAs	122
Physicochemical properties	Domiati cheese	Jojoba oil	Reduction in the cholesterol profile; improvement in the sensory scores; enhancement in the shelf stability	6
Techno-functionality	Shortbread cookies	Extra virgin olive oil emulsion gel	50% replacement of shortening exhibited desirable fracture and microstructure; caused 39% reduction in saturated fats and 19% reduction in total lipid content	46

and moisture barrier due to high hydrophobicity. These findings have opened the doors to explore many other novel functionalities of lipids to address the present-day challenges in both nutritive and technical terms.

KEYWORDS

- **chronic risks**
- **conjugated linoleic acid**
- **food functionality**
- **functional lipids**
- **medium chain fatty acid (MCFA)**
- **polyunsaturated fatty acid (PUFA)**

REFERENCES

1. Abedi, E.; Sahari, M. A. Long-Chain Polyunsaturated Fatty Acid Sources and Evaluation of their Nutritional and Functional Properties. *Food Sci. Nutr.* **2014,** *2*(5), 443–463.
2. Ackman, R. G. Marine Lipids and Omega-3 Fatty Acids. In *Handbook of Functional Lipids*; Akoh, C. C., Ed.; CRC Press: Boca Raton, FL, 2005; pp 314–327.
3. Ahmed, A.; Ali, S. W.; Imran, A.; Afzaal, M.; Arshad, M. S.; Nadeem, M.; Mubeen, Z.; Ikram, A. Formulation of Dates-Pit Oil-Based Edible Wax Coating for Extending the Storage Stability of Guava Fruit. *J. Food Process. Preserv.* **2020,** *44*(2), E-article 14336.
4. Alakhras, R.; Bellou, S.; Fotaki, G.; Stephanou, G.; Demopoulos, N. A.; Papanikolaou, S.; Aggelis, G. Fatty Acid Lithium Salts from *Cunninghamella Echinulata* have Cytotoxic and Genotoxic Effects on HL-60 Human Leukemia Cells. *Eng. Life Sci.* **2015,** *15*(2), 243–253.
5. Alejandre, M.; Poyato, C.; Ansorena, D.; Astiasarán, I. Linseed Oil Gelled Emulsion: A Successful Fat Replacer in Dry Fermented Sausages. *Meat Sci.* **2016,** *121*, 107–113.
6. Ali, R. F.; El-Anany, A. M.; Mousa, H. M. The Effect of Partially or Totally Replacing Milk Fat by Jojoba Oil in Cheese Production on Some Nutritional and Quality Properties. *Nutr Food Sci.* **2020,** *51*(1), 191–204.
7. Arabolaza, A.; Rodriguez, E.; Altabe, S.; Alvarez, H.; Gramajo, H. Multiple Pathways for Triacylglycerol Biosynthesis in *Streptomyces coelicolor*. *Appl. Environ. Microbiol.* **2008,** *74*(9), 2573–2582.
8. Atarés, L.; Chiralt, A. Essential Oils as Additives in Biodegradable Films and Coatings for Active Food Packaging. *Trends Food Sci. Technol.* **2016,** *48*, 51–62.
9. Athenaki, M.; Gardeli, C.; Diamantopoulou, P.; Tchakouteu, S. S.; Sarris, D.; Philippoussis, A.; Papanikolaou, S. Lipids from Yeasts and Fungi: Physiology, Production and Analytical Considerations. *J. Appl. Microbiol.* **2018,** *124*(2), 336–367.

10. Bahgat, K. A.; Elhady, M.; Abdel Aziz, A.; Youness, E. R.; Zakzok, E. Omega-6/Omega-3 Ratio and Cognition in Children with Epilepsy. *An. Pediatr. (Barc).* **2019,** *91*(2), 88–95.
11. BahramParvar, M.; Moghaddam, T. M.; Razavi, S. M. A. Effect of Deep-Fat Frying on Sensory and Textural Attributes of Pellet Snacks. *J. Food Sci. Technol.* **2014,** *51*(12), 3758–3766.
12. Béligon, V.; Christophe, G.; Fontanille, P.; Larroche, C. Microbial Lipids as Potential Source to Food Supplements. *Curr. Opin. Food Sci.* **2016,** *7*, 35–42.
13. Bell, J. A.; Griinari, J. M.; Kennelly, J. J. Effect of Safflower Oil, Flaxseed Oil, Monensin, and Vitamin-E on Concentration of Conjugated Linoleic Acid in Bovine Milk Fat. *J. Dairy Sci.* **2006,** *89*(2), 733–748.
14. Berasategi, I.; García-Íñiguez de Ciriano, M.; Navarro-Blasco, Í.; Calvo, M. I.; Cavero, R. Y.; Astiasarán, I.; Ansorena, D. Reduced-Fat Bologna Sausages with Improved Lipid Fraction. *J. Sci. Food Agric.* **2014,** *94*(4), 744–751.
15. Bernardo-Gil, M. G.; Grenha, J.; Santos, J.; Cardoso, P. Supercritical Fluid Extraction and Characterization of Oil from Hazelnut. *Eur. J. Lipid Sci. Technol.* **2002,** *104*(7), 402–409.
16. Berquin, I. M.; Edwards, I. J.; Chen, Y. Q. Multi-Targeted Therapy of Cancer by Omega-3 Fatty Acids. *Cancer Lett.* **2008,** *269*(2), 363–377.
17. Bourtoom, T. Edible Films and Coatings: Characteristics and Properties. *Int. Food Res. J.* **2008,** *15*(3), 237–248.
18. Brenna, J. T.; Akomo, P.; Bahwere, P.; Berkley, J. A.; Calder, P. C.; Jones, K. D.; Liu, L.; Manary, M.; Trehan, I.; Briend, A. Balancing Omega-6 and Omega-3 Fatty Acids in Ready-to-Use Therapeutic Foods (RUTF). *BMC Med.* **2015,** *13*(1), 117–121.
19. Brondz, I. Development of Fatty Acid Analysis by High-Performance Liquid Chromatography, Gas Chromatography, and Related Techniques. *Anal. Chim. Acta* **2002,** *465*(1–2), 1–37.
20. Câmara, A. K. F. I.; Okuro, P. K.; Santos, M.; de Souza Paglarini, C.; da Cunha, R. L.; Ruiz-Capillas, C.; Herrero, A. M.; Pollonio, M. A. R. Understanding the Role of Chia (*Salvia Hispanica* L.) Mucilage on Olive Oil-Based Emulsion Gels as a New Fat Substitute in Emulsified Meat Products. *Eur. Food Res. Technol.* **2020,** *246*(5), 909–922.
21. Castro-Gómez, P.; Garcia-Serrano, A.; Visioli, F.; Fontecha, J. Relevance of Dietary Glycerophospholipids and Sphingolipids to Human Health. *Prostaglandins Leukot. Essent. Fatty Acids* **2015,** *101*, 41–51.
22. Cescut, J.; Severac, E.; Molina-Jouve, C.; Uribelarrea, J. L. Optimizing Pressurized Liquid Extraction of Microbial Lipids using the Response Surface Method. *J. Chromatogr. A* **2011,** *1218*(3), 373–379.
23. Chakravorty, J.; Ghosh, S.; Jung, C.; Meyer-Rochow, V. B. Nutritional Composition of *Chondacris rosea* and *Brachytrupes orientalis*: Two Common Insects used as Food by Tribes of Arunachal Pradesh, India. *J Asia Pac. Entomol.* **2014,** *17*(3), 407–415.
24. Chang, C.; Wu, G.; Zhang, H.; Jin, Q.; Wang, X. Deep-Fried Flavor: Characteristics, Formation Mechanisms, and Influencing Factors. *Crit. Rev. Food Sci. Nutr.* **2020,** *60*(9), 1496–1514.
25. Chang, G.; Gao, N.; Tian, G.; Wu, Q.; Chang, M.; Wang, X. Improvement of Docosahexaenoic Acid Production on Glycerol by *Schizochytrium* Sp. S31 with Constantly High Oxygen Transfer Coefficient. *Bioresour. Technol.* **2013,** *142*, 400–406.
26. Charles, C. N.; Msagati, T.; Swai, H.; Chacha, M. Microalgae: An Alternative Natural Source of Bioavailable Omega-3 DHA for Promotion of Mental Health in East Africa. *Sci. Afr.* **2019**, *6*, E-article 00187.

27. Chen, D. W.; Balagiannis, D. P.; Parker, J. K. Egg Yolk Phospholipids: A Functional Food Material to Generate Deep-Fat Frying Odorants. *J. Sci. Food Agric.* **2019**, *99*(14), 6638–6643.
28. Chin, N. L.; Rahman, R. A.; Hashim, D. M.; Kowng, S. Y. Palm Oil Shortening Effects on Baking Performance of White Bread. *J. Food Process Eng.* **2010,** *33*(3), 413–433.
29. Choe, E.; Min, D. B. Chemistry of Deep-Fat Frying Oils. *J. Food Sci.* **2007,** *72*(5), R77–R86.
30. Colussi, G.; Catena, C.; Novello, M.; Bertin, N.; Sechi, L. A. Impact of Omega-3 Polyunsaturated Fatty Acids on Vascular Function and Blood Pressure: Relevance for Cardiovascular Outcomes. *Nutr. Metab. Cardiovasc. Dis.* **2017,** *27*(3), 191–200.
31. Couto, R. M.; Simoes, P. C.; Reis, A., Da Silva, T. L.; Martins, V. H.; Sánchez-Vicente, Y. Supercritical Fluid Extraction of Lipids from the Heterotrophic Microalga *Cryptheco-dinium cohnii*. *Eng. Life Sci.* **2010,** *10*(2), 158–164.
32. de Azeredo, H. M. C. Edible Coatings. In *Advances in Fruit Processing Technologies*; Rodrigues, S., Fernandes, F. A. N., Eds.; CRC Press: Boca Raton, FL, 2012; pp 345–362.
33. Dehghani, S.; Hosseini, S. V.; Regenstein, J. M. Edible Films and Coatings in Seafood Preservation: A Review. *Food Chem.* **2018,** *240*, 505–513.
34. Den-Hartigh, L. J.; Han, C. Y.; Wang, S.; Omer, M.; Chait, A. 10E, 12Z-Conjugated Linoleic Acid Impairs Adipocyte Triglyceride Storage by Enhancing Fatty Acid Oxidation, Lipolysis, and Mitochondrial Reactive Oxygen Species. *J. Lipid Res.* **2013,** *54*(11), 2964–2978.
35. Devi, A.; Khatkar, B. S. Physicochemical, Rheological and Functional Properties of Fats and Oils in Relation to Cookie Quality: A Review. *J. Food Sci. Technol.* **2016,** *53*(10), 3633–3641.
36. Eckel, R. H.; Hanson, A. S.; Chen, A. Y.; Berman, J. N.; Yost, T. J.; Brass, E. P. Dietary Substitution of Medium-Chain Triglycerides Improves Insulin-Mediated Glucose Metabolism in NIDDM Subjects. *Diabetes* **1993,** *41*(5), 641–647.
37. Economou, Ch. N.; Aggelis, G.; Pavlou, S.; Vayenas, D. V. Single Cell Oil Production from Rice Hulls Hydrolysate. *Bioresour. Technol.* **2011,** *102*(20), 9737–9742.
38. El-Sayed, E.; Ibrahim, K. Effect of the Types of Dietary Fats and Non-Dietary Oils on Bone Metabolism. *Crit. Rev. Food Sci. Nutr.* **2017,** *57*(4), 653–658.
39. Erickson, M. C. Chemistry and Function of Phospholipids. In *Food Lipids: Chemistry, Nutrition and Biotechnology*; Akoh, C. A., Min, D. B., Eds.; CRC Press: Boca Raton, FL, 2008; pp 39–62.
40. Ethier, S.; Woisard, K.; Vaughan, D.; Wen, Z. Continuous Culture of the Microalgae *Schizochytrium limacinum* on Biodiesel-Derived Crude Glycerol for Producing Docosahexaenoic Acid. *Bioresour. Technol.* **2011,** *102*(1), 88–93.
41. Fakas, S.; Papanikolaou, S.; Galiotou-Panayotou, M.; Komaitis, M.; Aggelis, G. Organic Nitrogen of Tomato Waste Hydrolysate Enhances Glucose Uptake and Lipid Accumulation in *Cunninghamella echinulata*. *J. Appl. Microbiol.* **2008,** *105*(4), 1062–1070.
42. Finco, A. M. O.; Mamani, L. D. G.; Carvalho, J. C.; de Melo Pereira, G. V.; Thomaz-Soccol, V.; Soccol, C. R. Technological Trends and Market Perspectives for Production of Microbial Oils Rich in Omega-3. *Crit. Rev. Biotechnol.* **2017,** *37*(5), 656–671.
43. Forsyth, S.; Gautier, S.; Salem, Jr. N. Global Estimates of Dietary Intake of Docosahexaenoic Acid and Arachidonic Acid in Developing and Developed Countries. *Ann. Nutr. Metab.* **2016,** *68*(4), 258–267.
44. Galante, J. H.; Tenore, R. R. Medium-Chain Triglycerides. In *Handbook of Functional Lipids*; Akoh C. C., Ed.; CRC Press: Boca Raton, FL 2005; pp 184–190.

45. García-Ayuso, L. E.; Velasco, J.; Dobarganes, M. C.; Luque de Castro, M. D. Accelerated Extraction of the Fat Content in Cheese Using a Focused Microwave-Assisted Soxhlet Device. *J. Agric. Food Chem.* **1999,** *47*(6), 2308–2315.
46. Giarnetti, M.; Paradiso, V. M.; Caponio, F.; Summo, C.; Pasqualone, A. Fat Replacement in Shortbread Cookies Using an Emulsion Filled Gel Based on Inulin and Extra Virgin Olive Oil. *LWT Food Sci. Technol.* **2015,** *63*(1), 339–345.
47. Givens, D. I.; Gibbs, R. A. Very Long Chain n-3 Polyunsaturated Fatty Acids in the Food Chain in the UK and the Potential of Animal-Derived Foods to Increase Intake. *Nutr. Bull.* **2006,** *31*(2), 104–110.
48. Griinari, J. M.; Bauman, D. E. Biosynthesis of Conjugated Linoleic Acid and its Incorporation into Meat and Milk of Ruminants. In *Advances in Conjugated Linoleic Acid Research*; Yurawecz, M. P., Mossoba, M. M., Kramer, J. K. G., Pariza, M. W., Nelson, G. J., Eds.; AOCS Press: Champaign, IL, 1999; pp 180–200.
49. Gustafsson, H. C.; Holton, K. F.; Anderson, A. N.; Nousen, E. K.; Sullivan, C. A.; Loftis, J. M.; Nigg, J. T.; Sullivan, E. L. Increased Maternal Prenatal Adiposity, Inflammation, and Lower Omega-3 Fatty Acid Levels Influence Child Negative Affect. *Front. Neurosci.* **2019,** *13*, 1035–1040.
50. Harnack, K.; Andersen, G.; Somoza, V. Quantitation of Alpha-Linolenic Acid Elongation to Eicosapentaenoic and Docosahexaenoic Acid as Affected by the Ratio of n6/n3 Fatty Acids. *Nutr. Metab.* **2009,** *6*(1), 10–18.
51. Hartmann, P.; Szabó, A.; Erős, G.; Gurabi, D.; Horváth, G.; Németh, I.; Ghyczy, M.; Boros, M. Anti-Inflammatory Effects of Phosphatidylcholine in Neutrophil Leukocyte-Dependent Acute Arthritis in Rats. *Eur. J. Pharmacol.* **2009,** *622*(1–3), 58–64.
52. Hassan, B.; Chatha, S. A. S.; Hussain, A. I.; Zia, K. M.; Akhtar, N. Recent Advances on Polysaccharides, Lipids and Protein Based Edible Films and Coatings: A Review. *Int. J. Biol. Macromol.* **2018,** *109*, 1095–1107.
53. Hassani, F.; Garousi, F.; Javanmard, M. Edible Coating Based on Whey Protein Concentrate-Rice Bran Oil to Maintain the Physical and Chemical Properties of the Kiwifruit (*Actinidia Deliciosa*). *Trakia J. Sci.* **2012,** *10*(1), 26–34.
54. Iemitsu, M.; Shimojo, N.; Maeda, S.; Irukayama-Tomobe, Y.; Sakai, S.; Ohkubo, T.; Tanaka, Y.; Miyauchi, T. The Benefit of Medium-Chain Triglyceride Therapy on the Cardiac Function of SHRs Is Associated with a Reversal of Metabolic and Signaling Alterations. *Am. J. Physiol. Heart Circ. Physiol.* **2008,** *295*(1), H136–H144.
55. Ishii, F.; Nii, T. Properties of Various Phospholipid Mixtures as Emulsifiers or Dispersing Agents in Nanoparticle Drug Carrier Preparations. *Colloids Surf. B Biointerfaces* **2005,** *41*(4), 257–262.
56. Jacome-Sosa, M. M.; Lu, J.; Wang, Y.; Ruth, M. R.; Wright, D. C.; Reaney, M. J.; Shen, J.; Field, C. J.; Vine, D. F.; Proctor, S. D. Increased Hypolipidemic Benefits of Cis-9, Trans-11 Conjugated Linoleic Acid in Combination with Trans-11 Vaccenic Acid in a Rodent Model of the Metabolic Syndrome, the JCR:LA-Cp Rat. *Nutr. Metab.* **2010,** *7*(1), 60–68.
57. Kadhum, A. A. H.; Shamma, M. N. Edible Lipids Modification Processes: A Review. *Crit. Rev. Food Sci. Nutr.* **2017,** *57*(1), 48–58.
58. Kataria, A.; Ahluwalia, P.; Sharma, S.; Singh, B. Cannabis Infused Foods. In *Trends & Prospects in Food Science & Processing Technology*; Prasad, V. M., Gupta, A., Singh, B., Misra, N., Mani, A., Eds.; Satish Serial Publishing House: Delhi, 2020; pp 145–172.

59. Khanvilkar, A. M.; Kamble, A. B.; Ranveer, R. C.; Ghosh, J. S.; Sahoo, A. K. Effect of Frying Media and Primary Packaging Material on Shelf Life of Banana Chips. *Int. Food Res. J.* **2016,** *23*(1), 284–288.
60. Kim, B. H.; Akoh, C. C. Recent Research Trends on the Enzymatic Synthesis of Structured Lipids. *J. Food Sci.* **2015,** *80*(8), C1713–C1724.
61. Kim, J. Y.; Lim, J.; Lee, J.; Hwang, H.-S.; Lee, S. Utilization of Oleogels as a Replacement for Solid Fat in Aerated Baked Goods: Physicochemical, Rheological, and Tomographic Characterization. *J. Food Sci.* **2017,** *82*(2), 445–452.
62. Komaiko, J.; Sastrosubroto, A.; McClements, D. J. Encapsulation of Omega-3 Fatty Acids in Nanoemulsion-Based Delivery Systems Fabricated from Natural Emulsifiers: Sunflower Phospholipids. *Food Chem.* **2016,** *203*, 331–339.
63. Krokida, M. K.; Oreopoulou, V.; Maroulis, Z. B.; Marinos-Kouris, D. Color Changes during Deep Fat Frying. *J. Food Eng.* **2001,** *48*(3), 219–225.
64. Lai, H. M.; Lin, T. C. Bakery Products: Science and Technology. Chapter 1; In *Bakery Products: Science and Technology*; Hui, Y. H., Corke, H., De Leyn, I., Nip, W. K., Cross, N. A., Eds.; Blackwell Publishing: Boston, MA, 2006; pp 3–65.
65. Lee, H. S.; Kim, K. H.; Park, S. H.; Hur, S. W.; Auh, J. H. Amylose-Lipid Complex as a Fat Replacement in the Preparation of Low-Fat White Pan Bread. *Foods* **2020,** *9*(2), 194–199.
66. Lee, J. Y.; Yoo, C.; Jun, S. Y.; Ahn, C. Y.; Oh, H. M. Comparison of Several Methods for Effective Lipid Extraction from Microalgae. *Bioresour. Technol.* **2010,** *101*(1), S75–S77.
67. Lee, S.; Kim, S.; Inglett, G. E. Effect of Shortening Replacement with Oatrim on the Physical and Rheological Properties of Cakes. *Cereal Chem.* **2005,** *82*(2), 120–124.
68. Lemoine, C.; Brigham, E.; Woo, H.; Koch, A.; Hanson, C.; Romero, K.; Putcha, N.; McCormack, M.; Hansel, N. Relationship Between Omega-3 and Omega-6 Fatty Acid Intake and Chronic Obstructive Pulmonary Disease Morbidity. *Ann. Am. Thorac. Soc.* **2020,** *17*(3), 378–381.
69. Ma, L.; Lengi, A. J.; McGilliard, M. L.; Bauman, D. E.; Corl, B. A. Effect of Trans-10, Cis-12 Conjugated Linoleic Acid on Activation of Lipogenic Transcription Factors in Bovine Mammary Epithelial Cells. *J. Dairy Sci.* **2014,** *97*(8), 5001–5006.
70. Maczek, C.; Böck, G.; Jürgens, G.; Schönitzer, D.; Dietrich, H.; Wick, G. Environmental Influence on Age-Related Changes of Human Lymphocyte Membrane Viscosity using Severe Combined Immunodeficiency Mice as an *In-Vivo* Model. *Exp. Gerontol.* **1998,** *33*(5), 485–498.
71. Man, Y. B. C.; Manaf, M. A. Medium-Chain Triacylglycerols. In *Nutraceutical Science and Technology*; Manaf, M. A., Man, Y. C., Eds.; CRC Press: Boca Raton, FL, 2006; pp 27–56.
72. Martínez-Monteagudo, S. I.; Saldaña, M. D. A. Retention of Bioactive Lipids in Heated Milk: Experimental and Modelling. *Food Bioprod. Process.* **2015,** *94*, 290–296.
73. Martins, D.; Custódio, L.; Barreira, L.; Pereira, H.; Ben-Hamadou, R.; Varela, J.; Abu-Salah, K. Alternative Sources of n-3 Long-Chain Polyunsaturated Fatty Acids in Marine Microalgae. *Mar. Drugs* **2013,** *11*(7), 2259–2281.
74. Mattice, K. D.; Marangoni, A. G. Fat Crystallization and Structure in Bakery, Meat, and Cheese Systems. Chapter 10; In *Structure-Function Analysis of Edible Fats*, 2nd ed.; Marangoni, A. Ed.; Elsevier: London, 2019; pp 287–311.
75. Mba, O. I.; Dumont, M.-J.; Ngadi, M. Deterioration Kinetics of Crude Palm Oil, Canola Oil and Blend during Repeated Deep-Fat Frying. *J Am. Oil Chem. Soc.* **2016,** *93*(9), 1243–1253.

76. McClements, D. J. Lipid-Based Emulsions and Emulsifiers. Chapter 3; In *Food Lipids: Chemistry, Nutrition and Biotechnology*, 3rd ed.; Akoh, C. C., Min, D. B., Eds.; Marcel Dekker, Inc.: Basel, 2008; pp 63–98.
77. McClements, D. J.; Gumus, C. E. Natural Emulsifiers—Biosurfactants, Phospholipids, Biopolymers, and Colloidal Particles: Molecular and Physicochemical Basis of Functional Performance. *Adv. Colloid Interface Sci.* **2016,** *234*, 3–26.
78. Mensink, R. P.; Sanders, T. A.; Baer, D. J.; Hayes, K.; Howles, P. N.; Marangoni, A. The Increasing Use of Interesterified Lipids in the Food Supply and Their Effects on Health Parameters. *Adv. Nutr.* **2016,** *7*(4), 719–729.
79. Miranda, D. T. S. Z; Batista, V. G.; Grando, F. C. C.; Paula, F. M.; Felício, C. A.; Rubbo, G. F. S.; Fernandes, L. C.; Curi, R.; Nishiyama, A. Soy Lecithin Supplementation Alters Macrophage Phagocytosis and Lymphocyte Response to Concanavalin A: Study in Alloxan-Induced Diabetic Rats. *Cell Biochem. Funct.* **2008,** *26*(8), 859–865.
80. Mohamed, S. A.; El-Sakhawy, M. Polysaccharides, Protein and Lipid-Based Natural Edible Films in Food Packaging: A Review. *Carbohydr. Polym.* **2020,** *238*, E-article 116178.
81. Mota, D. A.; Rajan, D.; Heinzl, G. C.; Osório, N. M.; Gominho, J.; Krause, L. C.; Soares, C. M.; Nampoothiri, K. M.; Sukumaran, R. K.; Ferreira-Dias, S. Production of Low-Calorie Structured Lipids from Spent Coffee Grounds or Olive Pomace Crude Oils Catalyzed by Immobilized Lipase in Magnetic Nanoparticles. *Bioresour. Technol.* **2020,** *307*, E-article 123223.
82. Nagao, K; Yanagita, T. Medium-Chain Fatty Acids: Functional Lipids for the Prevention and Treatment of the Metabolic Syndrome. *Pharmacol. Res.* **2010,** *61*(3), 208–212.
83. Ochsenreither, K.; Glück, C.; Stressler, T.; Fischer, L.; Syldatk, C. Production Strategies and Applications of Microbial Single Cell Oils. *Front. Microbiol.* **2016,** *7*, 1539–1542.
84. Osborn, H. T.; Akoh, C. C. Structured Lipids-Novel Fats with Medical, Nutraceutical, and Food Applications. *Compr. Rev. Food Sci. Food Saf.* **2002,** *1*(3), 110–120.
85. Pahlavani, M.; Razafimanjato, F.; Ramalingam, L.; Kalupahana, N. S.; Moussa, H.; Scoggin, S.; Moustaid-Moussa, N. Eicosapentaenoic Acid Regulates Brown Adipose Tissue Metabolism in High-Fat-Fed Mice and in Clonal Brown Adipocytes. *J. Nutr. Biochem.* **2017,** *39*, 101–109.
86. Pareyt, B.; Brijs, K.; Delcour, J. A. Impact of Fat on Dough and Cookie Properties of Sugar-Snap Cookies. *Cereal Chem.* **2010,** *87*(3), 226–230.
87. Parry, J.; Yu, L. Fatty Acid Content and Antioxidant Properties of Cold-Pressed Black Raspberry Seed Oil and Meal. *J. Food Sci.* **2004,** *69*(3), FCT189–FCT193.
88. Paul, A.; Frederich, M.; Megido, R. C.; Alabi, T.; Malik, P.; Uyttenbroeck, R.; Francis, F.; Blecker, C.; Haubruge, E.; Lognay, G. Insect Fatty Acids: Comparison of Lipids from Three Orthopterans and Tenebrio Molitor L. Larvae. *J. Asia Pac. Entomol.* **2017,** *20*(2), 337–340.
89. Perego, P.; Sordi, A.; Guastalli, R.; Converti, A. Effects of Changes in Ingredient Composition on the Rheological Properties of a Biscuit Industry Dough. *Int. J. Food Sci. Technol.* **2007,** *42*(6), 649–657.
90. Pino Moreno, J. M.; Ganguly, A. Determination of Fatty Acid Content in Some Edible Insects of Mexico. *J. Insects Food Feed* **2016,** *2*(1), 37–42.
91. Pintado, T.; Cofrades, S. Quality Characteristics of Healthy Dry Fermented Sausages Formulated with a Mixture of Olive and Chia Oil Structured in Oleogel or Emulsion Gel as Animal Fat Replacer. *Foods* **2020,** *9*(6), 830–838.

92. Pleissner, D.; Lam, W. C.; Sun, Z.; Lin, C. S. K. Food Waste as Nutrient Source in Heterotrophic Microalgae Cultivation. *Bioresour. Technol.* **2013,** *137*, 139–146.
93. Poerschmann, J.; Carlson, R. New Fractionation Scheme for Lipid Classes Based on "In-Cell Fractionation" using Sequential Pressurized Liquid Extraction. *J. Chromatogr. A* **2006,** *1127*(1–2), 18–25.
94. Rashid, R. M. S.; Barqi, A. J. A. Nutritional and Biological Evaluation of Low–Calorie Cake Prepared by Beeswax-Sunflower Oil Formulated Fat. *Indian J. Public Health Res. Dev.* **2020,** *11*(4), 1646–1651.
95. Regubalan, B.; Pandit, P.; Maiti, S.; Nadathur, G. T.; Mallick, A. Potential Bio-Based Edible Films, Foams, and Hydrogels for Food Packaging. In *Bio-Based Materials for Food Packaging*; Ahmed, S., Ed.; Springer: Singapore, 2018; pp 105–123.
96. Ren, L. J.; Li, J.; Hu, Y. W.; Ji, X. J.; Huang, H. Utilization of Cane Molasses for Docosahexaenoic Acid Production by *Schizochytrium* Sp. CCTCC M209059. *Korean J. Chem. Eng.* **2013,** *30*(4), 787–789.
97. Rinaldi, M.; Dall'Asta, C.; Paciulli, M.; Guizzetti, S.; Barbanti, D.; Chiavaro, E. Innovation in the Italian Ice Cream Production: Effect of Different Phospholipid Emulsifiers. *Dairy Sci. Technol.* **2014,** *94*(1), 33–49.
98. Rodrigues, D. C.; Cunha, A. P.; Brito, E. S.; Azeredo, H. M.; Gallão, M. I. Mesquite Seed Gum and Palm Fruit Oil Emulsion Edible Films: Influence of Oil Content and Sonication. *Food Hydrocoll.* **2016,** *56*, 227–235.
99. Ryu, B.-G.; Kim, K.; Kim, J.; Han, J. I.; Yang, J. W. Use of Organic Waste from the Brewery Industry for High-Density Cultivation of the Docosahexaenoic Acid-Rich Microalga, *Aurantio chytrium* Sp. KRS101. *Bioresour. Technol.* **2013,** *129*, 351–359.
100. Sahena, F.; Zaidul, I. S. M., Jinap, S.; Karim, A. A.; Abbas, K. A.; Norulaini, N. A. N.; Omar, A. K. M. Application of Supercritical CO_2 in Lipid Extraction: A Review. *J. Food Eng.* **2009,** *95*(2), 240–253.
101. Saini, R. K.; Keum, Y. S. Omega-3 and Omega-6 Polyunsaturated Fatty Acids: Dietary Sources, Metabolism, and Significance—A Review. *Life Sci.* **2018,** *203*, 255–267.
102. Salehi, F. Color Changes Kinetics during Deep Fat Frying of Carrot Slice. *Heat Mass Transf* **2018,** *54*(11), 3421–3426.
103. Santhi, D.; Kalaikannan, A.; Sureshkumar, S. Factors Influencing Meat Emulsion Properties and Product Texture: A Review. *Crit. Rev. Food Sci. Nutr.* **2017,** *57*(10), 2021–2027.
104. Scalfi, L.; Coltorti, A.; Contaldo, F. Postprandial Thermogenesis in Lean and Obese Subjects after Meals Supplemented with Medium-Chain and Long-Chain Triglycerides. *Am. J. Clin. Nutr.* **1991,** *53*(5), 1130–1133.
105. Schenkel, P.; Samudrala, R.; Hinrichs, J. Thermo-Physical Properties of Semi-Hard Cheese Made with Different Fat Fractions: Influence of Melting Point and Fat Globule Size. *Int. Dairy J.* **2013,** *30*(2), 79–87.
106. Sharma, R.; Ghoshal, G. Emerging Trends in Food Packaging. *Nutr. Food Sci.* **2018,** *48*(5), 764–779.
107. Sheng, J.; Vannela, R.; Rittmann, B. E. Evaluation of Methods to Extract and Quantify Lipids from Synechocystis PCC 6803. *Bioresour. Technol.* **2011,** *102*(2), 1697–1703.
108. Shinohara, H.; Wu, J.; Kasai, M.; Aoyama, T. Randomly Interesterified Triacylglycerol Containing Medium-and Long-Chain Fatty Acids Stimulates Fatty Acid Metabolism in White Adipose Tissue of Rats. *Biosci. Biotechnol. Biochem.* **2006,** *70,* 2919–2926.
109. Sivakanthan, S.; Madhujith, T. Current Trends in Applications of Enzymatic Interesterification of Fats and Oils: A Review. *LWT* **2020,** *132*, E-article 109880.

110. Solaesa, Á. G.; Sanz, M. T.; Falkeborg, M.; Beltrán, S.; Guo, Z. Production and Concentration of Monoacylglycerols Rich in Omega-3 Polyunsaturated Fatty Acids by Enzymatic Glycerolysis and Molecular Distillation. *Food Chem.* **2016,** *190*, 960–967.
111. Sun, G. Y.; Simonyi, A.; Fritsche, K. L.; Chuang, D. Y.; Hannink, M.; Gu, Z.; Greenlief, C. M.; Yao, J. K.; Lee, J. C.; Beversdorf, D. Q. Docosahexaenoic Acid (DHA): An Essential Nutrient and a Nutraceutical for Brain Health and Diseases. *Prostaglandins Leukot. Essent. Fatty Acids* **2018,** *136*, 3–13.
112. Sun, N.; Chen, J.; Wang, D.; Lin, S. Advance in Food-derived Phospholipids: Sources, Molecular Species and Structure as well as their Biological Activities. *Trends Food Sci. Technol.* **2018,** *80*, 199–211.
113. Surojanametakul, V.; Karnasuta, S.; Satmalee, P. Effect of Oil Type and Batter Ingredients on the Properties of Deep-Frying Flakes. *Food Sci. Technol.* **2020,** *40*(2), 592–596.
114. Takeuchi, H.; Noguchi, O.; Sekine, S.; Kobayashi, A.; Aoyama, T. Lower Weight Gain and Higher Expression and Blood Levels of Adiponectin in Rats Fed Medium-Chain TAG Compared with Long-Chain TAG. *Lipids* **2006,** *41*(2), 207–212.
115. Thomsen, B. R.; Yesiltas, B.; Sørensen, A. D. M.; Hermund, D. B.; Glastrup, J.; Jacobsen, C. Comparison of Three Methods for Extraction of Volatile Lipid Oxidation Products from Food Matrices for GC–MS Analysis. *J. Am. Oil Chem. Soc.* **2016,** *93*(7), 929–942.
116. Treede, I.; Braun, A.; Sparla, R.; Kühnel, M.; Giese, T.; Turner, J. R.; Anes, E.; Kulaksiz, H.; Füllekrug, J.; Stremmel, W.; Griffiths, G. Anti-Inflammatory Effects of Phosphatidylcholine. *J. Biol. Chem.* **2007,** *282*(37), 27155–27164.
117. Turley, J.; Thompson, J. *Nutrition: Your Life Science*, 2nd ed.; Cengage Learning: Boston, MA, 2015; p 560.
118. Ullah, R.; Nadeem, M.; Imran, M. Omega-3 Fatty Acids and Oxidative Stability of Ice Cream Supplemented with Olein Fraction of Chia (*Salvia hispanica* L.) Oil. *Lipids Health Dis.* **2017,** *16*(1), 34.
119. Utama, D. T.; Jeong, H. S.; Kim, J.; Barido, F. H.; Lee, S. K. Fatty Acid Composition and Quality Properties of Chicken Sausage Formulated with Pre-Emulsified Perilla-Canola Oil as an Animal Fat Replacer. *Poultry Sci.* **2019,** *98*(7), 3059–3066.
120. Vargas, M.; Albors, A.; Chiralt, A. Application of Chitosan-Sunflower Oil Edible Films to Pork Meat Hamburgers. *Procedia Food Sci.* **2011,** *1*, 39–43.
121. Venegas-Calerón, M.; Sayanova, O.; Napier, J. A. An Alternative to Fish Oils: Metabolic Engineering of Oil-Seed Crops to Produce Omega-3 Long Chain Polyunsaturated Fatty Acids. *Prog. Lipid Res.* **2010,** *49*(2), 108–119.
122. Vieira, F. G. K.; De Salles, R. K.; Mannes, P.; Kami, A. A.; Búrigo, T.; Geraldo, A. P. G.; Moreno, Y. M. F. Development and Acceptance of an Ice Cream as Food Alternative for Cancer Patients. *J. Culin. Sci. Technol.* **2020,** *18*(2), 89–97.
123. Wang, H.; Deng, Z.; Zhou, S.; Zhang, X. Using Potato Starch Wastewater to Produce GLA with *Cunninghamella echinulata. China Oils Fats* **2007,** *32*(8), 49–51.
124. Wang, Y.; Jacome-Sosa, M. M.; Proctor, S. D. The Role of Ruminant Trans Fat as a Potential Nutraceutical in the Prevention of Cardiovascular Disease. *Food Res. Int.* **2012,** *46*(2), 460–468.
125. Whitehurst, R. J. Ed.; *Emulsifiers in Food Technology*; John Wiley & Sons: Oxford, UK, 2008; p 264.
126. Xie, W.; Yang, X.; Zang, X. Interesterification of Soybean Oil and Methyl Stearate Catalyzed by Guanidine-Functionalized SBA-15 Silica. *J. Am. Oil Chem. Soc.* **2015,** *92*(6), 915–925.

127. Yang, B., Chen, H., Stanton, C., Ross, R. P., Zhang, H., Chen, Y. Q.; Chen, W. Review of the Roles of Conjugated Linoleic Acid in Health and Disease. *J. Funct. Foods* **2015,** *15*, 314–325.
128. Yang, L. G.; Song, Z. X.; Yin, H.; Wang, Y. Y.; Shu, G. F.; Lu, H. X.; Wang, S. K.; Sun, G. J. Low n-6/n-3 PUFA Ratio Improves Lipid Metabolism, Inflammation, Oxidative Stress and Endothelial Function in Rats Using Plant Oils as n-3 Fatty Acid Source. *Lipids* **2016,** *51*(1), 49–59.
129. Yu, K. S.; Cho, H.; Hwang, K. T. Physicochemical Properties and Oxidative Stability of Frying Oils during Repeated Frying of Potato Chips. *Food Sci. Biotechnol.* **2018,** *27*(3), 651–659.
130. Zambrano-Zaragoza, M. L.; Quintanar-Guerrero, D.; Del Real, A.; González-Reza, R. M.; Cornejo-Villegas, M. A.; Gutiérrez-Cortez, E. Effect of Nano-Edible Coating Based on Beeswax Solid Lipid Nanoparticles on Strawberry's Preservation. *Coatings* **2020,** *10*(3), 253–270.
131. Zheng, J. S., Hu, X. J., Zhao, Y. M., Yang, J.; Li, D. Intake of Fish and Marine n-3 Polyunsaturated Fatty Acids and Risk of Breast Cancer: Meta-Analysis of Data from 21 Independent Prospective Cohort Studies. *BMJ* **2013,** *346*, f3706-f3716.
132. Zhu, D.; Damodaran, S. Dairy Lecithin from Cheese Whey Fat Globule Membrane: Its Extraction, Composition, Oxidative Stability, and Emulsifying Properties. *J. Am. Oil Chem. Soc.* **2013,** *90*(2), 217–224.
133. Züge, L. C. B.; Maieves, H. A.; Silveira, J. L. M.; da Silva, V. R.; de Paula Scheer, A. Use of Avocado Phospholipids as Emulsifier. *LWT Food Sci. Technol.* **2017,** *79*, 42–51.

CHAPTER 12

ENCAPSULATION TECHNOLOGIES IN THE FOOD INDUSTRY

NADIRA ANJUM, INSHA ZAHOOR, MEENAKSHI TRILOKIA, and ANJU BHAT

ABSTRACT

The technique of encapsulation protects food components from volatile losses and prevents loss of color, taste, texture, and aroma from a particular food. Many techniques are being used for encapsulation of food ingredients and components, but spray drying, freeze drying, and coacervation have been mostly used in the field of food technology. The encapsulation technique has a wide range of applications in the field as it finds its way in meat and poultry products, beverages, dairy products, bakery products, and many more. Moreover, research is being carried out to use this technique for the development of functional ingredients like preservatives, colorants, and flavors for use in food products. It provides an important advantage of decreasing production and transportation cost by converting liquids into solids. Selection of an appropriate core material and encapsulation technique play a key role in the process as encapsulation efficiency and stability are determined by these factors. In future, research needs to be done on selection of wall material depending on the physiochemical characteristics of the material to be encapsulated and on the use of coencapsulation techniques for reduction in cost.

12.1 INTRODUCTION

Dried form of any food product is easy to handle, transport, and store than its fresh form. Dehydration on the other hand may result in unfavorable effects

Novel and Alternative Methods in Food Processing: Biotechnological, Physicochemical, and Mathematical Approaches. N. Veena, Megh R. Goyal, & Ritesh B. Watharkar (Eds.)

on texture, aroma, flavor, and other properties of the food product. Nowadays, the focus of food process engineers is toward the use of encapsulation, which is a process of entrapment of an active substance within a protective membrane called as wall material. This technique has been originally used in the field of biotechnology to ensure that the producer cells are efficiently separated from the metabolites. The technique of encapsulation has been developed about 60 years before for the purpose of effective drug delivery in the pharmaceutical industry.[12]

The increasing interest of consumers in functional foods and nutraceuticals has made researchers and scientists to shift their focus toward the bioactive components obtained from plant sources like fruits, vegetables, cereals, pulses, etc.[31] Also, the awareness of consumers about the foods with health benefits is increasing day by day; as a result, scientists and food manufacturers follow a trend of developing functional food ingredients to be used in the food products. These functional ingredients are unstable under the environmental as well as processing conditions.

Various studies have confirmed the antioxidant activity of various bioactive components present in several natural food products. The antioxidant activity of these components makes them suitable for treatment of various diseases in human body. But these components are known to be unstable and prone to oxidation. Upon oxidation, these components can generate highly reactive species called as free radicals that may cause many harmful effects on the body.[5] The components responsible for aroma and flavor may react with other ingredients in food during storage or processing resulting in undesirable changes in aroma and flavor, the health benefits or nutraceutical potential of such components may also be lost during such conditions. The bioactive components due to the limited solubility and bioavailability have limited applications in food products. To slow down the negative changes or prevent the components from the deteriorative changes and to increase their applications in food industry, encapsulation is considered as an efficient approach as the protective material used is meant to confine or restrict the interactions of the core with the environment surrounding it.[23]

Encapsulation involves proper mixing of the encapsulating material with the active ingredient. The mixture then formed is dried by selecting an appropriate drying method. Encapsulates vary in form and size according to the material as well as method of encapsulation. The encapsulates obtained can be simple or complex in shape. Irregular or spherical encapsulates can also be obtained. Sometimes the encapsulates obtained may contain multiple walls made either of same material or of different materials. Methods used for encapsulating any active ingredient can be physical, chemical, or physicochemical. Encapsulation efficiency can be obtained from the amount of any active agent encapsulated to its concentration initially present in the encapsulating mixture.

Depending on the size of the encapsulated material obtained, the process is categorized into three basic types like macroencapsulation, microencapsulation, and nanoencapsulation. Either any ingredient of food is encapsulated or sometimes the whole food product is encapsulated that leads to the categorization of encapsulation into various types. When the dimensions are small, the functionality of the core material will be higher as the surface area of smaller encapsulates will be higher. Because of the increased size obtained in macroencapsulation, it is used to a lesser extent than microencapsulation and nanoencapsulation techniques, and nanoencapsulation offers better results in terms of protection, bioavailability, and stability of bioactive components.

Classification of the encapsulated powder from any technique is done through a vibrating system with different sizes of sieves attached to it. Encapsulation efficiency of particles obtained even from a same technique can differ from each other.[18] Uses as well as the range of encapsulated components in food products is increasing very rapidly, which has increased the scope of value addition of food products in the food industry sector. Nowadays, focus of researchers is toward the use of coencapsulation methods where two or more components can be encapsulated together to have synergistic effects.[26] Size of encapsulates obtained in different techniques is given in Table 12.1.

TABLE 12.1 Types of Encapsulations Based on Sizes.

Technique	Size
Macroencapsulation	>5000 μm
Microencapsulation	0.2–5000 μm
Nanoencapsulation	2000 Å–0.2 μm

In food industry, the encapsulation can be used for the purpose of masking unpleasant taste and aroma that eventually helps in masking the unpleasant flavor of food constituents. Usually, polyphenols possess antioxidant activities, while on the other hand polyphenols are astringent that is not acceptable in some food products. To reduce the astringency to desired level, the process of encapsulation can prove to be an effective measure. It has been reported that the encapsulated material has better and improved properties than the unencapsulated materials.[14]

Since the techniques of encapsulation were known to be specific and expensive that limited the applications of encapsulation for large-scale production. Nowadays, the use of better as well as cost-effective materials has increased the production of food products from encapsulated materials. The number of products prepared by incorporating encapsulated ingredients is increasing day by day because the products possess better properties in terms of stability over extended periods, nutritional value, and antioxidant activities. A variety of materials including microbial cells, enzymes, minerals, vitamins, and antioxidants are commonly encapsulated.[21] Also encapsulated nutraceuticals are the focus of researchers. While designing any technique of encapsulation various factors are to be taken into consideration. Focus of researchers these days is to find effective ways to improve productivity of encapsulation techniques without affecting the quality of the product obtained.

This chapter focusses on scope of encapsulation technologies in food industry.

12.2 STAGES OF ENCAPSULATION

The process of encapsulation is considered as a multiple-stage process and usually involves three stages or steps. The process starts from mixing the active component with the material to the formation of the final encapsulate using various techniques depending on the nature of the active component and the encapsulating material. The steps involved are

- In the first step a solution, suspension, or an emulsion of an active component with the encapsulating material is prepared.
- The second step involves the use of any mechanical operation. The type of mechanical operation selected depends on the nature of feed mixture (whether solid or liquid matrix) obtained in step first.
- The last stage of encapsulation involves the use of chemical, physico-chemical, or physical processes for stabilization/solidification of both active ingredient and coating solution.

12.3 ENCAPSULATING MATERIALS

Before selecting any material for encapsulation, various factors are to be considered because some physical and chemical properties of the encapsulated material obtained depend on the type of material used for encapsulation. The material used should not react with the core as well as should be compatible with the substance to be encapsulated. The material used should be regarded as safe by the concerned regulatory bodies. Moreover, the cost of material should not be high.[1] Polysaccharides are the first and foremost choice to be used as an encapsulating agent. Proteins as well as lipids are also used as wall materials. In case of proteins, plant proteins have proven to give beneficial results as compared with animal proteins because the film-forming property of plant proteins is better. Also, plant proteins are known to be nonallergic in nature. Sometimes a combination of materials is used for encapsulation like combinations of maltodextrin and gums have given promising results. Moreover, combination of polysaccharides and proteins can also be used.[16] Various encapsulating agents and their properties are given in Table12.2.

TABLE 12.2 Various Wall Materials for Encapsulation.

Encapsulating agent	Properties	Examples	Reference
Carbohydrates	• Commonly used encapsulating agents as these are readily available and are less costly than other materials. • Increased solubility in water as well as the solutions with very low viscosities justifies their use as most readily used encapsulating agents.	Starch, cellulose, and other polymeric substances like pectin, alginates, agar, carrageenan, gums, etc.	29
Lipids	• The use of such materials in encapsulation is a recent advancement in this field. • When a lipid is used for encapsulation, the process is termed as liposome technology.	Fatty acids and phospholipids are mostly used. Glycerol and waxes can also be used.	24
Proteins	• Proteins are used in encapsulation due to their ability to bind several substances. • They also possess better emulsifying as well as film-forming properties.	Soy protein isolates, whey proteins (both isolates and concentrates), casein, gluten, and gelatin.	19

12.4 APPLICATIONS OF ENCAPSULATION IN FOOD INDUSTRY

The use of encapsulation in the field of food technology started in 1930s. In 1930s, there was a need for production of dry flavor powders so spray drying technique was used for such purposes. Since then, this technique is making a recognizable growth in development of products incorporated with encapsulated ingredients. Products with microencapsulated materials were formulated in the year 1954. Research on newer techniques of encapsulation was initiated by the pharmaceutical sector. The growth of encapsulation in the food industry sector can be seen in terms of the increased research being carried out in this field. Scopus database also showed that the number of articles published in this field from the past 40 years was 7000 in the year 2012.[13]

Food and encapsulation are very closely related to each other, in some cases these two are inseparable from each other. We know that a food system is always composed of several ingredients that differ from each other in terms of properties like stability in any environment, sensitivity, and degree of interaction with the surrounding matter and hence contribute to the complexity of the system. Also, the food system may sometimes consist of ingredients that are unstable or volatile but important for product to prevent the loss of such ingredients the technique of encapsulation is used. Encapsulation allows ease in handling by immobilizing the ingredient. The process also increases the surface area of the ingredient as a result its bioavailability as well as the effectiveness is increased to a higher extent. To ensure a targeted delivery, this process has provided promising results. Some of the food components especially those with health benefits possess an astringency that is not appetizing to the consumers, in order to mask that encapsulation is widely used.[30] Applications of encapsulation in beverage industry as well as in fermentation processes are also very well noticed and are gaining attention these days.

Various researchers have proved that when the food additives are encapsulated properly, their stability is enhanced. Apart from the increased stability of the food additives, the properties are also enhanced by the process of encapsulation and the improved properties help us to justify the use of this technique in the food industry. Encapsulated food ingredients can be used in a range of food products including animal foods, baby foods, bakery products, meat and meat products, dairy products, etc. In beverage industry for production of beer and wine, immobilized yeasts can be used to carry out the process of fermentation to make the process economical and help in reduction of cost of production. Labor requirement is reduced, and the materials can very well withstand the harsh industrial conditions.[10]

Use of encapsulation in the food industry has led to the development of such products, which were technically impossible. The growing concern of food industry is toward the development of such stabilization techniques that aim to preserve the functionality of bioactive components and also facilitates easier handling by improving physical properties of the components to be encapsulated. Apart from encapsulation of nutrients like vitamins and minerals, various carotenoids like lutein and lycopene and various sterols of plant origin (phytosterols) are successfully encapsulated. Pharmaceutical as well as cosmetic industries can even afford the use of encapsulation-like techniques usually of high cost. But in case of food industries, the main focus remains to reduce the production costs as a result resistance to implementation of newer techniques like encapsulation also is higher. Moreover, the newer techniques involved require trainings to be provided to the staff dealing with them that in turn leads to increase in the cost of implementation. The encapsulated materials used in foods should be well adapted to the gastrointestinal tract of humans.[15]

The development of cost-effective materials as well as techniques with time has increased the applications of encapsulation in food industries. Care must be taken to check that the sensory characteristics of any particular food product are not compromised by incorporation of any encapsulated ingredient. In addition, leakage through the wall material should also be prevented during encapsulation. Nowadays, the focus of researchers is to develop products by combination of the innovative techniques with the traditional methods available. The concept of beneficial microorganisms called as probiotics forces the researchers to develop functional foods, particularly in case of dairy products. Since the beneficial microorganisms present in various food products need to withstand the conditions during processing, storage as well as intake.[20]

12.4.1 PROBIOTICS

World Health Organization describes the term probiotics as live organisms, which provide several health benefits to the human body when taken in required amounts. Food products contain mostly *Lactobacillus* spp. and *Bifidobacterium* spp. as beneficial organisms. Probiotics help in prevention of various gastrointestinal disorders including diarrhea and lactose intolerance. But the conditions during manufacturing as well as in storage of the products results in decrease in the viable number of these beneficial or good bacteria in food products. Among the different techniques of encapsulation, microencapsulation is used effectively for probiotic bacteria during various stages of production.[7]

12.4.2 HIGHLY UNSATURATED OILS

Highly unsaturated oils are known to contain enough essential fatty acids that offer several health benefits to human body. Usage of essential fatty acids in food products has increased in past few years to prevent several diseases. The highly unsaturated oils are also known to contain a variety of fat-soluble bioactive components like vitamin E, polyphenols, carotenoids, etc., these bioactive components possess antioxidant activity.[6] On the other hand, these unsaturated fatty acids are prone to oxidation that may lead to undesirable changes in sensory as well as nutritive properties of these oils. Hence the process of encapsulation is widely used to prevent the oxidation of oils with high levels of unsaturation. The highly unsaturated oils are mostly found in flaxseeds, chia seeds, olives, fish, etc.

12.4.3 FLAVORS

Consumption of any food product is influenced by factors like flavor and aroma of that product. To improve the stability of flavors during processing, micro, and nanoencapsulation are widely used. Among the techniques of encapsulation, spray drying is frequently used for encapsulation of flavors in food industry. It has been studied that this technique of encapsulation results in higher retention of the components responsible for flavor, and minimum temperature exposure is experienced by the core material that prevents the component from heat-induced damage.

12.4.4 FRUIT JUICE

For the production of fruit juice powders, encapsulation is widely used. Among the drying methods spray drying has been successfully used to a large extent for production of fruit juice powders. Yield as well as other physicochemical parameters of the powder obtained is influenced by factors like type and concentration of encapsulating agent, inlet temperature, flow rate of air as well as encapsulation mixture.

12.4.5 FOOD COLOR

As compared with the synthetic colorants, natural colors are considered to have various health benefits but at the same time, they are unstable under various processing conditions. The techniques of encapsulation are mainly used to prevent the food colorants from degradation by light, oxygen, high temperature, etc. Mostly carbohydrates and proteins are used for the encapsulation of food colorants. For microencapsulation, spray drying is most suitable. On the other hand, electrospinning results in production of both micro as well as nanoparticles. Other technique called as antisolvent precipitation produces encapsulated particles dispersed in a liquid that can be further dried by any drying method.[17]

12.5 TECHNIQUES OF ENCAPSULATION

The techniques mostly used for encapsulation include spray drying, freeze drying, coacervation, extrusion, emulsification, spray chilling, fluid bed coating, and molecular inclusion. Among the different techniques of encapsulation, spray drying is mostly used in the food industries. These are discussed in this section.

12.5.1 SPRAY DRYING

Encapsulation by means of spray drying in food industry is both cost-effective and efficient technique and it is used in large-scale production of encapsulated flavors, oils, pigments, and fragrances. The most important characteristics are its flexible, continuous, and economical operation. Moreover, its production costs lesser than any other encapsulation techniques. It produces particles having size of <40 μm. In this process, the core material is dispersed into a polymer solution (such as protein, carbohydrate) that leads to formation of an emulsion or dispersion and atomization of the mixture into hot dry chamber. Due to the improvement of new wall material, new sprayer dryer, and atomizer design, the advancement in spray dried microcapsule materials have been possible. For example, the patented atomizer design facilitate particle with a narrow size distribution. Another new type of dryer known as "Leaf lash spray dryer" used high air velocity (300–400°C) suitable for encapsulation of citral and linalyl acetate devoid of degradation.[4]

The main advantages of spray drying technique are: useful for heat-labile materials, great retention of volatiles, good stability of the end-product, readily available, usually affords great barrier to the core material, and wide variety of wall materials available. On the other hand, disadvantages of spray drying are that produce nonuniformly sized materials, complexity of the equipment, inconsistent environment in the drying chamber, and loss of low-boiling point aromatics. Examples of spray drying are encapsulation of sumac flavor by spray drying in sodium chloride in salt cookies, salads, and crackers[2] and encapsulation of lipids into potato starch, tapioca, and corn by spray drying[11].

12.5.2 COACERVATION

This technique involves:

- Precipitation or separation of a liquid phase of coating material from a polymeric solution.
- Uniform layer coating of that phase around suspended core particles or material to be encapsulated.
- Solidification of the coating material via heat, crosslinking, or solvent removal techniques.

There are two available methods for coacervation, namely, simple process and complex process. In simple process, only a single macromolecule present, only one colloidal state, such as gelatin and gum acacia. On the other hand, in complex coacervation process, two or more molecules of opposite charges present, forms a complex concentrate due to ionic interaction (protein and polysaccharide) at the interface. Most commonly it is used in encapsulation of essential oils.[27] Examples of encapsulation are aspartame and peppermint oil etc.[28]

12.5.3 FREEZE DRYING

Freeze drying is also referred to as lyophilization. It can prolong the shelf-life of sensitive food products. Generally, it is suitable for the encapsulation of heat-sensitive bioactive compounds, probiotics, enzymes, and peptides. Encapsulation by means of freeze drying is attained by dispersion, dissolution, or emulsification of core materials in wall material matrix systems. Afterward colyophilization results in a poriferous, nonshrunken, and uncertain structure. Examples of freeze drying are β-carotene encapsulation within the maltodextrin via freeze drying had greater retention amount.[9] Encapsulation of anthocyanins obtained from black carrot juice had maximum retention.[25]

12.5.4 EXTRUSION

Extrusion technique is based on immobilization of the core material with polysaccharide gel that interacts with a multivalent ion. In other words, it involves amalgamation of the core in an aqueous solution of polymer (like sodium alginate solution) accompanied by mixture that undergoes dropwise extrusion through a reduced caliber pipette or syringe into hardening solution, that is, calcium chloride.[32] Usually, this technique is intended for vitamin C and colors encapsulation. This extends the shelf life by >2 years.

12.5.5 EMULSIFICATION

The emulsification is usually used for encapsulation of lipophilic components that usually are insoluble in water. In this technique, the active ingredient acts as a dispersed phase and the encapsulating material acts as a continuous phase. Apart from the dispersed and the continuous phase, a surface-active agent is also used. Depending on the arrangement of the hydrophilic and lipophilic phases, emulsions obtained can be of oil-in-water or water-in-oil types. Sometimes double emulsions are also formed. For example, carotenoids form oil-in-water emulsion on encapsulation while bioactive components like polyphenols form water-in-oil emulsion. Double emulsions are known to have several applications for encapsulation of various components in food industry. Like any component responsible for flavor of food can be encapsulated or an off-flavor can be masked by double emulsion technique.[34] This technique provides higher stability to components susceptible to oxidation. Mostly flavor components and omega-3 fatty acids are encapsulated by this advanced form of double emulsions.[22] The increased stability of encapsulated technique is due to the increased thickness obtained due to the multiple layers of encapsulating medium around the core. When these emulsions are to be converted into a stable dry form, drying by either spray, or freeze drying can be done.

12.5.6 SPRAY CHILLING OR SPRAY COOLING

In spray chilling or spray cooling, the active ingredient is coated with lipids. These two techniques of encapsulation differ from each other in terms of melting point of lipids used. Spray cooling is used at higher melting temperatures of lipids as compared with spray chilling, where it is between 34°C and 48°C. In contrast to other encapsulation techniques, the air used in these techniques is first cooled to ambient or refrigerated conditions. Various vitamins, acidulants, and flavors are usually encapsulated by these techniques.[33]

12.5.7 FLUIDIZED BED COATING

This technique involves the use of a chamber into which the feed mixture of core ingredient and the coating solution is poured. The core material usually is in the form of solid particles. Various factors like humidity and air velocity are controlled to obtain optimum encapsulating conditions. The mixture is atomized with the help of an atomizer. The time taken by the particles in the chamber decides the amount of encapsulating material used to coat the active ingredient. The particles thus obtained possess particle size >50 μm. The bigger particles may have a size up to 500 μm and difference between the size of the particles obtained should not vary too much. If the coating is of hot melt type (vegetable oil), hardening of the encapsulating material is obtained by employing cold air and the core material from the capsule is released on physical breakage. On the other hand, evaporation of the solvent is done by hot air when the coating is solvent based and the capsule releases the constituents upon dissolution in water. Salt added to various bakery products and various food additives like sorbic acid, citric acid, etc., can be encapsulated by this technique.[8]

12.5.8 MOLECULAR INCLUSION

Molecular inclusion has been a less exploited technique. Due to the amphiphilic behavior, β-cyclodextrin is considered as the most suitable wall-material in this technique. This technique usually involves an interaction between a substance bearing a cavity and a ligand. This molecule possesses a cavity in the center (oleophilic inner pouch) surrounded by hydrophilic surface. The cavity has a very small diameter of about 0.65 nm. On formation of the inclusion complex, water molecules present in the molecule cavity is replaced by any less polar component. During the final stage, the formed molecule is precipitated from the solution. The ligand is bound to the cavity-bearing substance by hydrogen bonding, Vander Waal's forces. Precipitated molecule is finally separated and dried. Entrapment of active ingredient by this technique is considered as reversible.

A change in pH of solution leads to the breakage of bonds between the active core and the cavity-bearing substance resulting in the release of the component to the target. The active component encapsulated is also released when the linkages between the monomers of encapsulating material are cleaved by the activity of specific enzymes or by increasing temperature. The linkages of the β-cyclodextrin are of $\alpha(1\rightarrow4)$ type. The most important advantage of this technique is the unique structure of the obtained encapsulates. The major drawback of this technique is its higher cost. Only a particle of specific size can be loaded into the cavity and the smaller size of the cavity of molecule also affects the capacity of the molecule for loading the active ingredients.[3]

12.6 SUMMARY

The encapsulation is important in the field of food industry to improve nutritive value, health benefits, and also the market value of food products. Since the techniques of encapsulation are newer to the field of food industry and application of encapsulation at industrial level has various difficulties that are to be solved first. So, ways to increase the effectiveness needs to be further studied. Scientists and researchers should focus on increasing applications of encapsulation for production of various products on large scale. The encapsulated components can be used in food products for the purpose of fortification to improve nutritive value, provide health benefits, and other functional properties. Further, newer techniques, materials, and conditions should be chosen for encapsulation.

ACKNOWLEDGMENT

The authors are thankful to the Indian Council of Medical Research, Ministry of Health and Family Welfare, GOI, New Delhi for providing financial support in the form of ICMR-SRF.

KEYWORDS

- **active ingredient**
- **carriers**
- **encapsulation**
- **Indian Council of Medical Research**
- **stability**

REFERENCES

1. Agnihotri, N.; Mishra, R.; Goda, C.; Arora, M. Microencapsulation: Novel Approach in Drug Delivery: A Review. *Indo Glob. J. Pharm. Sci.* **2012,** *2*(1), 1–20.
2. Bayram, O. A.; Byram, M.; Tekin, A. R. Spray Drying of Sumac Flavor Using Sodium Chloride, Sucrose, Glucose and Starch as Carriers. *J. Food Eng.* **2005,** *69*(2), 253–269.
3. Becket, G.; Schep, L. J.; Tan, M. Y. Improvement of the In Vitro Dissolution of Praziquantel by Complexation with Alpha-, Beta- and Gamma - Cyclodextrins. *Int. J. Pharm.* **1999,** *179*(1), 65–71.
4. Bhandari, B. R.; Dumoulin, E. D.; Richard, H. M. J.; Noleau, I.; Lebert, A. M. Flavor Encapsulation of Spray Drying: Application to Citral and Linalyl Acetate. *J. Food Sci.* **1992,** *57*(1), 217–221.
5. Chen, W.; Xu, Y.; Zhang, L.; Su, H.; Zheng, X. Blackberry Subjected to In Vitro Gastrointestinal Digestion Affords Protection against Ethyl Carbamate-induces Cytotoxicity. *Food Chem.* **2016,** *212*, 620–627.
6. Chew, S. C.; Tan, C. P.; Long, K.; Nyam, K. L. Effect of Chemical Refining on the Quality of Kenaf (*Hibiscus cannabinus*) Seed Oil. *Ind. Crops Prod.* **2016,** *89*, 59–65.
7. Das, A.; Ray, S.; Raychaudhuri, U.; Chakraborty, R. Microencapsulation of Probiotic Bacteria, and Its Application in Food Technology. *Int. J. Agric. Environ. Biotechnol.* **2014,** *7*(1), 47–53.
8. De Zarn, T. J. Food Ingredient Encapsulation. Chapter 7; In *Encapsulation and Controlled Release of Food Ingredients*; Risch, S. J., Reineccius, G. A., Eds.; American Chemical Society (ACS) Publications: Washington, DC, 1995; pp 74–86.
9. Desobry, S. A.; Netto, F. M.; Labuza, T. P. Comparison of Spray-Drying, Drum-Drying and Freeze-Drying for β-Carotene Encapsulation and Preservation. *J. Food Sci.* **1997,** *62*(6), 1158–1162.
10. Diviès, C.; Cachon, R.; Cavin, J. F.; Prévost, H. Theme 4: Immobilized Cell Technology in Wine Production. *Crit. Rev. Biotechnol.* **1994,** *14*(2), 135–153.
11. Drusch, S.; Serfert, Y.; Heuvel, A. V. D.; Schwarz, K. Physicochemical Characterization and Oxidative Stability of Fish Oil Encapsulated in an Amorphous Matriz Containing Trehalose. *Food Res. Int.* **2006,** *39*(7), 807–815.
12. Dziezak, J. D. Microencapsulation and Encapsulated Food Ingredients. *Food Technol.* **1988,** *2*(4), 136–151.

13. Estevinho, B. N.; Rocha, F.; Santos, L.; Alves, A. Microencapsulation with Chitosan Spray Drying for Industry Applications—A Review. *Trends Food Sci. Technol.* **2013,** *31*(2), 138–155.
14. Fang, Z.; Bhandari, B. Encapsulation of Polyphenols—A Review. *Trends Food Sci. Technol.* **2010,** *21*(10), 510–523.
15. Favaro-Trindade, C. S.; De Pinho, S. C.; Rocha, G. A. Review: Microencapsulation of Food Ingredients. *Braz. J. Food Technol.* **2008,** *11*(2), 103–112.
16. Gibbs, B. F.; Kermasha, S.; Alli, I.; Mulligan, C. N. Encapsulation in the Food Industry: A Review. *Int. J. Food Sci. Nutr.* **1999,** *50*(3), 213–224.
17. Idham, Z.; Muhamad, I. I.; Sarmidi, M. R. Degradation Kinetics and Color Stability of Spray Dried Encapsulated Anthocyanins from *Hibiscus sabdariffa* L. *J. Food Process Eng.* **2012,** *35*(4), 522–542.
18. Jafari, S. M.; He, Y.; Bhandari, B. Role of Powder Particles on Encapsulation Efficiency of Oils during Spray Drying. *Dry. Technol.* **2007,** *25*(6), 1081–1089.
19. Karthik, P.; Anandharamakrishnan, C. Microencapsulation of Docosahexaenoic Acid by Spray-Freeze-Drying Method and Comparison of its Stability with Spray-Drying and Freeze-Drying Methods. *Food Bioproc. Technol.* **2013,** *6*(10), 2780–2790.
20. Kim, K. I.; Yoon Y. H.; Baek Y. J. Effects of Rehydration Media and Immobilization in Ca-Alginate on the Survival of *Lactobacillus casei* and *Bifidobacterium bifidum*. *Korean J. Dairy Sci.* **1996,** *18*(3), 193–198.
21. Kirby, C. J. Microencapsulation and Controlled Delivery of Food Ingredients. *Food Sci. Technol. Today* **1991,** *5*(2), 74–80.
22. Klinkesorn, U.; Sophanodora, P.; Chinachoti, P.; McClements, D. J.; Decker, E. A. Increasing the Oxidative Stability of Liquid and Dried Tuna Oil-in-Water Emulsions with Electrostatic Layer-by-Layer Deposition Technology. *J. Agric. Food Chem.* **2005,** *53*(11), 4561–4566.
23. Lesmes, U.; McClements, D. J. Structure-Function Relationships to Guide Rational Design and Fabrication of Particulate Food Delivery Systems. *Trends Food Sci. Technol.* **2009,** *20*(10), 448–457.
24. Mozafari, M. R. Liposomes: An Overview of Manufacturing Techniques. *Cell. Mol. Biol. Lett.* **2005,** *10*(4), 711–719.
25. Murali, S.; Kar, A.; Mohapatra, D.; Kalia, P. Encapsulation of Black Carrot Juice Using Spray and Freeze Drying. *Food Sci. Technol. Int.* **2014,** *21*(8), 604–612.
26. Nedovic, V.; Kalušević, A.; Manojlovic, V.; Levic, S.; Bugarski, B. An Overview of Encapsulation Technologies for Food Applications. *Procedia Food Sci.* **2011,** *1*, 1806–1815.
27. Oliveira, M. N.; Sivieri, K.; Henrique, J.; Alegro, A.; Saad, S. M. I. Technological Aspects of Functional Foods Containing Probiotics. *Braz. J. Pharm. Sci.* **2002,** *38*(1), 1–21.
28. Pakzad, H.; Alemzadeh, I.; Kazemi, A. Encapsulation of Peppermint Oil with Arabic Gum-Gelatin by Complex Coacervation Method. *Int. J. Eng. Trans. B Appl.* **2013,** *26*(8), 807–814.
29. Poornima, K.; Sinthya, R. Application of Various Encapsulation Techniques in Food Industries. *Int. J. Latest Eng. Res. Appl.* **2017,** *2*(10), 37–41.
30. Shahidi, F.; Han, X. Q. Encapsulation of Food Ingredients. *Crit. Rev. Food Sci. Nutr.* **1993,** *33*(6), 501–547.
31. Shishir, M. R. I.; Xie, L.; Sun, C.; Zheng, X.; Chen, W. Advances in Micro- and Nano Encapsulation of Bioactive Compounds Using Biopolymer and Lipid Based Transporters. *Trends Food Sci. Technol.* **2018,** *78*, 34–60.

32. Swarbrick, J., Ed.; *Encyclopedia of Pharmaceutical Technology*, 2nd ed.; Informa Healthcare: New York, 2004; p 5536.
33. Taylor, A. H. Encapsulation System and their Applications in the Flavor Industry. *Food Flavor Ingredients Process Packag.* **1983,** *4*, 48–52.
34. Van Der Graaf, S.; Schroën, C. G. P. H.; Boom, R. M. Preparation of Double Emulsions by Membrane Emulsification–A Review. *J. Membr. Sci.* **2005,** *251*(1–2), 7–15.

CHAPTER 13

TECHNICAL CHARACTERISTICS FOR ENCAPSULATION IN THE FOOD INDUSTRY

CAROLINA E. DEMAMAN ORO, MARCELO L. MIGNONI, GIOVANI L. ZABOT, GECIANE T. BACKES, ROGÉRIO M. DALLAGO, and MARCUS V. TRES

ABSTRACT

Encapsulation has received broad interest in several applications due to its capacity of preventing interactions that do not contribute to the food matrix and are also being able to improve the delivery of target compounds. In this chapter, this technique is reviewed to present different characteristics to preserve several bioactive compounds. The chapter discusses the selection of micro or nanoencapsulation processes, but also the shell and core physicochemical properties in order to improve its final application. We also present the broadly used wall materials and encapsulated food compounds, and also highlight the process' promising results noticed in recent researches. At the end of the chapter, the major technological challenges and advantages and future outlooks on the subject of encapsulation technology are provided.

13.1 INTRODUCTION

Encapsulation may be defined as a process that covers active agent compounds with some kind of caps or coats made of carrier materials, resulting, thus, in micro or nanometer scale particles.[43,50] Several techniques have been

Novel and Alternative Methods in Food Processing: Biotechnological, Physicochemical, and Mathematical Approaches. N. Veena, Megh R. Goyal, & Ritesh B. Watharkar (Eds.)

developed to encapsulate compounds and these techniques include liposomes, hydrogel, solvent evaporation, molecular inclusion, emulsions, freeze drying, spray drying, electrospinning, supercritical antisolvent (SAS), nanoemulsions, coacervation, and spray cooling, among others. This technology here presented is widely applied in food industry where the objective is to encapsulate bioactive compounds.[50]

The several health benefits from the nutraceuticals, natural, and medicinal properties have enhanced its market demand,[51] which are leading to product development in the industries.[29] Although there are numerous food components encapsulation techniques, most of them are based on drying techniques. It can be explained because most of the time, the food components that undergo encapsulating process are commonly in a liquid state.[11]

It has been suggested the several advantages of the natural substances encapsulation,[25] since the capsule is a protective barrier against light, oxygen, pH, moisture, heat, shear, and consequently prevents degradation of the sensible compounds.[50] Furthermore, food encapsulation systems are favorable to human digestive systems, since it allows controlled delivery and stability against disruptive conditions, such as temperature, bile salt composition, pH, electrolyte balance, and mechanical stress. Thus, the encapsulation technology has a broad application, since it is able to protect the gastric environment for vitamins and also deliver drugs or probiotics to the colon.[10]

This book chapter presents a critical review on the characteristics and main applications of encapsulation techniques in food-related areas.

13.2 GENERAL PROPERTIES AND CHARACTERISTICS OF ENCAPSULATED MATERIALS

The bioactive compounds application faces several challenges, such as low levels of solubility, considerable instability, uncontrolled release, and low levels of bioavailability.[44] For these reasons, encapsulation has received considerable interest from researchers. It is agreed that controlled release behavior maintains the stability of powders for longer periods. Also, encapsulation can eliminate unfavorable flavor and taste of some compounds of interest when required,[21,44] reduce storage space,[25] improve the bioavailability and stability, in the same time retention is maximized, and the release in the human body is controlled.[16] Encapsulation also reduces the losses caused by evaporation or volatility loss. It provides the enhancement of reactivity barrier, the improvement of biological activity, physical stability, and also of bioactive compounds shelf life.[50]

An illustration of successful applications and reviews has been reported in the literature (Figure 13.1) with the criteria for search as "encapsulation," "encapsulated food," and "encapsulated bioactive compounds." The research survey for articles and reviews was done by searching for information in the last few years (2015–2019) in the main international scientific databases and portals of the proposed subject.

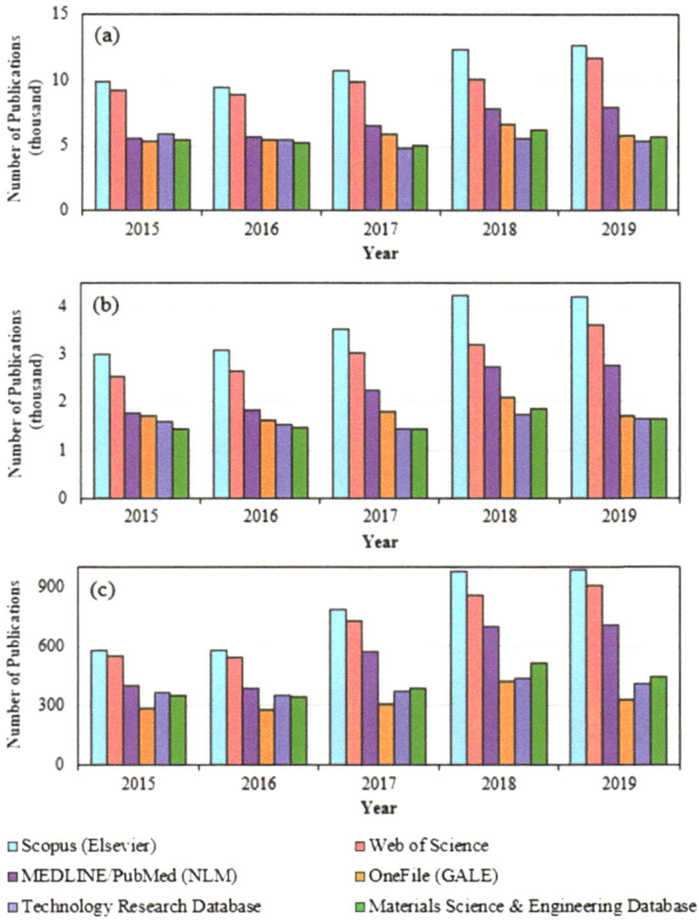

FIGURE 13.1 Number of publications between 2015 and 2019 for (a) encapsulation, (b) encapsulated food, and (c) encapsulated bioactive compounds.

Encapsulation is a subject that attracts attention because of the potential of application in different fields of knowledge. Figure 13.1 shows the increase in the number of publications about encapsulation between 2015 and 2019, with an increase of 15.8%, 23.1%, and 32.8% in publications for encapsulation, encapsulated food, and encapsulated bioactive compounds, respectively. The major number of publications was found in 2018 ("encapsulated food" and "encapsulated bioactive compounds") and 2019 ("encapsulation").

During the analyzed period, there was a rise in the number of publications of encapsulated bioactive compounds, which are mainly related to chemistry, engineering, antioxidants, plant extracts, agriculture, diet and clinical nutrition, fruits, flavonoids, and phenols. Although the encapsulation of bioactive compounds provides increased stability, the selection of a suitable encapsulation process is needed.

13.3 SELECTION OF MICRO OR NANOENCAPSULATION PROCESSES

Food industry is broadly benefited by the microencapsulation and nanoencapsulation-based system and techniques. The proper selection of encapsulation materials and methods is important when choosing between micro or nanostructured encapsulate systems.[50] It is possible to find several encapsulation techniques (Figure 13.2) in order to preserve them from environmental stress and improve its physical-chemical functionalities, stability, and bioavailability.[32] The development of physical, physicochemical, and chemical encapsulation techniques allows for obtaining encapsulated compounds[37] with different sizes. The micro or nanoencapsulation process selection is determined by the used materials, physicochemical properties, and its final application.[5]

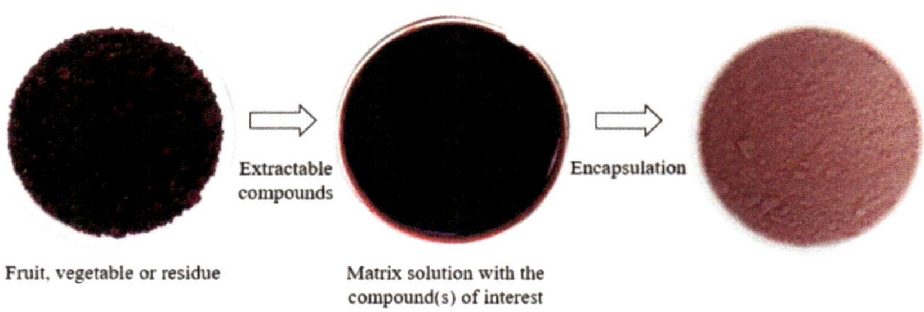

FIGURE 13.2 Schematic encapsulation process.

Microencapsulation can be found in the most used methods. This can be achieved by physical methods but also by chemical methods. According to their sizes, nanoparticles can range from 1 nm to 1 μm, and microparticles from 1 to 800 μm. This way, it is hard to produce nanoparticles by the microparticles techniques.[9]

Figure 13.3 shows how many works about encapsulation techniques were published in the literature until 2020. The research survey for only articles was done by searching for information in the main international scientific databases and portals (Scopus, Web of Science, MEDLINE/PubMed, OneFile (GALE), Technology Research Database, and Materials Science and Engineering Database, among others). The survey was carried out with the keywords for search as "encapsulation + technique."

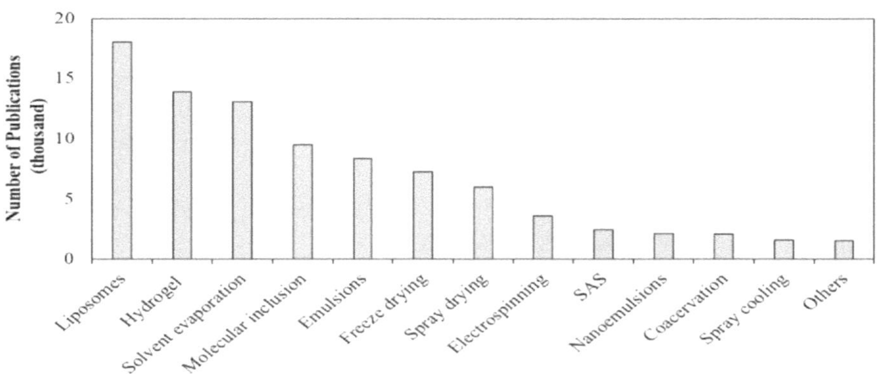

FIGURE 13.3 Number of publications in the literature about the main encapsulation techniques.

Most of the published articles are about encapsulation by the liposome technique (20.2% of total publications). Liposomes simulate cell membranes and have a high encapsulation and loading capacity with outstanding biocompatibility and safety,[22] which explain their demand. They are prepared from nontoxic and natural materials, being safety for health and having high biodegradability and biocompatibility.[17,22] Besides, liposomes also entrap and deliver hydrophilic, hydrophobic, and amphiphilic substances.[17] However, they probably present weak physical and chemical stability, which continues to have low encapsulation efficiency. The same occurs with fusion and aggregation of lipid vesicles, lipid oxidation or burst release, and instability.[17,43]

Aytekin et al.[6] succeeded to encapsulate propolis-loaded liposome formulations by the modified ethanol injection method. Propolis sustains its lipophilic character and because of this, high loading capacity (66.5%)

and encapsulation efficiency (57.3%) could be obtained. The same could be observed with propolis-loaded liposome formulations, which maintained the high antioxidant activity that is observed in free propolis.[6] Quan et al.[41] investigated the internalization of the anthocyanin cyanidin-3-O-glucoside liposomes across the gastric epithelial cell monolayer (GES-1 cells). The authors achieved 75% of encapsulation efficiency and could notice both regulatory and antioxidant effects on the biological activity of the cells.

The second most used technique is hydrogel (15.5%), which can encapsulate both lipophilic and hydrophilic bioactive compounds.[50] Hydrogels are composed by 3D networks designed by cross-linked hydrophilic polymers. They are also composed by water, which fills the space between macromolecules, and are originated from proteins or polysaccharides, which are entirely biodegradable.[28] However, synthetic hydrogels are also employed in many important applicative fields. For example, hydrogels are excellent reusable materials for water purification by contaminant adsorption.[59] Hydrogel-based solar steam generator is a desirable and efficient desalination process.[46]

Another application that shows the absorbent potential of hydrogel was described by Suner and Sahiner.[52] They synthesized different guar gum hydrogels through chemical crosslinking via bulk polymerization and reverse micelle microemulsion techniques to be used for protein absorption. The authors demonstrated that protein absorption increased with a decrease in pH. Also, they concluded that the hydrogels were blood compatible and noncytotoxic on fibroblast cells up to 1 mg/mL concentration.

Solvent evaporation is a much used technique in several industries such as pharmaceutical ones. Poly(lactic-co-glycolic acid) (PLGA) is broadly used as encapsulation material and is a biodegradable polymer. The oil-in-water method is usually used for insoluble or poorly water-soluble substances of interest. In summary, the method consists in the dissolution of the drug or compound of interest in a solvent containing the polymer, emulsification, extraction and solvent evaporation, and recovery and drying of the spheres/capsules to eliminate any residual solvent.[24] Due to the characteristics of the method, the solvent evaporation is usually used as a single or double emulsion-solvent evaporation method.

Ospina-Villa et al.[36] used a single emulsion solvent evaporation method to encapsulate proteins from *Leishmania panamensis* into PLGA particles. Villicaña-Molina et al.[58] prepared microspheres through an emulsion and solvent evaporation method and loaded with acetaminophen, which is an anti-inflammatory drug. They used chitosan and pectin as coating and glutaraldehyde as a cross-linking agent. These authors demonstrated that the two biopolymers are suitable for this method, but chitosan showed the best results and interaction at the molecular level with the acetaminophen.

Molecular inclusion consists of a generally hydrophilic shell that has a cavity inside, which can accommodate the hydrophobic core compound,[39] and the method can be combined with other method to achieve better results. Plenagl et al.[38] improved antimicrobial photodynamic therapy by improving liposomal formulations as carriers for hypericin. The authors noticed the possibility of liposomal encapsulation in the case of parenteral application.

Xiao et al.[61] noticed in watermelon flavor, the incorporation of seven characteristic aroma compounds into γ-cyclodextrin without the same proportion as in free flavor and exhibited different release behavior. The functional groups and the interaction force of the inclusion complexes can explain the achieved results. Both flavor and aroma can be encapsulated with molecular inclusion to utilization in food industries to enhance the sensorial properties of products. Also, cyclodextrins eliminate or reduce unwanted components, such as unwanted taste or odor, hygroscopicity, and consequently microbial contamination.[23]

Different types of emulsions can be used for encapsulation, such as single emulsions (water-in-oil and oil-in-water) and double emulsions (oil-in-water-in-oil and water-in-oil-in-water).[43,50] These different types are one of the reasons for the popularity (9.3%, Figure 13.3) of the method. Double emulsions have many food applications such as reducing fat content, increasing fatty acid profile, encapsulation of functional and active components, and reducing sugar/salt content of foods by enhancing taste intensity. Emulsions have controlled release and chemical stability,[19] in which the interface layer is designed to control the release and/or digestion.[10]

Pulit-Prociak et al.[40] reported the encapsulation of active compounds (birch leaf extract and tannic acid) in biopolymer micelles suspended in emulsion formulations. According to the authors, the encapsulation of tannic acid with controlled release demonstrated satisfactory results, and lecithin as an emulsifier was sufficient to obtain the desired formulation. Although, birch leaf extract did not show controlled release, kinetically stable emulsions were successfully obtained with 80–90% of encapsulation efficiency.[40] Besides emulsions, one of the lipid-based nanocarrier as oil-in-water emulsion is nanoemulsion with lipid droplet size between 10 and 100 nm.[18]

Freeze drying or lyophilization is a technique useful due to porous structure for thermally sensitive compounds,[25] such as essential oils, pigments, proteins, and microorganisms, and consists of drying the sample by sublimating its water content.[9] Lyophilization is a process widely used to produce heat-sensitive fruit juice powders to maintain its quality and stability.[53] Core materials and matrix solutions are homogenized and then colyophilized to make dried materials. Although is a simple technique to encapsulate water-soluble essences and natural aromas, that has a long dehydration period,[26] it is expensive and time consuming[63] and shows irregular, cracked, and porous surface morphology that might be resulted from the powder grinding process.[29]

Mar et al.[29] encapsulated bioactive compounds of *Clidemia japurensis* and *Clidemia hirta* juices by freeze drying in maltodextrin containing different dextrose equivalents and observed that the lowest dextrose equivalent carrier presented the best retention and stability of the bioactive compounds. A similar result was achieved by Yamashita et al.[62] In their study, lyophilization efficiently encapsulated an anthocyanin-rich extract obtained from blackberry by-products, with high anthocyanin retention (76% for maltodextrins with 10 dextrose equivalents; 68% for maltodextrins with 20 dextrose equivalents).

Spray drying, atomization, or pulverization produces microcapsules in a relatively simple and continuous processing operation. In this sense, spray drying is a unit operation in which the feed solution is transformed in dried particulate form by spraying the feed solution towards a hot drying medium. Although spray drying technique has been widely used for drying heat-sensitive foods, pharmaceuticals, and other substances because of the rapid solvent evaporation.[27] Spray drying is not indicated for thermally sensitive compounds due to high-temperature air and is limited in terms of the choice of coating materials.[25,63]

Talón et al.[55] encapsulated eugenol by spray drying using different wall materials (whey protein, soy lecithin, and maltodextrin in combination with oleic acid and chitosan). They had a higher encapsulation efficiency (95–98%) and antibacterial effect against *Escherichia coli* and *Listeria innocua*. These authors reported bigger particles and large agglomerates of chitosan and particles spherical in systems containing oleic acid. Also, in the tested aqueous simulants, all powders released substantially their total content of eugenol at equilibrium.[55] The encapsulation process can be optimized by response surface methodology, as described by Singh et al.[51] when optimizing Jamun pulp powder by spray drying.

To reduce the unwanted effects of a particular method, a combination of two or more techniques can be used. For example, spray-freeze drying is a combination of spray drying and freeze drying techniques. The method consists of spraying the feed solution into a liquid nitrogen medium and then drying the particles under vacuum in the freeze dryer. As a result, a reduction of drying time of the freeze drying method is achieved as well as overcomes the high-temperature effect of the spray drying process.[63] Another example is spray-dried coated and uncoated liposomes described by Guldiken et al.[15] when evaluating different encapsulation systems for black carrot extract.

Spray cooling or spray chilling is a similar technique to spray drying, and the difference is the use of a cooling chamber instead of a drying chamber. This technique is indicated for thermally sensitive and can overcome the disadvantages of spray drying.[7,45]

Electrospinning is a method to produce submicron or nanoscale polymer fibers. The technique is conducted by applying an electrical field to continuously draw the droplet of polymer solution or melt polymer into a fine fiber,[60] forming structures with a high specific surface at low cost and with a large diversity of structures.[54] The fibers can be used to coat support materials useful, for example, in the development of multilayer active packaging.[56] The effect of the addition of different surfactants on the characteristics of electrospun zein nanofibers and the release of cinnamaldehyde essential oil as an antibacterial package has been reported.[48]

SAS precipitation is a technique based on supercritical fluids applied to precipitate and encapsulate several compounds. Although supercritical CO_2 is a great antisolvent, most of the works describing this method used toxic and environmentally unfriendly solvents, such as dichloromethane, hexane, and ethyl acetate. Consequently, the final product is not applicable in food products.[25] SAS consists of atomization by supercritical CO_2 of a solution containing the compound of interest solubilized in a solvent and the wall material, with subsequent precipitation in a vessel.[47] Machado et al.[25] studied the use of CO_2 as antisolvent and ethanol as solvent of the organic solution (extract + polyvinylpyrrolidone) to encapsulate anthocyanin-rich extract from blackberry residues.

Coacervation is an expensive technique and it appears to be employed only for food grade. The method consists of three steps that must be carried under continuous agitation, which leads to formation of three immiscible phases, deposition, and formation of particles of the liquid polymer coating.[64] Onbas et al.[34] obtained the encapsulation efficiency of propolis extract by complex coacervation of 98.77% with similar cytotoxic activity in comparison to the free extract. Tavares et al.[57] also succeeded in encapsulating garlic extract by complex coacervation using whey protein isolate, chitosan, and gum Arabic as wall materials. Other techniques, such as polycondensation, nanoliposomes, fluidized bed, droplet extrusion, and electrospraying, can be employed to achieve the desired encapsules of bioactive compounds.

13.3.1 PHYSICOCHEMICAL PROPERTIES OF WALL MATERIALS

Different substances can be employed as coating materials to encapsulate solids, liquids, or gases of different types and properties. However, it must be taken into account that the regulations for food additives are more stringent than for other applications. In this sense, the materials used as a protective coating must be recognized as "generally recognized as safe," biodegradable, and capable of forming a barrier between the internal phase and its surroundings.[5] It is also important to consider the functionality, encapsulate concentration, target release, and stability requirement.[50]

Generally, food biopolymers, such as proteins and polysaccharides, are employed as wall material in the encapsulation process.[43] Lipids and other organic and inorganic materials also can be employed as wall material.[50] The choice of wall/shell material will depend on the nature of the compound to be encapsulated, functional characteristics of the particles (for example, size, charge, polarity, load capacity, permeability, degradability, release profile, polarity, and solubility), the method and the expected final result.[35] For example, to produce free-flowing fruit juice powders, agents like maltodextrin, gum Arabic, κ-carrageenan, whey protein isolates, etc., have been employed.[51] Some characteristics of wall materials are shown in Table 13.1.

Natural carriers also can be used to improve functional food. Pumpkin protein isolate from pumpkin oil cake was successfully reported for encapsulation of beetroot juice using freeze and spray drying techniques.[8]

The wall material must be a good emulsifier, should have low viscosity at high concentration, and possess good dissolution and network-forming characteristics. Finally, the wall material must preserve the bioactive compound at different conditions of processing throughout the food chain.[50]

13.4 APPLICATIONS OF ENCAPSULATED COMPOUNDS IN FOOD-RELATED AREAS

Currently, there are two main applications of bioactive ingredients in the food industry: (1) mainly as value-added ingredients to improve the beneficial properties of foods; (2) for the development of smart packaging that resists food spoilage. However, most of these bioactive compounds are very unstable and easily oxidizable under various conditions, especially when exposed to a combination of oxygen, light, and/or heating during processing and/or storage. The oxidative degradation of bioactive compounds can lead to the formation of free radicals, unpleasant taste, or odor, in addition to affecting their biological activity.[32,60] Likewise, encapsulation contributes to the coating of undesirable flavors, the formation of solid particles, reduced evaporation or loss of volatility, and improved physical stability, biological activity, and shelf life of bioactive compounds.[1,12,20,50]

TABLE 13.1 Characteristics of Wall Materials for Encapsulation.

Material	Characteristics	Reference
Casein	Amphiphilic molecule, high nutritional value and absorbability, negligible toxicity, ability to self-assemble as micelle and bind ions, and small molecules, excellent emulsifying, and stabilizing properties	14
Chitosan	Natural polysaccharide, ability to control the release, cationic biopolymer, low solubility in water, and electro-potential responsible for its stability	21,42
Cyclodextrins	Safe, environmentally friendly, biodegradable properties, satisfactory loading, and release performance	61
Gum hydrocolloids or gum polysaccharides	Sustainable, biocompatible, biodegradable, good emulsifying characteristics, and high stability during processing	42,2
Maltodextrin	Low cost, good barrier, neutral aroma and taste, low viscosity, low hygroscopicity, water-solubility, and good compatibility with various core materials	29,11
Oligochitosan	Water-solubility, low viscosity, film formation, immunoregulation, antimicrobial, and free-radical scavenging effect	21
Pectin	Biocompatible, biodegradable, water-soluble, and nontoxic	43
Polyvinyl alcohol	Synthetic polymer, water-soluble, completely biodegradable, colorless and odorless, good mechanical properties, high oxygen barrier capacity, and biocompatibility	56
Polyvinylpyrrolidone	Synthetic polymer, water-soluble, nontoxic, biocompatible, adhesive power, and amphiphilic nature	65

The application of pure bioactive compounds is limited in biological formulations due to some particular characteristics, that is, rapid release, low bioavailability, low solubility, as well as easy deterioration in the presence of environmental stresses.[3,50] Nowadays, there is a high need to create food products with positive health effects, such as antidiabetic, anti-inflammatory, antioxidant properties, etc.[33] In fact, people are looking for products that are free of artificial and synthetic additives and that can promote their health. Thus, there is a growing interest in natural food ingredients by both consumers and the food industry.[4]

Fruits and vegetables are generally rich in bioactive compounds, which are heat-sensitive components. The rate of deterioration may vary with the process and storage conditions applied and vary again with product composition and packaging material.[49] Because of these drawbacks, it is important to preserve the active components of fruits and vegetables using encapsulation techniques. The encapsulated material can even be used to restore components lost during processing. Several authors have described successful techniques in encapsulating different plant sources, such as curcumin,[3] black carrot,[15,31] blackberry,[25] among others.

The most well-known application of encapsulation in foods is the preservation of aroma and flavor compounds, which can be lost through evaporation, oxidation, or interactions with other compounds. In vitro studies showed that the release of phenolic compounds from microcapsules was higher in the simulated intestinal fluid than in gastric medium. This study demonstrated the use of four types of wall materials: maltodextrin, gum Arabic, alginate, and pectin for the microencapsulation of golden berry juice by freeze drying.[11]

Fortification of foods with encapsulated vitamins is desirable for health. For example, fortification of vitamin D has been a challenge for the food industry because of the poor water-soluble nature of vitamin D, instability, and heterogeneous distribution in the food matrix.[30] This behavior justifies the need to use effective encapsulation techniques. Iron encapsulation has been reported and has promising use in salads, ice cream, and whipped cream.[19] Also, these vitamins and nutrients to be encapsulated can come from residues from fruit industries, which generates the reuse of residues with high active power and to confer the food additional health advantages besides their nutritional value.[13]

13.5 ADVANTAGES OF USING ENCAPSULATED MATERIALS IN FOOD PROCESSING

Encapsulation leads to a more stable compound and protects it from unwanted interactions. Therefore, the compost has no interaction with the product, packaging, and surroundings, which protects it from all factors that can lead to degradation or inactivation. Because of this, a high load of compounds with active properties can be added to the capsule, which demonstrates higher yield compared with the free substance, and small amounts of encapsulated can be added to obtain the desired effect. The main advantages are

- Food fortification with different micronutrients encapsulated in simple products;
- Addition or replacement of natural flavors, aromas, and vitamins;
- Controlled release of the compound, either during storage or during digestion;
- Use of encapsulated compounds in active and intelligent packaging; and
- Elimination of unwanted flavor and taste.

In addition, active packaging design with differentiated properties and consequent longer shelf life is easily achieved with the use of appropriate wall materials and techniques. In conclusion, nutritional value and consequent health promotion are provided with the use of encapsulated compounds in food.

13.6 FUTURE CHALLENGES

Bioactive compounds have unique characteristics and behaviors. Therefore, a better understanding of the physicochemical interaction of these compounds with wall materials is needed. Furthermore, the study of the delivery of the encapsulated compounds by the human body, as well as the mechanisms in each part of the digestion, are necessary for each class of compounds and their combinations with the wall materials.

The biggest challenge for the food industry is the cost of equipment and the viability of the wall materials that will be employed. Several studies are emerging to make encapsulation techniques more accessible and also more widespread.

13.7 SUMMARY

The encapsulation of compounds of interest to add in different food products has several benefits and is an emerging technology. Research on different encapsulation techniques is increasing each year, showing its potential for use in food-related areas. This chapter presented the encapsulation techniques most employed in food products and presented details of some of the most common wall materials found in the literature. More research regarding the interaction of encapsulated compounds and wall materials is needed, as well as the absorption mechanisms in the human body.

KEYWORDS

- **bioactive compounds**
- **biopolymers**
- **encapsulation**
- **food applications**
- **microencapsulation**
- **nanoencapsulation**
- **physicochemical properties**
- **wall materials**

REFERENCES

1. Aguiar, J.; Estevinho, B. N.; Santos, L. Microencapsulation of Natural Antioxidants for Food Application—The Specific Case of Coffee Antioxidants—A Review. *Trends Food Sci. Technol.* **2016,** *58*, 21–39.
2. Ahmad, S.; Ahmad, M.; Manzoor, K.; Purwar, R.; Ikram, S. A Review on Latest Innovations in Natural Gums Based Hydrogels: Preparations & Applications. *Int. J. Biol. Macromol.* **2019,** *136*, 870–890.
3. Ariyarathna, I. R.; Karunaratne, D. N. Microencapsulation Stabilizes Curcumin for Efficient Delivery in Food Applications. *Food Packag. Shelf Life*, **2016,** *10*, 79–86.
4. Assadpour, E.; Mahdi Jafari, S. A Systematic Review on Nanoencapsulation of Food Bioactive Ingredients and Nutraceuticals by Various Nanocarriers. *Crit. Rev. Food Sci. Nutr.* **2018,** *59*(19), 3129–3151.
5. Atay, E.; Fabra, M. J.; Martínez-Sanz, M.; Gomez-Mascaraque, L. G.; Altan, A.; Lopez-Rubio, A. Development and Characterization of Chitosan/Gelatin Electrosprayed Microparticles as Food Grade Delivery Vehicles for Anthocyanin Extracts. *Food Hydrocoll.* **2018,** *77*, 699–710.
6. Aytekin, A. A.; Tuncay Tanrıverdi, S.; Aydın Köse, F.; Kart, D.; Eroğlu, İ.; Özer, Ö. Propolis Loaded Liposomes: Evaluation of Antimicrobial and Antioxidant Activities. *J. Liposome Res.* **2020,** *30*(2), 107–116.

7. Bampi, G. B.; Backes, G. T.; Cansian, R. L.; de Matos, F. E.; Ansolin, I. M. A.; Poleto, B. C.; Corezzolla, L. R.; Favaro-Trindade, C. S. Spray Chilling Microencapsulation of *Lactobacillus acidophilus* and *Bifidobacterium animalis* subsp. *lactis* and Its Use in the Preparation of Savory Probiotic Cereal Bars. *Food Bioproc. Technol.* **2016,** *9*(8), 1422–1428.
8. Čakarević, J.; Šeregelj, V.; Tumbas Šaponjac, V.; Ćetković, G.; Čanadanović Brunet, J.; Popović, S.; Kostić, M. H.; Popović, L. Encapsulation of Beetroot Juice: A Study on the Application of Pumpkin Oil Cake Protein as New Carrier Agent. *J. Microencapsul.* **2020,** *37*, 121–133.
9. Canizales, J. R.; Velderrain Rodríguez, G. R.; Domínguez Avila, J. A.; Preciado Saldaña, A. M.; Alvarez Parrilla, E.; Villegas Ochoa, M. A.; González Aguilar, G. A. Encapsulation to Protect Different Bioactives to be Used as Nutraceuticals and Food Ingredients. In: *Bioactive Molecules in Food*; Mérillon, J. M., Ramawat, K. G., Eds.; Springer International Publishing: Switzerland, 2019; pp 1–20.
10. Corstens, M. N.; Berton-Carabin, C. C.; de Vries, R.; Troost, F. J.; Masclee, A. A. M.; Schroën, K. Food-Grade Micro-Encapsulation Systems that may Induce Satiety via Delayed Lipolysis: A Review. *Crit. Rev. Food Sci. Nutr.* **2017,** *57*(10), 2218–2244.
11. Dag, D.; Kilercioglu, M.; Oztop, M. H. Physical and Chemical Characteristics of Encapsulated Goldenberry (*Physalis Peruviana* L.) Juice Powder. *LWT Food Sci. Technol.* **2017,** *83*, 86–94.
12. de Souza Simões, L.; Madalena, D. A.; Pinheiro, A. C.; Teixeira, J. A.; Vicente, A. A.; Ramos, Ó. L. Micro- and Nano Bio-Based Delivery Systems for Food Applications: In Vitro Behavior. *Adv. Colloid Interf. Sci.* **2017,** *243*, 23–45.
13. Gaber Ahmed, G. H.; Fernández-González, A.; Díaz García, M. E. Nano-Encapsulation of Grape and Apple Pomace Phenolic Extract in Chitosan and Soy Protein via Nanoemulsification. *Food Hydrocoll.* **2020,** *108*, E-article 105806.
14. Ghatak, D.; Iyyaswami, R. Selective Encapsulation of Quercetin from Dry Onion Peel Crude Extract in Reassembled Casein Particles. *Food Bioprod. Process.* **2019,** *115*, 100–109.
15. Guldiken, B.; Linke, A.; Capanoglu, E.; Boyacioglu, D.; Kohlus, R.; Weiss, J.; Gibis, M. Formation and Characterization of Spray Dried Coated and Uncoated Liposomes with Encapsulated Black Carrot Extract. *J. Food Eng.* **2019,** *246*, 42–50.
16. Guo, J.; Giusti, M. M.; Kaletunç, G. Encapsulation of Purple Corn and Blueberry Extracts in Alginate-Pectin Hydrogel Particles: Impact of Processing and Storage Parameters on Encapsulation Efficiency. *Food Res. Int.* **2018,** *107*, 414–422.
17. Hassane Hamadou, A.; Huang, W. C.; Xue, C.; Mao, X. Formulation of Vitamin C Encapsulation in Marine Phospholipids Nanoliposomes: Characterization and Stability Evaluation during Long Term Storage. *LWT Food Sci. Technol.* **2020,** *127*, E-article 109439.
18. Hemmatkhah, F.; Zeynali, F.; Almasi, H. Encapsulated Cumin Seed Essential Oil-Loaded Active Papers: Characterization and Evaluation of the Effect on Quality Attributes of Beef Hamburger. *Food Bioproc. Technol.* **2020,** *13*(3), 533–547.
19. Ilyasoglu Buyukkestelli, H.; El, S. N. Development and Characterization of Double Emulsion to Encapsulate Iron. *J. Food Eng.* **2019,** *263*, 446–453.
20. Jia, Z.; Dumont, M. J.; Orsat, V. Encapsulation of Phenolic Compounds Present in Plants Using Protein Matrices. *Food Biosci.* **2016,** *15*, 87–104.

21. Jiang, G. L.; Zhu, M. J. Preparation of Astaxanthin-Encapsulated Complex with Zein and Oligochitosan and Its Application in Food Processing. *LWT Food Sci. Technol.* **2019,** *106*, 179–185.
22. Kalam, M. A.; Alkholief, M.; Badran, M.; Alshememry, A.; Alshamsan, A. Co-Encapsulation of Metformin Hydrochloride and Reserpine into Flexible Liposomes: Characterization and Comparison of in Vitro Release Profile. *J. Drug Deliv. Sci. Technol.* **2020,** *57*, E-article 101670.
23. Krishnaswamy, K.; Orsat, V.; Thangavel, K. Synthesis and Characterization of Nano-Encapsulated Catechin by Molecular Inclusion with Beta-Cyclodextrin. *J. Food Eng.* **2012,** *111*(2), 255–264.
24. Li, M.; Rouaud, O.; Poncelet, D. Microencapsulation by Solvent Evaporation: State of the Art for Process Engineering Approaches. *Int. J. Pharm.* **2008,** *363*(1–2), 26–39.
25. Machado, A. P. da F.; Alves Rezende, C.; Alexandre Rodrigues, R.; Fernández Barbero, G.; de Tarso Vieira e Rosa, P.; Martínez, J. Encapsulation of Anthocyanin-Rich Extract from Blackberry Residues by Spray-Drying, Freeze-Drying and Supercritical Antisolvent. *Powder Technol.* **2018,** *340*, 553–562.
26. Mahdavee-Khazaei, K.; Jafari, S. M.; Ghorbani, M.; Hemmati-Kakhki, A. Application of Maltodextrin and Gum Arabic in Microencapsulation of Saffron Petal's Anthocyanins and Evaluating Their Storage Stability and Color. *Carbohydr. Polym.* **2014,** *105*(1), 57–62.
27. Mahdavi, S. A.; Jafari, S. M.; Ghorbani, M.; Assadpoor, E. Spray-Drying Microencapsulation of Anthocyanins by Natural Biopolymers: A Review. *Dry. Technol.* **2014,** *32*(5), 509–518.
28. Mao, L.; Lu, Y.; Cui, M.; Miao, S.; Gao, Y. Design of Gel Structures in Water and Oil Phases for Improved Delivery of Bioactive Food Ingredients. *Crit. Rev. Food Sci. Nutr.* **2020,** *60*(10), 1651–1666.
29. Mar, J. M.; Silva, L. S.; Rabelo, M. D. S.; Muniz, M. P.; Nunomura, S. M.; Correa, R. F.; Kinupp, V. F.; Campelo, P. H.; Bezerra, J. D. A.; Sanches, E. A. Encapsulation of Amazonian Blueberry Juices: Evaluation of Bioactive Compounds and Stability. *LWT Food Sci. Technol.* **2020,** *124*, E-article 109152.
30. Maurya, V. K.; Aggarwal, M. A Phase Inversion Based Nanoemulsion Fabrication Process to Encapsulate Vitamin D3 for Food Applications. *J. Steroid Biochem. Mol. Biol.* **2019,** *190*, 88–98.
31. Murali, S.; Kar, A.; Mohapatra, D.; Kalia, P. Encapsulation of Black Carrot Juice Using Spray and Freeze Drying. *Food Sci. Technol. Int.* **2015,** *21*(8), 604–612.
32. Niu, B.; Shao, P.; Luo, Y.; Sun, P. Recent Advances of Electrosprayed Particles as Encapsulation Systems of Bioactives for Food Application. *Food Hydrocoll.* **2020,** *99*, E-article 105376.
33. Oancea, A. M.; Aprodu, I.; Ghinea, I. O.; Barbu, V.; Ioniţă, E.; Bahrim, G.; Râpeanu, G.; Stănciuc, N. A Bottom-up Approach for Encapsulation of Sour Cherries Anthocyanins by Using β-Lactoglobulin as Matrices. *J. Food Eng.* **2017,** *210*, 83–90.
34. Onbas, R.; Kazan, A.; Nalbantsoy, A.; Yesil-Celiktas, O. Cytotoxic and Nitric Oxide Inhibition Activities of Propolis Extract along with Microencapsulation by Complex Coacervation. *Plant Foods Hum. Nutr.* **2016,** *71*(3), 286–293.
35. Ortiz-Basurto, R. I.; Rubio-Ibarra, M. E.; Ragazzo-Sanchez, J. A.; Beristain, C. I. Microencapsulation of *Eugenia uniflora* L. Juice by Spray Drying Using Fructans with Different Degrees of Polymerisation. *Carbohydr. Polym.* **2017,** *175*, 603–609.

36. Ospina-Villa, J. D.; Gómez-Hoyos, C.; Zuluaga-Gallego, R.; Triana-Chávez, O. Encapsulation of Proteins from Leishmania Panamensis into PLGA Particles by a Single Emulsion-Solvent Evaporation Method. *J. Microbiol. Methods* **2019,** *162*, 1–7.
37. Ozkan, G.; Franco, P.; De Marco, I.; Xiao, J.; Capanoglu, E. A Review of Microencapsulation Methods for Food Antioxidants: Principles, Advantages, Drawbacks and Applications. *Food Chem.* **2019,** *272*, 494–506.
38. Plenagl, N.; Seitz, B. S.; Duse, L.; Pinnapireddy, S. R.; Jedelska, J.; Brüßler, J.; Bakowsky, U. Hypericin Inclusion Complexes Encapsulated in Liposomes for Antimicrobial Photodynamic Therapy. *Int. J. Pharm.* **2019,** *570*, E-article 118666.
39. Potdar, S. B.; Landge, V. K.; Barkade, S. S.; Potoroko, I.; Sonawane, S. H. Flavor Encapsulation and Release Studies in Food. Chapter 16; In *Encapsulation of Active Molecules and Their Delivery System*; Sonawane, S. H., Bhanvase, B. A., Sivakumar, M., Eds.; Elsevier: Amsterdam, Netherlands, 2020; pp 293–321.
40. Pulit-Prociak, J.; Kabat, M.; Węgrzyn, E.; Zielina, M.; Banach, M. Encapsulation of Antioxidant Compounds in Biopolymer Micelles. *Chem. Eng. Commun.* **2020,** *207*(3), 393–412.
41. Quan, Z.; Guan, R.; Huang, H.; Yang, K.; Cai, M.; Meng, X. Antioxidant Activity and Absorption of Cyanidin-3-O-Glucoside Liposomes in GES-1 Cells in Vitro. *Biosci. Biotechnol. Biochem.* **2020,** *84*(6), 1239–1249.
42. Raeisi, S.; Ojagh, S. M.; Quek, S. Y.; Pourashouri, P.; Salaün, F. Nano-Encapsulation of Fish Oil and Garlic Essential Oil by a Novel Composition of Wall Material: Persian Gum-Chitosan. *LWT Food Sci. Technol.* **2019,** *116*, E-article 108494.
43. Rehman, A.; Ahmad, T.; Aadil, R. M.; Spotti, M. J.; Bakry, A. M.; Khan, I. M.; Zhao, L.; Riaz, T.; Tong, Q. Pectin Polymers as Wall Materials for the Nano-Encapsulation of Bioactive Compounds. *Trends Food Sci. Technol.* **2019,** *90*, 35–46.
44. Rezaei Erami, S.; Raftani Amiri, Z.; Jafari, S. M. Nanoliposomal Encapsulation of Bitter Gourd (Momordica Charantia) Fruit Extract as a Rich Source of Health-Promoting Bioactive Compounds. *LWT Food Sci. Technol.* **2019,** *116*, E-article 108581.
45. Saifullah, M.; Shishir, M. R. I.; Ferdowsi, R.; Tanver Rahman, M. R.; Van Vuong, Q. Micro and Nano Encapsulation, Retention and Controlled Release of Flavor and Aroma Compounds: A Critical Review. *Trends Food Sci. Technol.* **2019,** *86*, 230–251.
46. Salehi, A. A.; Ghannadi-maragheh, M.; Torab-, M.; Torkaman, R.; Asadollahzadeh, M. Hydrogel Materials as an Emerging Platform for Desalination and the Production of Purified Water. *Sep. Purif. Rev.* **2020,** *50*(4), 380–399.
47. Santana, Á. L.; Macedo, G. A. Challenges on the Processing of Plant-Based Neuronutraceuticals and Functional Foods with Emerging Technologies: Extraction, Encapsulation and Therapeutic Applications. *Trends Food Sci. Technol.* **2019,** *91*, 518–529.
48. Shao, P.; Liu, Y.; Ritzoulis, C.; Niu, B. Preparation of Zein Nanofibers with Cinnamaldehyde Encapsulated in Surfactants at Critical Micelle Concentration for Active Food Packaging. *Food Packag. Shelf Life* **2019,** *22*, E-article 100385.
49. Shinwari, K. J.; Rao, P. S. Stability of Bioactive Compounds in Fruit Jam and Jelly during Processing and Storage: A Review. *Trends Food Sci. Technol.* **2018,** *75*, 181–193.
50. Shishir, M. R. I.; Xie, L.; Sun, C.; Zheng, X.; Chen, W. Advances in Micro and Nano-Encapsulation of Bioactive Compounds Using Biopolymer and Lipid-Based Transporters. *Trends Food Sci. Technol.* **2018,** *78*, 34–60.
51. Singh, C. S.; Paswan, V. K.; Rai, D. C. Process Optimization of Spray Dried Jamun (*Syzygium Cumini* L.) Pulp Powder. *LWT Food Sci. Technol.* **2019,** *109*, 1–6.

52. Suner, S. S.; Sahiner, N. Biocompatible Macro, Micro and Nano Scale Guar Gum Hydrogels and Their Protein Absorption Capacity. *J. Macromol. Sci. A* **2020,** *57*(12), 810–818.
53. Suravanichnirachorn, W.; Haruthaithanasan, V.; Suwonsichon, S.; Sukatta, U.; Maneeboon, T.; Chantrapornchai, W. Effect of Carrier Type and Concentration on the Properties, Anthocyanins and Antioxidant Activity of Freeze-Dried Mao [*Antidesma Bunius* (L.) Spreng] Powders. *Agric. Natur. Resourc.* **2018,** *52*(4), 354–360.
54. Taheri, A.; Jafari, S. M. Gum-Based Nanocarriers for the Protection and Delivery of Food Bioactive Compounds. *Adv. Colloid Interf. Sci.* **2019,** *269*, 277–295.
55. Talón, E.; Lampi, A. M.; Vargas, M.; Chiralt, A.; Jouppila, K.; González-Martínez, C. Encapsulation of Eugenol by Spray-Drying Using Whey Protein Isolate or Lecithin: Release Kinetics, Antioxidant and Antimicrobial Properties. *Food Chem.* **2019,** *295*, 588–598.
56. Tampau, A.; González-Martínez, C.; Chiralt, A. Polyvinyl Alcohol-Based Materials Encapsulating Carvacrol Obtained by Solvent Casting and Electrospinning. *React. Funct. Polym.* **2020,** *153*, E-article 104603.
57. Tavares, L.; Barros, H. L. B.; Vaghetti, J. C. P.; Noreña, C. P. Z. Microencapsulation of Garlic Extract by Complex Coacervation Using Whey Protein Isolate/Chitosan and Gum Arabic/Chitosan as Wall Materials: Influence of Anionic Biopolymers on the Physicochemical and Structural Properties of Microparticles. *Food Bioproc. Technol.* **2019,** *12*(12), 2093–2106.
58. Villicaña-Molina, E.; Pacheco-Contreras, E.; Aguilar-Reyes, E. A.; León-Patiño, C. A. Pectin and Chitosan Microsphere Preparation via a Water/Oil Emulsion and Solvent Evaporation Method for Drug Delivery. *Int. J. Polym. Mater. Polym. Biomater.* **2020,** *69*(7), 467–475.
59. Weerasundara, L.; Gabriele, B.; Figoli, A.; Ok, Y. S.; Bundschuh, J. Hydrogels: Novel Materials for Contaminant Removal in Water—A Review. *Crit. Rev. Environ. Sci. Technol.* **2021,** *51*(17), 1970–2014.
60. Wen, P.; Zong, M. H.; Linhardt, R. J.; Feng, K.; Wu, H. Electrospinning: A Novel Nano-Encapsulation Approach for Bioactive Compounds. *Trends Food Sci. Technol.* **2017,** *70*, 56–68.
61. Xiao, Z.; Hou, W.; Kang, Y.; Niu, Y.; Kou, X. Encapsulation and Sustained Release Properties of Watermelon Flavor and Its Characteristic Aroma Compounds from γ-Cyclodextrin Inclusion Complexes. *Food Hydrocoll.* **2019,** *97*, E-article 105202.
62. Yamashita, C.; Chung, M. M. S.; dos Santos, C.; Mayer, C. R. M.; Moraes, I. C. F.; Branco, I. G. Microencapsulation of an Anthocyanin-Rich Blackberry (Rubus Spp.) by-Product Extract by Freeze-Drying. *LWT Food Sci. Technol.* **2017,** *84*, 256–262.
63. Yoha, K. S.; Moses, J. A.; Anandharamakrishnan, C. Effect of Encapsulation Methods on the Physicochemical Properties and the Stability of Lactobacillus Plantarum (NCIM 2083) in Synbiotic Powders and in-Vitro Digestion Conditions. *J. Food Eng.* **2020,** *283*, E-article 110033.
64. Yousuf, B.; Gul, K.; Wani, A. A.; Singh, P. Health Benefits of Anthocyanins and Their Encapsulation for Potential Use in Food Systems: A Review. *Crit. Rev. Food Sci. Nutr.* **2016**, *56*(13), 2223–2230.
65. Zakharov, A. G.; Voronova, M. I.; Bazanov, A. V.; Surov, O. V. Porous Composites of Water-Soluble Polymers with Cellulose Nanocrystals. *J. Solgel Sci. Technol.* **2019,** *92*(2), 484–495.

INDEX

A

Agro-industrial waste, 205
bioactive ingredients, 207–208
classification, 206–207
fermentation, effect, 221–223
fermented, 218–221
food products, 213
fortified commercial food products, 208, 212, 214
functional valorized, 219
fungal, types of
solid-state fermentation (SSF), 217
submerged fermentation (SF), 217–218
limitations, 214–215
nutritional and functional attributes, 224–225
potential health benefits, 209–211
processing, 206–207
valorization, 215–217
Airborne power, 8
Assembling
molecular glues, 156–157

B

Biotechnological approaches, applications, 177
benefits, 197–198
concerns, 197–198
food industries
biodegradable packaging, 191–192
biopreservatives, 185–186
biosensors, 193–194
biosurfactants, 189–191
enzymes, 182–185
fermented foods, 181–182
food ingredients, 189
functional foods, 188–189
GM food, 192–193
prebiotics, 186–187
probiotics, 186–187
single-cell proteins (SCPS), 187–188
synbiotics, 186–187
foodomics
nutrition, 194–197
safety, 194–197
history, 180
methods
current status, 180–181
scope, 180–181
significance, 180–181

C

Citrus in Meghalaya
species, 238–239
Citrus indica, 239–240
Citrus latipes, 240
Citrus limetta, 241–244
Citrus limon, 246
Citrus macroptera, 241
Citrus medica, 245
Citrus reticulata, 247
Citrus indica, 239–240
medicinal properties, 240
Citrus latipes, 240
applications, 240–241
Citrus limetta
sweet lime, effect, 245
Citrus limon
compounds, 246–247
health benefits, 247
Citrus macroptera
use, 241
Citrus medica
pharmalogical properties, 245–246
Citrus reticulata
Khasi mandarin, medicinal properties, 248
physicochemical composition, 248

D

Deactivation of microorganisms, 37
Dielectric properties, 26–27

Directed self-assembly, 159
Drying
  infrared (IR), 33
  microwave-assisted freeze drying (MAFD), 31
  microwave-assisted vacuum drying (MAVD), 31–32
  ohmic heating (OH), 32–33
  osmotic dehydration (OD), 32
  ultrasonication (US), 33

E

Electroporation
  pulsed electric field (PEF), 53
    in foods, 55
    frequency, 53–54
    microbial inactivation, 54–56
    polarization, 53–54
    shape, pulse characteristics, 53–54
  treated foods, 52
  ultra-high voltage electric fields (UHVEF), 51
    applications, 51–52
Emulsification, 15–16
Enzymatic activity, 10–11
Enzymes, 182–185
Excluded volume (EV), 155–156
Extraction of compounds, 13
Extraction time, 40–41

F

Food industries
  biodegradable packaging, 191–192
  biopreservatives, 185–186
  biosensors, 193–194
  biosurfactants, 189–191
  enzymes, 182–185
  fermented foods, 181–182
  food ingredients, 189
  functional foods, 188–189
  GM food, 192–193
  prebiotics, 186–187
  probiotics, 186–187
  single-cell proteins (SCPS), 187–188
  synbiotics, 186–187
Food irradiation
  recent research
    cereals, 61
    fresh vegetables, 59–60
    fruits, 58–59
    juices, 58–59
    meat products, 60
    pulses, 61
    spices, 59–60
Food structural designing (FSD)
  native, destabilization, 161–162
  new phases, development, 161–162
  stabilization, 162
Food structure (FS), 153–154
  assembling, 156
  food structural designing (FSD)
    steps, 159–161
  formation, 157
  mapping techniques, 162–163
  principles, 157–158
  structural stability and incompatibility, 154
Foodomics
  nutrition, 194–197
  safety, 194–197
Foods and food products, 3. *See also* Ultrasound processing
  applications, 9–10
    cleaning, 16
    cutting, 13–14
    drying, 12–13
    emulsification, 15–16
    extraction of compounds, 13
    filtration, 14
    food properties, modification, 16–17
    freezing, 11–12
    homogenization, mixing, 14–15
    microbial, enzymatic activity, 10–11
  generation, 5
    liquid driven, 5–6
    magnetostrictive, 5
    piezoelectric transducer, 5
  history, 4–5
  importance, 17
  principle, 6
  ultrasound, combined technologies, 8
    manosonication, 9
    manothermosonication, 9
    ultrasonication (US), 8
  ultrasound system, 6–7
    airborne power, 8
    parallel vibrating plates, 7

probes, 7
radial vibrating system, 8
ultrasonic baths, 7
Free volume (FV), 155
Fungal
types of
solid-state fermentation (SSF), 217
submerged fermentation (SF), 217–218

G

Glass transition temperature (GTT), 164–165

H

Heat pump drying, 85
applications, 86
classification, 86
principle, 85–86
working, 85–86
High-pressure processing (HPP)
applications, 56–58
process description, 56
Hybrid drying, 76–78
multiple innovations, 78

I

Ice cream crystallization, 142–143
Irradiation of foods and food products, 93
advantages, 104–105
applications, 101
high-dose, radappertization, 103–104
around globe, 106–107
consumer acceptability, 106
disadvantages, 105–106
ionizing radiation, 95–96
principle, 96
labeling of, 100–101
low-dose, radicidation
germination inhibition, 101
pest disinfestations, 102
medium dose, radurization
pathogens in food, 102
shelf-life extension, 102–103
spices, 103
milestones in, 94–95
historic indicators, 95
process of, 96–97
components, 97–98
dosage, 98–99
food material after treatment, 99
microorganisms, effects on, 99–100
sources, 98
prospects, 107
safety of radioactivity, 100
regarding toxicity, 100

K

Khasi mandarin
medicinal properties, 248

L

Leavening reaction, 126–128
Low-dose, radicidation
germination inhibition, 101
pest disinfestations, 102

M

Mathematical modeling in food processing, 115
first-order rate constants, 125
fundamental and empirical, 120–122
cooking of vegetables, 123–126
fermentation, 128–129
leavening reaction, 126–128
kinetic, 119–123
mathematical modeling, 117–119
nomenclature, composition, 137
product quality optimization, 140
ice cream crystallization, 142–143
ingredients for snack bars, 140–142
thermal resistance, 126
unified models, 129–132
retort processing, 132–135
wheat flour dough, 136–139
Maximum freeze concentration (MFC), 167
Medium dose, radurization
pathogens in food, 102
shelf-life extension, 102–103
spices, 103
Microfiltration (MF), 48
advantages, 49
application, 50
characteristics, 49–50
food industry, 48–49
Microwave drying
distribution, 30

mechanism, 30
migration, 30
quality characteristics, 31
Microwave heating, 23
assisted, 29–30
deactivation of microorganisms, 37
dielectric properties, 26–27
drying
infrared (IR), 33
microwave-assisted freeze drying (MAFD), 31
microwave-assisted vacuum drying (MAVD), 31–32
ohmic heating (OH), 32–33
osmotic dehydration (OD), 32
ultrasonication (US), 33
factors affecting
dependence on density, 29
effect of composition, 28–29
moisture content, 28
physical structure, 29
temperature, 28
mechanism, 25
microwave drying
distribution, 30
mechanism, 30
migration, 30
quality characteristics, 31
microwave-assisted extraction (MAE), 38–39
cycle, 40–41
extraction time, 40–41
factors affecting, 39
nature of solvent, 39–40
plant matrix, 41
stirring effect, 41
temperature, 40
pasteurization technique, 33–34
fluid state food, 34–35
mechanism, 34
solid state food, 35
sensory characteristics, 26
sterilization technique, 35
benefits, 38
mechanism, 35–36
microwave-assisted thermal sterilization, 36–37
Microwave-assisted extraction (MAE), 38–39
cycle, 40–41
extraction time, 40–41
factors affecting, 39
nature of solvent, 39–40
plant matrix, 41
stirring effect, 41
temperature, 40
Microwave-assisted freeze drying (MAFD), 31
Microwave-assisted vacuum drying (MAVD), 31–32

N

Nutraceutical food, 237
Citrus in Meghalaya, 238–239
Citrus indica, 239–240
Citrus latipes, 240
Citrus limetta, 241–244
Citrus limon, 246
Citrus macroptera, 241
Citrus medica, 245
Citrus reticulata, 247
Citrus indica
medicinal properties, 240
Citrus latipes
applications, 240–241
Citrus limetta
sweet lime, effect, 245
Citrus limon
compounds, 246–247
health benefits, 247
Citrus macroptera
use, 241
Citrus medica
pharmalogical properties, 245–246
Citrus reticulata
Khasi mandarin, medicinal properties, 248
physicochemical composition, 248
nutritional aspects, 243–244
species, 239

O

Ohmic heating (OH), 32–33
Osmotic dehydration (OD), 32

P

Pasteurization technique, 33–34
fluid state food, 34–35
mechanism, 34
solid state food, 35
Potential of nonthermal, 47
food irradiation, recent research
cereals, 61
fresh vegetables, 59–60
fruits, 58–59
juices, 58–59
meat products, 60
pulses, 61
spices, 59–60
high-pressure processing (HPP)
applications, 56–58
process description, 56
irradiation, 58
microfiltration (MF), 48
advantages, 49
application, 50
characteristics, 49–50
food industry, 48–49
Pretreatments or predrying treatments, 70–71
acid, 73
alkaline, 72–73
blanching, 71–72
freezing, 74
salting, 72
sulfiting, 73–74
sulfur dioxide, 73–74
sulfuring, 73–74
Pulsed electric field (PEF) (electroporation), 53
in foods, 55
frequency, 53–54
microbial inactivation, 54–56
polarization, 53–54
shape, pulse characteristics, 53–54

R

Radio frequency drying (RFD), 83
applications, 83–85
principle, 83
working, 83
Refractance window (RWTM), 80–81
working system, 81–82

S

Single-cell proteins (SCPS), 187–188
Soft condensed matter (SCM), 151
Solid-state fermentation (SSF), 217
State diagram (SD), 163–164
applications, 167–168
construction, 164
Sterilization technique, 35, 36
benefits, 38
mechanism, 35–36
microwave-assisted thermal sterilization, 36–37
Structural design of food products, 149
assembling
molecular glues, 156–157
construction
freezing, 166–167
glass transition temperature (GTT), 164–165
maximum freeze concentration (MFC), 167
solid-crystal melting temperature, 166
data of food products, 169–171
food structural designing (FSD)
native, destabilization, 161–162
new phases, development, 161–162
stabilization, 162
food structure (FS), 153–154
assembling, 156
food structural designing (FSD), 159–161
formation, 157
mapping techniques, 162–163
principles, 157–158
structural stability and incompatibility, 154
principles
directed self-assembly, 159
phase separation, 158
self-assembly, 159
spontaneous self-assembly, 159
soft condensed matter (SCM), 151
stability and incompatibility
excluded volume (EV), 155–156
free volume (FV), 155
state diagram (SD), 163–164
applications, 167–168
construction, 164

Submerged fermentation (SF), 217–218
Superheated steam drying, 86–87

T

Thermophysical and physicochemical technologies, 69
- drying characteristics, 74–75
- heat pump drying, 85
  - applications, 86
  - classification, 86
  - principle, 85–86
  - working, 85–86
- hybrid drying, 76–78
  - multiple innovations, 78
- IR drying, 78–80
- kinetics, 74–75
- need for new developments, 75–76
- pretreatments or predrying treatments, 70–71
  - acid, 73
  - alkaline, 72–73
  - blanching, 71–72
  - freezing, 74
  - salting, 72
  - sulfiting, 73–74
  - sulfur dioxide, 73–74
  - sulfuring, 73–74
- radio frequency drying (RFD), 83
  - applications, 83–85
  - principle, 83
  - working, 83
- refractance window (RWTM), 80–81
  - working system, 81–82
- superheated steam drying, 86–87

U

Ultra-high voltage electric fields (UHVEF), 51
- applications, 51–52

Ultrasonication (US), 8, 33
Ultrasound processing
- combined technologies
  - manosonication, 9
  - manothermosonication, 9
  - ultrasonication (US), 8
  - ultrasound system, 6–7
- foods and food products, 3, 9–10
  - cleaning, 16
  - cutting, 13–14
  - drying, 12–13
  - emulsification, 15–16
  - extraction of compounds, 13
  - filtration, 14
  - food properties, modification, 16–17
  - freezing, 11–12
  - homogenization, mixing, 14–15
  - microbial, enzymatic activity, 10–11
- generation, 5
  - combined technologies, 8
  - history, 4–5
  - importance, 17
  - liquid driven, 5–6
  - magnetostrictive, 5
  - piezoelectric transducer, 5
  - principle, 6
- system
  - airborne power, 8
  - parallel vibrating plates, 7
  - probes, 7
  - radial vibrating system, 8
  - ultrasonic baths, 7

Unified models, 129–132
- retort processing, 132–135
- wheat flour dough, 136–139